Friendship

ORIGINS OF HUMAN BEHAVIOR AND CULTURE

Edited by Monique Borgerhoff Mulder and Joe Henrich

Friendship

Development, Ecology, and Evolution of a Relationship

DANIEL J. HRUSCHKA

University of California Press

BERKELEY   LOS ANGELES   LONDON

University of California Press, one of the most distinguished university presses in the United States, enriches lives around the world by
advancing scholarship in the humanities, social sciences, and natural
sciences. Its activities are supported by the UC Press Foundation and by
philanthropic contributions from individuals and institutions. For more
information, visit www.ucpress.edu.

University of California Press
Berkeley and Los Angeles, California

University of California Press, Ltd.
London, England

Library of Congress Cataloging-in-Publication Data

Hruschka, Daniel J., 1972–
    Friendship : development, ecology, and evolution of a relationship /
by Daniel J. Hruschka.
        p.   cm. — (Origins of human behavior and culture ; v.5)
    Includes bibliographical references and index.
    ISBN 978-0-520-26546-2 (cloth : alk. paper) — ISBN 978-0-520-
26547-9 (pbk. : alk. paper)
        1. Friendship—Social aspects.   2. Kinship.   3. Human behavior.
    4. Interpersonal relations.   I. Title.
    GN486.3.H78   2010
    302.3′4—dc22   2010008844

Manufactured in the United States of America

19 18 17 16 15 14 13 12 11 10
10 9 8 7 6 5 4 3 2 1

This book is printed on Cascades Enviro 100, a 100% post consumer
waste, recycled, de-inked fiber. FSC recycled certified and processed
chlorine free. It is acid free, Ecologo certified, and manufactured by
BioGas energy.

To Dee, Ella, and Emily

Contents

Illustrations

TABLES

Boxes

Acknowledgments

This book was written on the shoulders of friends—those who shared their own views of friendship through conversations and deeds and especially those who kindly waded through early versions of the manuscript and provided valuable suggestions for improving it. For those few hearty souls who read the entire manuscript, I owe a great debt of gratitude. Kenny Maes is likely responsible for any prose that makes you laugh. Frank Scott untwisted my convoluted phrasing and pushed for more figures. Tim Taylor helped me with style and was a font of knowledge about friendship in literature. I would also like to extend special thanks to Peter Hruschka, who has been an unswerving companion on this project from my first dissertation chapters to the final proofs of the book, and for constantly reminding me that words matter. Thanks to Ryan Brown, Ronda Butler-Villa, Napoleon Chagnon, Nathan Collins, Lee Cronk, Dee Hruschka, Judy Hruschka, Ting Jiang, Brandon Kohrt, Shirley Lindenbaum, Monique Borgerhoff Mulder, Thor Veen, Laura Ware, and Jon Wilkins for valuable comments on portions of the manuscript. And thanks to passersby at the Santa Fe Institute, especially Aaron Clauset, for taking "just one more look" at the illustrations. Each in his or her unique way made this a much better book.

I would not have followed the path to this book without the thoughtful work of other scholars. Joan Silk, Rita Smaniotto, John Tooby, and Leda Cosmides are some of the first scholars to have questioned standard evolutionary accounts for altruism among friends. Joe Henrich pointed me to their work and helped hone my understanding of evolutionary approaches to cooperation and friendship. Robert Brain's cross-cultural survey of friendship inspired my own study of friendship around the world, and the work of anthropologists who dared to study friendship when kinship ruled anthropology provided important insights and necessary raw material for the cross-cultural component of the study.

The Santa Fe Institute generously granted necessary time and encouragement to complete the manuscript. The National Science Foundation, Emory University's Department of Anthropology, the Center for Myth and Ritual in American Life at Emory University, and my advisors, Dan Sellen, Joe Henrich, Carol Worthman, and Bradd Shore, provided valuable support during the completion of my dissertation, which ultimately became the foundation for the present book. University of California Press, and Blake Edgar especially, have been tremendously supportive in making the transition from a thesis to a work that might be of interest beyond my dissertation advisors. Jacqueline Volin, Jimmée Greco, Celeste Newbrough, and Jean McAneny deserve special thanks for their thorough work in shepherding the book through final production and copyediting, indexing, and proofreading.

My deepest thanks go to my family—Dee, Ella, Emily, Tom, Carolyn, Mom, and Poppa—for their enduring patience and support and for showing me that family and friendship are not naturally divided, but rather flourish best together.

Introduction

The Adaptive Significance of Friendship

I have not as yet mentioned a circumstance which influenced my
whole career more than any other. This was my friendship with
Professor Henslow.

CHARLES DARWIN, *Recollections of the
Development of My Mind and Character*

In the late spring of 1876, nearly seventeen years after his first publication
of *On the Origin of Species* and following decades of careful description of
the natural world, Charles Darwin sat down to write a sketch of his life. He
devoted only sixty pages to the topic, detailing his early encounters with
the natural world, his compulsive beetle collecting, his lackluster attempt
at earning a medical degree, and his five years of voyaging on H.M.S.
Beagle. However, when Darwin described the circumstance that most
influenced his intellectual career, he focused not on his encounters with
books or the natural world, but rather on a friendship—his intimate bond
with his Cambridge mentor and fellow naturalist John Henslow. Grounded
in a shared passion for the natural world, the friendship between Darwin
and Henslow developed at Cambridge over frequent walks, country expeditions, and home visits, as the two pondered questions in religion and
natural science. Their friendship lasted from 1828 until Henslow's death
in 1861, and over the years, Henslow played a singular role in Darwin's
intellectual development. In addition to introducing Darwin to the scientific study of geology, botany, and zoology, Henslow arranged Darwin's
position on the H.M.S. *Beagle*, where the young scientist would ultimately
make observations critical to his theory of natural selection.

An astute and meticulous observer of the natural world, Darwin recognized the importance of friendships everywhere in the story of his
personal development. Darwin's friends introduced him to new ideas,
provided academic opportunities, and supported his theories on evolution
in an atmosphere of vigorous academic debate.[1] Rarely, however, did these
friends provide the kind of material support bearing on the life-or-death
struggle for existence that figured so prominently in Darwin's theory of
evolution and natural selection. Therefore, it is not surprising that, in con-

trast to his recurring treatment of the subject in the short natural history of his own development, Darwin referred to friendship only a handful of times in the sum of his scientific works on human evolution.

The apparent discrepancy in Darwin's own writings—between the importance of friendships in his own life and the role that friendships might have played over the course of human evolution—reflects current thinking about friendship in the modern West.[2] Many of us have friends, and they reward us in diverse ways, engaging us with stimulating conversation, improving our mood, and relieving us from minor inconveniences by sharing a ride, lending a hand, or taking the time to think through problems. However, while friends make us happy and help us in small ways, it is not entirely clear that they are important in the high-stakes game of survival and reproduction.[3] As the twentieth-century social commentator C.S. Lewis wrote, "Friendship is unnecessary, like philosophy, like art. . . . It has no survival value." In line with this view, theories of human evolution have generally neglected the adaptive importance of friendships, instead focusing on exchange regulated by kin-biased altruism, pair-bonding, or strictly balanced give-and-take.

The purpose of this book is twofold. First, it brings to the foreground the unique ways that friendships, defined here as long-term relationships of mutual affection and support, have helped people deal with the struggles of daily life in a wide range of human societies.[4] Depending on the culture, friends share food when it is scarce, provide backup during aggressive disputes, lend a hand in planting and harvesting, and open avenues of exchange across otherwise indifferent or hostile social groups. And behavior among friends is not necessarily regulated in the same way as behavior in other relationships, such as those among biological kin or mates. Nor is it regulated in terms of strictly balanced, tit-for-tat exchange. Rather, I will argue that the help provided by friends is regulated by a system based on mutual goodwill that motivates friends to help each other in times of need. How humans are able to cultivate goodwill and successfully maintain friendships when the potential for exploitation is theoretically so great is a fascinating question, and one that will figure prominently in this book.

Beyond the basic unifying elements of mutual affection and support, friendships can be established and maintained in diverse ways across cultures, many of which are difficult to reconcile with ideals of friendship in the United States and Europe. People in other places and times have inherited friendships from parents and other family members, sanctified friendships through public wedding-like rituals, and entered friendships based on the wishes of family and community elders.[5] In many societies, close friends are sufficiently valuable that it is acceptable to violate the law to protect them. And some defining features of friendship in the U.S., such as a focus on emotional rather than material support, are of minor importance in other societies. Therefore, in addition to identifying core features of friendship, the book's second goal is to document and account for the recurring yet diverse ideals and behaviors associated with friendship in human societies.

I approach these goals from three perspectives—developmental, ecological, and evolutionary—each of which opens up complementary vistas on how friendships have emerged as a social form among humans and how they continue to arise in everyday life. The first perspective taken in the book is developmental and acknowledges that much of human behavior is fashioned through a process of social learning that takes place over a lifetime. Therefore, this book examines how people learn the rules of friendship in their natal cultures and how they cultivate friendships with one another over time. The second perspective is ecological and recognizes that a key human adaptation is the ability to adjust behavior to the vicissitudes of local environments. Thus, we might expect friendships to vary in their particular functions and developmental trajectories in different ecological settings.[6] For example, how do the friendships of foragers in harsh and highly variable environments differ from those of steadily employed middle-class citizens of a modern nation-state? Are there societies where friendships are unnecessary or indeed absent, as some scholars have proposed? The third, evolutionary, perspective asks how behaviors among friends ultimately influence survival and reproduction, why a capacity for something like friendship might have arisen and endured among humans, and what other animals might possess the capabilities necessary for the cultivation of friendship-like relationships. From these three perspectives on friendship's origins, I develop an account that ranges from ultimate evolutionary explanations of friendship's ubiquitous appearance in human social life to proximal descriptions of the psychological processes involved in learning and regulating behaviors among friends in changing and uncertain contexts.

Before proceeding further, it is worth considering in more detail what we mean by friendship, how it differs from other kinds of relationships, and how friendships uniquely aid in the struggles of daily life (box 1). Philosophy perhaps more than any other discipline has dealt with these issues, and I begin by reviewing how philosophers have defined friendship, not only in terms of which behaviors are observed among friends, but also

---

BOX 1 What Is a Friend?

Friend is a slippery concept. Among Lepcha farmers in eastern Nepal, the closest word for *friend* can be extended to many kinds of relationships, including trading partnerships with foreigners, relationships based on mutual aid, and childhood companions (Gorer 1938). In English, politicians use it to address masses of supporters, nation-states use it to declare economic and political alliances, and social networking sites use the term for any kind of mutually recognized tie. As a testament to its conceptual spread, the word *friend* is spoken and written more in English than any other relational term—even more than *mother* or *father* (Leech, Rayson, and Wilson 2001). In the midst of such ubiquitous and diverse usage, one aim of this book will be to identify what is meant by the word *friend* and how individuals who self-identify as friends, and especially close friends, feel about and behave toward each other. While this approach works well in English-speaking contexts, it poses serious problems when one travels to other cultures that use other words for friend-like relationships. I discuss in more detail how to deal with this issue of cross-cultural translation in chapter 2.

---

what underlying motivations guide such behaviors. Next, I briefly outline how the approach to regulating behaviors among friends differs from that used in other relationships, such as those between kin or between partners who exchange on the basis of quid pro quo (something for something). Finally, I propose that this way of regulating relationships provides one solution to a recurring problem in evolutionary biology and the social sciences: mutual aid in uncertain environments. These three questions—how is friendship regulated, how does friendship differ from other relationships, and why is friendship useful—will arise throughout the book.

PHILOSOPHERS DEFINING FRIENDSHIP

In contrast to its relative neglect by students of human evolution, friendship has been a recurring topic in philosophy.[7] Big names in Western thought, ranging from Aristotle to twentieth-century French philosopher Jacques Derrida, have attempted to identify the essential qualities of friendship, to define its place in the social order, and to give advice on dealing with friends. Aristotle devoted two of the ten books in his *Nicomachean Ethics* to the subject and laid out the necessary conditions for the relationship: a friend must wish well for the other, the other must share this goodwill, and both must recognize that these feelings are mutual. Predating many later treatments of friendship, Aristotle's work also made clear distinctions between friendships based purely on mutual utility and those based on mutual goodwill. Twenty-four centuries later, Jacques Derrida, the father of philosophical deconstructionism, wrote an entire book on the challenge of knowing whether someone is a friend or an enemy.

Non-Western intellectual traditions have also given friendship serious thought. In their advice on leading a proper life, the Buddha and followers of Confucius outlined the types of friendships that one should seek in daily life and those that one should avoid. Over three thousand years ago in present-day Punjab, Vedic hymns were written that enumerated the obligations of friends: friends should provide food and protect one another's honor, and foremost should not abandon one another in times of need.

These diverse traditions frequently define friendship in terms of rules and violations—how one should behave toward friends, what friends should do for one another, and examples of false friends who violate codes of good conduct. For example, in his advice to followers in the *Sigalovada Sutra*, the Buddha outlined five appropriate behaviors toward friends that closely reflect modern Western ideals: (1) be generous, (2) speak kindly, (3) provide care, (4) be equal, and (5) be truthful. According to the Buddha, friends will return the favor by offering protection and consolation in times of need. In the same text, the Buddha also illustrated four violations of friendship as "foes in the guise of friends": (1) the selfish friend who only fulfills his duty out of fear, (2) the friend who promises much but does not deliver when one is in need, (3) the flatterer who speaks ill behind one's back, and (4) the ruiner who leads one to intoxication, late-night revelry, idle entertainment, and gambling.[8]

Behaviors such as being truthful and providing care often play an important part in philosophers' definitions. However, behaviors alone are insufficient to define friendship. We also need to understand what makes people *want* to engage in these behaviors and how these expectations are enforced and encouraged. Consider drawing up a contract with a close friend stating the conditions under which each should help the other or resorting to small claims court to address a close friend's bad behavior. These measures would not conflict with most of the Buddha's rules, but they would likely violate our own notions of friendship. Though the Buddha focused mostly on the rules of friendship, he also recognized the importance of *how* the rules are followed, by stating, for example, that friends should not help out of fear but rather from feelings of compassion and loving-kindness. More broadly, people in a wide range of cultures carefully avoid certain kinds of accounting—such as strict give-and-take—when interacting with friends. A recurring theme of this book will be *how* friends follow and enforce the rules of friendship, and why this distinguishes friendship from other kinds of relationships, such as kin ties or trade relationships based on reciprocal exchange or barter.

FRIENDSHIP: A SPECIAL KIND OF RECIPROCAL ALTRUISM

Friendship is only one among many ways that humans—and other organisms—co-regulate one another's behavior.[9] Among cooperative relationships, for example, evolutionary theorists have generally focused on those regulated by kin-biased altruism, pair-bonding with mates, and strict tit-for-tat exchange.[10] How do human friendships differ from these kinds of relationships?

Observers have frequently noted similarities in the ways people behave toward close friends and closely related kin. In both cases, people often help for the sake of helping, rather than from fear of punishment or out of some expectation of return. People apply similar vocabularies, of love, loyalty, and goodwill, when talking about close family and friends. Indeed, they often explicitly incorporate non-kin friends into their families by calling them sister, brother, aunt, or uncle. For these reasons, some scholars have argued that friendship may be an application of the mechanisms regulating kin-biased altruism to non-kin individuals.[11] However, despite these superficial similarities, helping behaviors among friends differ in important ways from those among kin, depending in different ways on feelings of closeness and the costs of helping, a topic I will explore further in chapter 3.

Another possible foundation for friendship is pair-bonding between mating partners. Like biological kin, spouses and mates talk about love and loyalty, and they often help one another in unconditional ways. In the U.S. and other societies, many people refer to their spouse as their best friend. Indeed, friendships may recruit many of the same psychological and physiological processes involved in cultivating pair bonds. However, there are some problems with this explanation. Other mammals also form long-lasting pair bonds. For example, mouse-like prairie voles enter life-long monogamous unions that focus on common territory defense and pup rearing. However, these bonds *require* sexual activity (or human intervention to influence choice of mates) to form. Therefore, if human friendships are based on a template of pair-bonding, we must also explain how friendships can arise without the other trappings of pair-bonding, such as sexual desire, sexual behavior, and another common feature of human pair-bonds, single-minded, romantic obsession with a partner (chapter 4).

Finally, friendship also shares many similarities with reciprocally altruistic behavior whereby unrelated individuals help others depending on the quality of past exchanges and on the expectation of aid in the future. Such behavior is inherently risky, because one person may cheat by first enjoying the help of another but then failing to help in return. In his groundbreaking 1971 article "The Evolution of Reciprocal Altruism," Robert Trivers described how altruistic behaviors among non-kin could evolve by natural selection if costs and benefits were equally exchanged over sufficiently numerous interactions. Although not as common as kin-biased behaviors, such exchange relationships appear occasionally in the natural world. In coral reefs across the Pacific Ocean, bluestreak cleaner wrasses provide parasite-removal services to larger fish. In Central Mexico, vampire bats frequently regurgitate valuable blood-meals to share with hungry (non-kin) partners. And around the world, humans engage in all manner of reciprocal exchanges, whether we consider Nama pastoralists sharing water in the dry deserts of southern Africa, Tausug farmers of the Philippines rushing to the support of friends during feuds, or Ache foragers of South America sharing the fruits of their hunting and gathering.[12]

A decade after Trivers's account of the evolution of reciprocal altruism, political scientist Robert Axelrod and evolutionary biologist William Hamilton formalized (and dramatically simplified) the concept of reciprocally altruistic behavior in a game called the prisoner's dilemma. In the canonical prisoner's dilemma game, police have arrested two partners-in-crime, but without a confession from either of the conspirators the police can only make the case for a lesser charge. Hoping to divide and conquer, the police separate the prisoners into soundproof cell blocks, and they give each prisoner the opportunity to rat out his mate. If both prisoners keep quiet (thus cooperating amongst themselves), they both enjoy the much-reduced sentence of six months' jail time. If only one squeals, then he goes home scot-free, but the sucker faces a ten-year sentence. If both squeal, they both face a steep three-year sentence. If they know they'll never meet again, each prisoner does better alone by squealing. However, if both squeal on each other, then they get more time than if they had both kept quiet. The prisoner's dilemma game cleanly captures the trade-off between potential gains to be made by cooperating (in this case keeping quiet) and the possible risks of exploitation at the hands of a selfish partner.

Using a repeated version of this game, where the same players must face one another over many interactions, Axelrod and Hamilton showed how individuals following a simple cooperative strategy, popularly known as tit-for-tat, could avoid exploitation and outperform greedy defectors.[13] Tit-for-tat involved simply cooperating with a partner until that partner defected, at which point one refused to cooperate any further. The strategy only required knowing a partner's previous actions and opened up the possibility that organisms as simple as bacteria might have the capacity to cooperate. It also captured the kinds of quid pro quo exchanges often found in arm's-length commercial trades among humans.[14] The mathematical elegance of the repeated prisoner's dilemma game was also appealing, and over time reciprocal altruism became synonymous with tit-for-tat cooperation in the repeated prisoner's dilemma game.

The standard repeated prisoner's dilemma game, while elegantly capturing the tension between the temptation of immediate gratification and the promise of long-term cooperation, also represents a very limited view of the conditions in which cooperation might evolve. First, it assumes that the opportunities for helping a partner occur in lock-step alternation with uniform costs to helping, so that one could readily and immediately observe if a partner was cheating. In the real world, however, the opportunities to help a friend can be spaced over very long intervals in unknowable ways and involve vastly different costs and benefits. Needs can also become highly unbalanced. Due to a string of bad luck, for example, one friend may need a steady flow of help while the other friend needs none. Moreover, a friend may legitimately not be able to help when the need arises. The uncertain timing and size of needs and the uncertain ability of particular friends to help at a moment of need make the task of regulating reciprocal aid in such contexts very difficult. In such situations, a simple strategy based on keeping a strict balance of benefits and costs (e.g., tit-for-tat) would be very brittle. At the slightest failure of a partner, it would lead to the dissolution of friendship at best and recurring retaliation between partners at worst, with no possibility of repair. Over the past two decades researchers have dealt with some of these issues, such as the uncertain timing of needs, while leaving others relatively unexplored.[15] However, to deal with these added contingencies in exchange, one must often consider more complex strategies, raising questions about how humans could actually do the mental calculations required to enact such strategies.

In addition to these theoretical problems with tit-for-tat in regulating cooperation in real-life environments, there is an empirical problem. There is abundant evidence that human friends don't help one another in a tit-for-tat manner by responding directly to the balance of favors or a partner's past actions. Indeed, friends frequently avoid such strict accounting. Rather, when making decisions to help, they focus on the twin facts that so-and-so is a friend and she is in need.[16] In such cases, evaluations of friendship rather than accounting of past and possibly future exchanges are the most proximate reasons for the decision to help. This move, from choosing to help based on a tit-for-tat accounting system to helping because a friend is in need, also has implications for how people think and behave with friends. The question "Is Ella a friend?" requires new criteria to discern Ella's goodwill and feelings in the friendship. What are Ella's intentions toward me? Does she consider *me* a friend? Does she understand my needs and preferences? Does she pay too much attention to the balance of exchanges? These are important questions, because they bear indirectly on a partner's willingness to help in the future.

The addition of novel elements in decision making, in this case the task of evaluating the quality of one's friendship, opens up new potential for disruption of decision making and thus novel forms of exploitation. For example, unknown individuals, from panhandlers and con artists to politicians, often invoke the term *friend* to prime our helping behavior. In his famed book *How to Win Friends and Influence People*, Dale Carnegie described a number of tactics intended to make people feel that they are your friends so that they will help you in the future. It is also possible for chronic inequality to develop among friends as long as both feel that each still maintains goodwill. Such patterns of exploitation are a result of relying on friendship as the proximal reason for helping rather than focusing directly on the history of exchanges. A major question in this book will be how such attempts to divert and generalize the construct of friendship succeed (and fail) in altering real helping behavior, and what defenses people use to deter such manipulation.

WHY FRIENDSHIP, AND WHY HUMANS?

Friendship bonds bear some resemblance to bonds between closely related kin and mating partners and to ties based on quid pro quo exchanges. But a central part of this book will be to show that friendship involves a unique set of regulatory processes. Feelings of closeness are important predictors of help among friends but much less so among biological kin, suggesting that helping among friends is not due to a confusion of friends with kin (chapter 3). Friends do not need sexual attraction, sexual behavior, or the common rearing of offspring to cultivate their relationships, as occurs among mating pairs (chapter 4). And close friends violate many of the rules proposed for maintaining reciprocal altruism. Close friends eschew strict reciprocity, rather helping based on need. Friends are less sensitive to the balance of favors than are strangers and acquaintances and are more generous to one another, even when their partner won't find out whence the kind act came (chapter 1). From an evolutionary perspective, what selective pressures might have favored this need-based, low-monitoring form of reciprocal helping, when other, more basic modes of regulating cooperation and exchange were likely available?

I propose that the psychological systems underlying the ability and propensity to cultivate friendships were selected (or at least not rooted out by selection) because they uniquely addressed common adaptive problems of cooperation and mutual aid in uncertain contexts. In other words, friendship, as a system regulating altruistic behavior, solves a computational task in uncertain environments that cannot be met by simple reactive exchange strategies, such as tit-for-tat accounting.

Humans are relatively unique among animals in their capacity for cumulative cultural learning, whereby novel tools, activities, preferences, and artifacts can emerge and be preserved with some degree of fidelity over generations.[17] With this capacity for culture comes an explosion in the kinds of goods and favors that individuals can exchange, including food, knowledge of good foraging sites, child care, access to mates, shelter construction, sex, mentoring, guard duty against animal predators and other human groups, safe haven in other villages, support in disputes, grooming and parasite removal, labor, implements for hunting and food preparation, and manufactured goods, such as cloth, string, weapons, tools, and prestige items.[18] Compare this to the relative paucity of goods and services observed in exchanges among our closest relatives—chimpanzees.[19] The great diversity of possible exchanges among humans, as well as the uncertain timing of needs in each of these domains, drastically increases the complexity of strict accounting based purely on inputs and outputs.

One possible solution to this accounting problem would be to avoid it, and to instead rely exclusively on the goodwill of closely related kin for help in these domains. However, over the course of hominin evolution, some favors, such as access to mates, food sharing across ecological zones, and support in disputes with kin, would have been difficult if not impossible for close kin alone to provide. In the highlands of Papua New Guinea, Binumarien horticulturalists more than double the number of available gardening helpers by relying on biologically unrelated "social kin," who are as reliable as biological kin in providing aid. Ju/'hoansi foragers in the southern African desert invest in social insurance against hungry times by cultivating extensive webs of friendship outside of their circle of closely related kin. Yanomamo villagers in Venezuela rely on marital alliances, in addition to ties with close genetic kin, to build up coalitions that are sufficiently large to win in community-wide brawls. These examples are admittedly limited to contemporary human groups, but they also represent common forms of exchange—food sharing, labor exchange, and coalition support—that would likely have been important throughout human evolution.[20] The fact that friends are so reliably cultivated and recruited to engage in these kinds of exchange suggests that friendship plays an important role beyond genetic kinship in solving these problems of everyday life.

The question "Why humans?" also draws attention to the physiological mechanisms in humans that support the cultivation and maintenance of this low-monitoring, need-based form of mutual aid. How are the brain systems and neurotransmitters involved in other kinds of relationships, such as those among romantic partners or human parents and their offspring, recruited to promote the unconditional aid and long-term bonding observed among friends? What role do the neuropeptides involved in mammalian bonding, such as oxytocin and vasopressin, play in the development of friendship? And how do these systems operate differently in humans than in other animals? In terms of development, what physiological and psychological mechanisms mediate the unfolding of friendship, such as the often long courtship that leads from acquaintanceship to unconditional support, the increased forgiveness among friends that can preserve a relationship from premature death, and the transformation of thought from calculated help to knee-jerk altruism? Many of these questions do not have definitive answers yet, but I will do my best to review the growing body of research on physiological and psychological systems that likely underwrite the human capacity to make and keep friends.

THE BOOK

The general outline provided so far raises a number of questions that I will discuss in more detail in the chapters to follow. To what degree does something like friendship recur across human cultures, and are there core features that define friendship in these diverse settings? How does friendship differ from other kinds of relationships, such as those based on biological kinship or sexual attachment? How do people come to view others as friends, and what defenses do they use to avoid incorrectly assuming that someone has their well-being at heart? How do friends successfully regulate helping behaviors in uncertain environments? When does such regulation fail? And how is the regulation of friendships sensitive to the local cultural and social milieu? In this book, I bring together current work from a number of fields—including anthropology, psychology, sociology, and economics—to answer these questions. Figure 1 outlines these questions by chapter.

Chapter 1 asks, What is friendship? This question has been a source of debate from Plato to the present day. To tackle it, I start by reviewing the work of social psychologists and economists who have used surveys, experiments, and behavioral observation to understand the internal workings of friendships. I present evidence that friends do not help one another based on a careful balance of accounts or a concern about future payoffs. Moreover, I outline the key psychological and social processes—including feelings of closeness, love, and trust, as well as ways of communicating these feelings—involved in the everyday working of friendship. While this provides a solid starting point, most studies that permit such a fine-grained understanding of feelings, behavior, and communication among friends are concentrated in a narrow range of societies (i.e., the U.S. and other industrialized nations), making it difficult to extend these findings to understand what friendship might be like for the vast majority of humans living today and those who have lived in the past.

To remedy this narrow focus, chapter 2 turns to non-Western and small-scale societies to systematically examine friendship's place in human life. Specifically, is friendship a way of relating that arises across a wide array of human groups? Or is it particular to certain places and times? To answer these questions, chapter 2 explores relationships similar to friendship in societies ranging from small groups of hunter-gatherers to the densely populated cities of modern nation-states. This approach permits a view of the unity *and* diversity in the ways that humans cultivate and maintain friendships. It also leads to a core definition of friendship as a relationship involving support in times of need that is regulated by mutual affection between friends.

In chapters 3 and 4 I compare friendship with two other kinds of relationships: biological kinship and sexual attachment. Chapter 3 focuses on the similarities and differences between friends and close kin, weighing current theories about whether our feelings and behaviors toward friends simply extend psychological systems for kinship or rather reflect distinct psychological processes. Chapter 4 briefly deals with the relationship between friendship and the kinds of motivations, feelings, and behaviors

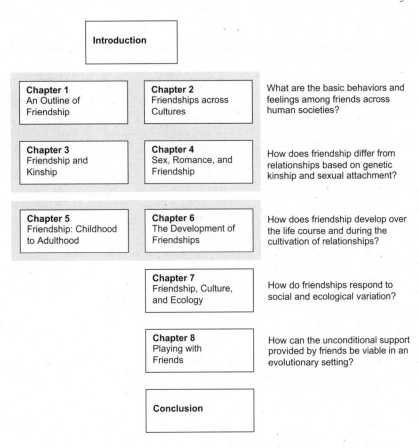

FIGURE 1.   Outline of the book, with relevant questions

involved in sexual attachment. Specifically, it differentiates among three systems involved in sexual attachment—sexual behavior, romantic obsession, and long-term attachment—and examines the relationship of these systems to friendship.

The capacity for friendship does not emerge instantly at birth, and an important part of childhood in many societies is learning how to be a friend. Chapter 5 reviews current research on how children's thinking about friendship changes in adolescence, from simple "liking" to abstract conceptualizations of trust, loyalty, and betrayal. Moreover, it examines how this general developmental trend is colored by particular factors, such as culture, personal predispositions, and gender.

Chapter 6 focuses on how friendships develop over time, as partners test one another's commitment and intentions, defend against false friends, and maintain a relationship in spite of occasional violations. It examines how friends send honest signals of empathy and goodwill through such behaviors as sharing secrets, disregarding the balance of exchange, and giving small gifts. It also reviews how such signals can be manipulated so as to exploit an individual's goodwill and how people defend themselves against such machinations.

Despite a common underlying structure to friendship in most human societies, cultural, social, and ecological conditions also influence friendships. Chapter 7 outlines key ways that friendship differs across societies, in terms of the relative importance of emotional and material support, the degree to which people help friends over other obligations and loyalties, and the kinds of help that friends provide. The chapter also reviews and critiques theories commonly proposed to account for cultural and ecological differences in friendship, such as the influence of resource uncertainty, geographic mobility, and changes in communication technology.

In chapter 8, I examine in more detail why the unconditional, need-based support among close friends can make economic and evolutionary sense. I formalize the argument proposed in this book, that friendship provides a way to regulate exchange and reduce the possibility of cheating, but also to avoid prematurely destroying a beneficial relationship in highly uncertain environments. Specifically, people who cultivate friendships by starting small and gradually raising the stakes ultimately create a mutually beneficial context where the best strategy for both friends is not to focus on past behaviors or to deliberate about future interactions, but rather to determine whether someone is a friend.

A SHORT NOTE ON METHODOLOGY

This book draws from work spanning a wide range of disciplines, including anthropology, economics, sociology, psychology, and biology. Each discipline has a preferred set of methods for exploring the world and testing claims about it, and so this book necessarily synthesizes a diverse set of methods, including ethnographic descriptions, behavioral experiments, hypothetical decision scenarios, self-report and observational data, longitudinal and cross-sectional designs, cross-cultural and cross-national comparisons, meta-analyses, and case studies. Each method has its own strengths and weaknesses and provides unique insights into the psychological and behavioral workings of friendship. To provide a background to these approaches, the book contains eight methods boxes, each of which

BOX 2 Cohen's D-statistic and Criteria for Reporting Studies

When describing the results of studies, I will generally use Cohen's d-statistic. The d-statistic captures the difference between two groups but also adjusts this for how different people are *within* groups. For example, a d-statistic of 0 means that the average values of two groups are identical, while a d-statistic of 2.0 indicates a very large difference between the groups. The bell curves in figure 2 show how much the distribution of heights between two groups (e.g., men and women) would overlap for a given d-statistic. For a d-statistic of 0.20, there is almost complete overlap, while the distributions are quite distinct when d increases to 2.0.

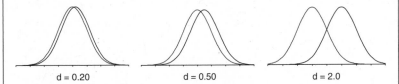

| d = 0.20 | d = 0.50 | d = 2.0 |

FIGURE 2. The overlap of two distributions given a particular d-statistic

If the d-statistic for sex differences in height were 2.0, then if we pulled a man and a woman off the street, the man would be taller than the woman 93 percent of the time. Therefore, sex tells us a great deal about who will be taller. If the d-statistic were zero, on the other hand, then the odds of predicting who was taller would be fifty-fifty—no better than chance. Any cutoff is necessarily arbitrary, but as a crude attempt to separate the wheat from the chaff, I focus on those results where the d-statistic is at a minimum 0.50. If this were the value for sex differences in height, then a man would be taller than a woman 64 percent of the time.

To calculate the d-statistic for two groups, one subtracts the first group mean from the second and then divides this difference by the pooled standard deviations for the two groups. When such data is not provided in an article, it is possible to estimate the d-statistic from other available information (Rosnow, Rosenthal, and Rubin 2000). Moreover, given a Pearson's correlation (r) for continuous data, there is also a straightforward way to calculate an equivalent d-statistic. Thus, the d-statistic provides one common metric for comparing the effect sizes reported in different studies.

describes a commonly used research method and what that method can and cannot tell us about friendship.

One important methodological challenge that arises when synthesizing so many studies is how to compare the relative importance of each study's claims. Social scientists frequently publish results that meet a criterion of "statistical significance," such as the claim that women share more personal details with their friends than do men or the assertion that partners who feel closer to each other are more likely to help each other in times of need. What such claims usually mean is that the difference between two groups (e.g., men and women) or the association between two variables (e.g., height and age) is probably not zero. They do not tell us how big such differences or associations are or to what degree they really matter. Indeed, even the most minute difference between two groups will become statistically significant when there is a sufficient number of observations in one's dataset. Therefore, statistical significance alone tells us very little about the practical importance of the difference. It is necessary but not sufficient. Of greater value for the purposes of this book is knowing how different two groups are or how much one variable can predict another. Therefore, in this book I focus attention on those published results that show moderate to large differences or associations, as described in box 2.

THE CHALLENGE OF DEFINING FRIENDSHIP

To identify friendship in the real world, and, more important, across different cultures, it will be necessary to examine how we might measure such abstract concepts as closeness, love, and trust, and how we might determine if friends are helping regardless of past behavior or future consequences. Moreover, we would hope that such a definition would be meaningful whether we applied it to humans living in the highlands of Papua New Guinea or the plains of Central Asia. In the next chapter, I tackle this issue by defining these concepts in more detail, thus providing a framework for understanding and comparing friendships as they exist in diverse cultural settings.

1 An Outline of Friendship

Den neie.
I should like to eat your intestines.

In the highlands of Papua New Guinea, a Wandeki man shouts this phrase as an old friend comes to visit. At first glance, the expression is startling, invoking gory images of cannibalism. Even in islands not far from New Guinea, the promise of eating someone's body parts is a sign of anger and aggression. However, in the presence of a Wandeki friend, the phrase means something quite the opposite—unbridled affection and happiness at seeing a companion after a long separation. The greeting continues as the two men wrap their arms around each other and the visitor responds in kind, *"A! Ene den neie!"*—"Yes, I too should like to eat your intestines."[1]

From the perspective of a European or American, the appropriate behaviors among friends in other cultures may appear bizarre and indeed unfriendly. Consider, for example, the obligation among Dogon farmers of Mali not only to attend a close friend's funeral, but also to dress in rags, overturn jars of millet porridge, and insult the generosity of the family. Among Bozo fishermen in the same region, friends demonstrate their love by making lewd comments about the genitals of one another's parents.[2] Given these diverse and frequently counterintuitive behaviors between friends, how can we hope to define the relationship? As the Wandeki example suggests, focusing exclusively on overt behavior is not enough. The intentions, feelings, and thoughts behind those behaviors also matter in differentiating a hostile act from a gesture of friendship.

Accordingly, I examine friendship as an integrated social and psychological system defined not only by behaviors, but also by underlying feelings and motivations. Figure 3 illustrates this multidimensional view of friendship. Behaviors between friends are the most visible parts of the system—both to participants in a relationship and to researchers who attempt to observe it. In the course of daily life, behaviors such as gift giving and kind acts and words are about the only observables for grasping the workings of friendship. Below this visible surface of behavior, psychological processes, such as perceptions, feelings, and motivations, play a role in steering actions among friends. For the last half-century, psychologists have tried to get a handle on this submerged system through individual self-reports and behavioral experiments. Deeper still are physiological mechanisms, including the activity of neurons, neurotransmitters, and hormones. Researchers have only recently begun to investigate activity at this physiological level by measuring the concentration of specific chemicals in the body and taking pictures of blood flow in the brain. Although each of these analytical levels is equally important, the relative weight I give to each of them is also a function of how much time and effort researchers have devoted to their study. Therefore, the fact that there are so few observations about the physiological underpinnings of friendship says more about the relatively short time period in which they have been studied than about their relative importance in the functioning of friendship.

The chapter is organized into three sections that focus on key aspects of the friendship system, as shown in figure 3. The first section focuses on behavior and describes how two important activities among self-described close friends—helping and sharing—are not observed to the same extent among strangers and acquaintances. Moreover, it describes why three standard explanations for friends' increased generosity—a norm of reciprocity, an urge to balance accounts, and a concern about the shadow of the future—do not fit empirical findings from observation and experiment. The second section focuses on psychological constructs commonly used to describe feelings among friends—including closeness, love, and trust—and how these relate to behavior. The last section brings the discussion full circle by examining how people display and communicate these internal psychological states through behavior, and why the mutual communication and recognition of these feelings and intentions is an important part of maintaining a friendship.

A caveat is due here. Comparing claims about why friends help one another often requires carefully controlled experiments that can parse the precise relationships between variables such as subjective closeness and helping (see box 3). Therefore, I devote considerable time to describing such experiments. However, while such experiments serve an important purpose in the scientific process, they also come with limitations. First, the more tightly controlled an experiment, the more artificial and oversimplified it becomes, raising questions about how much it can tell us about behavior in the real world. This is a necessary evil of experimental research, and one to

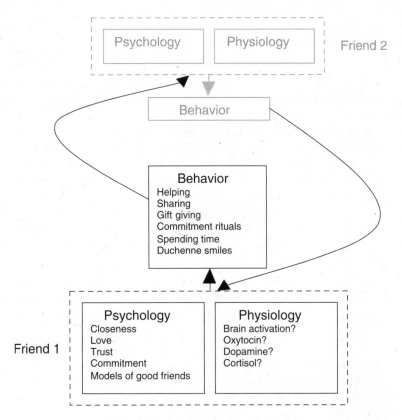

FIGURE 3. Behavioral, psychological, and physiological constructs involved in interactions among friends. One friend's behavior influences another's feelings, thoughts, and physiological processes, which in turn influence behavior.

consider when interpreting its results. Second, researchers have traditionally found it easier to conduct such experiments in the United States and Europe (and most frequently on college campuses). Therefore, there is a substantial Western (and collegiate) bias in this chapter. I will attempt to remedy this in chapter 2 by looking to descriptions of friendship, like that of the Wandeki men, found in a wider range of world cultures.

HELPING AND SHARING AMONG FRIENDS, AND WHY THREE COMMON MECHANISMS CAN'T EXPLAIN THEM

Among Trobriand sea voyagers off the coast of Papua New Guinea, friends who trade with one another also provide support and lodging for

---

BOX 3 Behavioral Experiments

Asking people how they should or would behave toward friends is an important first step in finding out what to expect of actual behavior. Due to the way that ethnographies are often written, such normative statements may be the closest we can get to understanding how friends actually act toward one another in a diverse range of societies. However, such normative statements also raise the question, do people really behave thus? Or are they simply deluding themselves?

Behavioral experiments provide one way of examining how people really act when confronted with a particular situation, and they have one major advantage over direct observation of behavior in natural settings. By permitting careful control of the social situation, such experiments can tease apart variables, such as the effects of the shadow of the future and a norm of reciprocity, by precisely defining the situation. For example, they can answer questions such as "How would someone behave if the shadow of the future vanished?" In this chapter, behavioral experiments are crucial for showing how friends do not help for the same reasons that strangers or acquaintances do.

Despite their usefulness, behavioral experiments also have several downsides. First, to facilitate control of the situation, they have often been conducted in a limited range of settings, most notably universities in the United States and Europe. Although this is changing (Henrich et al. 2006), the results I present in this chapter are exclusively from the United States. Second, the highly controlled nature of such experiments raises questions about the degree to which respondents actually behave as they would in the real world. For example, a recent study by anthropologists Michael Gurven and Jeffrey Winking among Tsimane gardeners in Bolivia suggests that how people share in experiments may stray quite dramatically from how much they share food and contribute to village feasts in the real world (Gurven and Winking 2008). Therefore, while behavioral experiments are useful for teasing apart how people behave in highly controlled situations, more work is necessary to understand how such behavior reflects the ways that people respond to situations in real-life settings.

---

one another when one of them is traveling. Among Baka Pygmy foragers and their farming neighbors in Central Africa, close friends (or *loti*) openly share and exchange material and social goods; male friends may even exchange wives. And in U.S. high schools, friends stand up for one another against verbal backstabbing, they keep important secrets, and they help talk one another through problems and conflicts.[3] These examples illustrate a recurring expectation about friendship in its diverse manifestations around the world. Whether one asks a Wandeki gardener in Papua New Guinea or a Turkana cattle herder in East Africa, the reply will be the same: Close friends should share what they have and help one another in times of need.

Despite the ubiquity of an ideal of mutual sharing and support among friends, little quantitative data exists in cross-cultural settings to determine whether this ideal reflects real acts. In the U.S., at least, carefully designed experiments comparing self-described friends with acquaintances and strangers have confirmed that these norms and expectations are consistent with actual behaviors. For example, when given the opportunity to share money or food in a laboratory setting, people are more willing to share with friends. And when asked to play a game in which partners can acquire more money by shirking than by cooperating, friends are more likely to cooperate.[4] Indeed, in many studies of support reported in everyday life, people help close friends at levels comparable to immediate kin (chapter 3).

Most economic and evolutionary analyses have proposed three mechanisms to explain such behaviors among non-kin friends: a norm of reciprocity, an urge to balance favors, and a concern about the shadow of the future.[5]

From an evolutionary or economic perspective, each of these three mechanisms is a way to ensure that an investment in helping will not be lost. The first two mechanisms deal with monitoring past behaviors, keeping accounts and withholding help from those who hold a deficit. The third relies on estimating how one's actions might influence a friend's behavior in the future. These mechanisms successfully explain much of the helping and sharing observed among strangers and acquaintances in experimental settings, an observation that has led researchers to assume that the same mechanisms are at work among self-described close friends. However, a growing body of literature in social psychology, sociology, and economics indicates that this is not the case. In the next few sections, I outline these different accounts and then describe the experimental evidence showing why they haven't been able to explain increased sharing and helping among friends.

The Norm of Reciprocity

People often try to reciprocate the good deeds (and misdeeds) of others. In the early twentieth century, French sociologist Marcel Mauss postulated that a norm of reciprocity underlay this tendency to return favors, and that it was a basic principle of gift giving in many of the world's cultures. In the 1960s, American sociologist Alvin Gouldner extended Mauss's argument by postulating that the norm of reciprocity was a human universal. According to this norm, people expect their favors to be reciprocated, and if a partner violates these expectations, then people react with fewer and smaller favors. Moreover, people have an urge to reciprocate kind acts. Verified in numerous experiments with strangers and acquaintances, the norm of reciprocity forms the basis of marketing techniques in which gifts, whether T-shirts, address labels, or Hare Krishna religious materials, are given to potential clients to encourage purchases or donations.[6] A tendency to reciprocate past behaviors is also the underlying principle of tit-for-tat strategies in the prisoner's dilemma game and is a common form of exchange in arm's-length commercial trade.

Behaviors consistent with the norm of reciprocity have been observed in numerous experimental and naturalistic settings. However, most of these experiments have focused on strangers and acquaintances. When researchers have examined how friends reciprocate, they have found something very different; friends are *less* rather than more likely to follow the norm of reciprocity than are strangers or acquaintances.[7] For example, in the 1990s investigators set up an experiment to see how giving a gift to a partner—either a stranger or a friend—influenced that partner's willingness to help later on. In the experiment's gift treatment, one student gave another student a can of soda before asking the student to buy some lottery tickets ($1 each). In the control treatment, no soda was offered before the request to buy lottery tickets. When the experiment was conducted among strangers, the students who received sodas bought nearly twice as many $1 lottery tickets from their benefactor (mean = 1.31 compared to 0.69, d = 1.03). This fits nicely with the argument that strangers follow a norm of reciprocity. However, among friends the findings were very different. The gift of a soda had no effect on the number of tickets a friend bought. Indeed, friends uniformly bought more tickets from one another, and when friends were given a soda, they agreed to buy slightly *fewer* tickets (2.63 compared to 2.94, d = -0.19).[8] This lack of short-term reciprocity among friends has been confirmed in a number of experiments, including those examining how young children share precious items such as candy, toys, and crayons and how members of non-Western societies behave in "trust games" (box 4) (five experiments, average d-statistic = 0.77; see appendix C).[9]

These experimental results fit an ideal expressed in many places around

BOX 4 Reciprocity and Formal Friendship among the Maasai

One study among Maasai herders in eastern Africa nicely demonstrates how friend-like relationships can attenuate a norm of reciprocity. Among Maasai herders, *osotua* (literally, umbilical cord) refers to a friend-like relationship based on feelings of mutual respect and responsibility. Requests for help among *osotua* should be based on genuine need, and gifts among *osotua* do not create debts or obligations. Indeed, Maasai consider it inappropriate to use words like *debt* (*sile*) and *pay* (*alak*) when discussing *osotua*. In short, *osotua* exchange is based on a norm of need rather than a norm of reciprocity.

Lee Cronk, an anthropologist who has worked with Maasai herders for more than twenty years, wanted to see if framing an exchange in terms of *osotua* might make Maasai less likely to follow a norm of reciprocity. He asked pairs of individuals to play what is called the investment game. Both players were given an endowment of 100 Kenyan shillings, and one player, the investor, was given the opportunity to send any amount of his endowment to the second player, the trustee. The trustee received *three times* what was sent (in addition to his endowment). So, if the investor sent 50 shillings, the trustee received 150 shillings to add to his original 100 shillings. Finally, the trustee could keep all of the surplus, or he could send back any amount to the investor. The sequence of moves that would make the largest, most equitable payoff for both parties would be for the investor to send all 100 shillings. This would give the trustee a total of 400 shillings (300 + 100 endowment). To be equal, he would then return 200 shillings to the investor. In the best of all possible worlds, each player could double his original endowment, but this would require following a norm of reciprocity.

Cronk framed the game in two ways. For some participants, he simply described the game. For others he referred to it with a Maasai word, calling it the *osotua* game. When individuals played the regular, unframed game, trustees who received more also returned more ($d = 0.62$). They appeared to follow a norm of reciprocity. However, when playing the "*osotua* game," trustees were less sensitive to how much the investor sent their way ($d = 0.37$). Interestingly, a later study among U.S. college students who simply had read about *osotua* showed similar results (non-*osotua* game $d = 0.75$ vs. *osotua* game $d = 0.29$). In short, simply framing a game as the *osotua* game made partners behave less reciprocally (Cronk 2007; Cronk and Wasielewski 2008).

the world that friends should eschew a norm of reciprocity, focusing rather on a friend's need. Chuuk islanders in the Pacific Ocean state that friends should not expect their favors to be returned. The main expectation among Tzeltal maize farmers in Mexico or Shluh barley farmers in Morocco is that any kind of repayment among friends is deferred. And Arapesh gardeners in Papua New Guinea regard strict accounting among friends with distaste.[10] Although a hard-nosed behavioralist may discount such expectations as mere ideals that poorly match behaviors, the previous experiments confirm that such ideals indeed reflect how friends behave toward one another (at least in tightly controlled experimental settings).

The Urge to Balance Favors

An urge to balance favors is like a norm of reciprocity but involves maintaining a balance of favors over the long term rather than responding to particular past deeds. According to one influential theory of relationships, equity theory, a partner in a relationship should be happiest when his or her inputs and outputs (however measured) balance those of the other partner. This theory predicts that people will act in ways that maintain equity in their relationship—by helping more when they have received an excess of help and helping less when the balance is perceived to be tipped in the other partner's favor.[11]

Like findings regarding the norm of reciprocity, however, several lines of evidence indicate that friends are actually *less* concerned about balance than are acquaintances and strangers. In ethnographic groups around the world, friends are expected to ignore the balance of accounts. Koryak reindeer herders in Siberia state that friends should not keep score. Thai farmers should not reckon help given by friends in their fields or at home. And Guarani maize farmers of southern Brazil should not weigh or balance their friends' help with clearing, tilling, or harvesting.[12]

These expectations are corroborated by several experiments and survey studies that have examined how individuals focus on the relative balance of inputs and outputs in their relationships. For example, in one experiment with college students, researchers compared how friends (and strangers) paid attention to the input of their partner during a cooperative task. Researchers separated a pair of friends (or strangers) into two rooms to take turns on a fifteen-minute exercise—searching a matrix of numbers for particular sequences. While one partner sifted through rows of numbers, the other waited in another room. A red light in the waiting room lit up every time the worker completed ten sequences, indicating how much he or she had contributed to the task. And behind a double-sided mirror, an experimenter recorded the number of times the waiting partner looked up to check the red light. Strangers glanced at the light much more than did friends, an observation that the researchers interpreted as a greater concern about a partner's inputs to the task. Interestingly, when the researchers changed the experiment so that the red light indicated that the worker was in need, friends glanced at the light much more than did strangers.[13]

Another experiment suggests that friends also care less about equality when splitting payoffs and more about their total group payoff. In this study, researchers asked ten- to twelve-year-old boys to choose between splitting a low group payoff (50 cents) equally or a high group payoff (90 cents) unequally. In this study, friends were more likely to agree on the higher, unequal payoff than were strangers ($d = 0.53$). In short, friends cared more about their total outcome as a pair rather than about maintaining equity.[14]

These studies indicate that friends are less concerned than acquaintances about short-term balance in their relationship or the inputs of their partner. A disregard for balance is also confirmed by studies of longer-term exchange among friends. When people are asked to rate or quantify the inputs and outputs in their close friendships, partners in balanced friendships are somewhat more satisfied with the relationship than those in unbalanced friendships (seven studies, average $d = 0.44$). However, this pales in comparison to the negative effect of imbalance in non-close relationships ($d = 1.34$). Underbenefited friends are no more angry about their situation than are overbenefited friends (two studies, average $d = 0.03$), and they are no less satisfied with their friendship (six studies, average $d = -0.10$). Moreover, inequity in either direction poorly predicts the probability of ending a friendship (d is less than 0.10).[15]

One weakness of such studies is a reliance on individuals' subjective assessments of how much they put into a relationship and how much they take out of it. Such assessments are prone to many kinds of error, both systematic and random, and so the lack of observed association may simply indicate poor measurement. Nonetheless, the findings from these experiments and observational studies present little evidence for the assertion that friends are more generous because they are concerned about balancing accounts between one another. Indeed, the limited evidence available indicates that if anything, friends care less about inequality than do strangers and acquaintances.

The Shadow of the Future

The shadow of the future is a metaphor for the influence that possible consequences of our behavior can have on how we choose to act today. A concern about future consequences guides many of the decisions we make in daily life—to save money, to be nice to our boss, or to forgo a drink at a workday lunch. It can also influence our decisions to help others. Specifically, the possibility of future interactions casts a shadow over present decisions as we estimate how our actions will influence our partners' reactions down the road. Numerous behavioral experiments have shown that increasing the likelihood of future interactions with a stranger or acquaintance also increases one's willingness to help and to cooperate. Conversely, making such behaviors anonymous, thus eliminating the possibility for future interactions and removing the shadow of the future, decreases the likelihood of sharing, helping, or cooperating.[16] The shadow of the future seems a plausible explanation for the increased levels of sharing and helping among friends, since friends expect to be together over a longer time horizon than do acquaintances or strangers. According to this view, one friend helps another because he expects that his actions will influence the other's behavior toward himself in the future.

Only recently have researchers designed experiments that can determine to what degree helping among friends depends on the perceived consequences of their actions for the friendship. An important part of such experiments is to ensure that the potential helper believes that the recipient will never know from whom the kind act came. With anonymity, a donor's decision to give should not depend on how the gift might affect a partner's future behavior toward the donor. In other words, the bright light of anonymity floods out any shadow that the future might cast on people's present decisions.

In one recent experiment that examined the effects of removing the shadow of the future, researchers made use of the popular networking site Facebook as one way to assure anonymity. Through the Facebook website, researchers asked Harvard College students to identify up to ten of their best friends, and the researchers counted only those friendships that were reciprocated. Some students were designated as "decision makers," and over the course of several days the researchers asked them to make several decisions about sharing with or helping particular partners (with real money provided by the experimenters).[17]

In some of the scenarios, the decision maker was asked to make a decision about sharing with someone whom they only knew distantly as a friend-

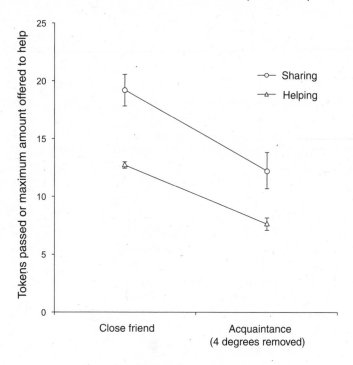

FIGURE 4. Increased sharing and helping toward close friends and acquaintances with anonymous interaction, from Leider et al. 2009. The sharing game involved a threefold increase in money when shared. Error bars are standard errors.

of-a-friend-of-a-friend-of-a-friend. In other cases, it was a close friend who would be informed about the donor's identity (the non-anonymous condition). In other cases, it was a close friend who would not find out about the donor's identity (the anonymous condition).

After studying thousands of decisions, the researchers found that students passed at least 50 percent more of their surplus to close friends than they did to distant acquaintances, even when the recipients would not discover their identities. Therefore, people gave much more to close friends than to acquaintances, even when their actions were completely anonymous. These findings suggest that much of the greater generosity among friends is not due to the so-called shadow of the future.[18] Of course, the shadow of the future had some relevance, and transfers to friends increased another 24 percent relative to strangers when the recipient would find out

BOX 5 Fear, Shame, and Reputation in Decisions to Help Friends

Close friends often help each other and share with little concern for past behaviors or future consequences. However, in any specific appeal for help, a wide range of emotions and concerns can creep in. For example, public rituals to consecrate a friendship often inject another emotion—fear of sanctions, either divine or social—into the decision-making process. Another concern is reputation. *Yïluñta* is the name Amharic speakers in Addis Ababa, Ethiopia, use to refer to a set of unspoken rules and expectations for fulfilling financial, material, and social obligations to friends and neighbors (and sometimes family). *Yïluñta* also refers to emotions, such as shame, felt when one's reputation is on the line. A person who faces a choice between appeasing a friend's request—for a loan or to spend time together when one is busy—and letting her down will be "caught" by *yïluñta* and will think, *"Ara! What if someone saw me? What if someone said something?"* By refusing a friend's request, people feel they risk losing that friend, and in turn risk jeopardizing their reputation, not to mention bringing some measure of shame to their whole family.

Urban residents of Addis Ababa say that *yïluñta* has both a good and bad side. It keeps people in line with social expectations, but it also involves a tinge of self-interest (as opposed to selfless generosity). Commentators say *yïluñta* is potentially bad because it makes some already materially impoverished people go out of their way to meet social obligations, sometimes to the point of unhealthy self-sacrifice, particularly when it comes to sharing food. Such ambivalence is an integral part of understanding friendship in Addis Ababa. Consider the following quotes from two women in Addis Ababa, who jokingly attempt to describe their culture by comparing it to the culture of Western foreigners: "We [Ethiopians] care for each other and love each other, not like you [foreigners]" and "You have become like [Ethiopian people]—forgetting your best friends!" The ambivalence revealed by these statements may be due to social life in a busy, crowded city, which makes it more difficult for people to fulfill all friends' requests for help and easier to shirk social obligations without disastrous outcomes for their reputation.

Yïluñta is a reminder that friends, even close friends, sometimes worry about the future consequences of their behavior toward one another—specifically what others will think about them. It also raises a number of interesting questions. Do qualities of a friendship, such as closeness, love, or trust, influence the degree to which *yïluñta* affects decisions to help friends? Fine-grained experiments in the United States show that close friends are less sensitive to the shadow of the future than are strangers, but does this extend to people living in other places and times? Cross-cultural research that combines deep ethnographic understanding with fine-grained observational and experimental studies will be necessary to answer these questions. (Text by Kenneth Maes, reproduced with permission.)

the decision maker's identity. However, the stronger result is that most of the greater sharing among friends could not be explained by any signaling between friends or by concerns about the consequences of the decision makers' actions on their friends' future behavior.[19]

In the experimental and observational studies described previously, researchers have found that much of the giving, helping, and sharing among friends cannot be explained in terms of three commonly proposed mechanisms for regulating exchange in evolutionary and economic frameworks: the norm of reciprocity, the urge to balance favors, and the shadow of the future. What, then, is motivating the increased helping and sharing among friends? Without evidence for the importance of past behaviors or future consequences on present behavior, the most parsimonious explanation is that friends are intrinsically motivated to share and help when the opportunity arises. Of course, this fits with how the members of many cultures explain the motivations that drive helping among friends. Ovimbundu farmers in Angola claim that they assist friends purely from motives of friendship, Iroquois farmers in the northeast U.S. report helping friends out of feelings of affection, and Greek herders state that the help of friends ideally derives from sentiments of friendship.[20] The experimental evidence described previously suggests that these claims are more than just ideals. Friends appear to disregard many of the signs that strangers and acquaintances cling to when making decisions to help one another (but see box 5).

Friends' relative insensitivity to past behaviors and future payoffs fits experimental and observational data. But it is puzzling from an evolutionary perspective that emphasizes survival of the fittest. In short, this insensitivity would make individuals vulnerable to exploitation at the hands of false friends. Before trying to resolve this puzzle, it will be necessary to understand the kinds of feelings and motivations that might underlie such unconditional behavior in more detail.

HOW WE THINK AND FEEL ABOUT FRIENDS

The words that people use in describing their feelings toward friends (e.g., warmth, closeness, love, liking, trust, and commitment) carry diffuse and imprecise meanings that can seem hopelessly unmeasurable from a scientific perspective. How does one quantify love, or compare levels of closeness or commitment? Research in the last three decades has focused on this challenge, attempting to clarify how these diffuse words reflect psychological and physiological processes underlying behavior among friends. Here I will focus on three words that people often use to express these feelings—closeness, love, and trust—and explore how the feelings they describe are linked to behavior.

Closeness

In the U.S., people often differentiate among friends based on an idiom of proximity, or closeness. Although someone may have many friends, he or she may consider only a handful to be close. Moreover, it is possible to extend this spatial metaphor in various ways, for example, by noting that one is "drifting apart" from a friend or that "we're like two peas in a pod." Other languages also use spatial proximity as a metaphor for the quality of friendships. In Russian, for example, one can call a close friend *blizkij drug* (close friend), in Nepali, *najikai saathi* (nearby friend), in Mongolian, *dotnii naiz* (inside friend), and in French, *ami proche* (close friend). In Korean, the closeness of the relationship between both friends and family members is captured by the word *cheong*, which refers to the melding of individual identities into a new collective unit and incorporates elements of unconditional acceptance, trust, and intimacy.[21]

Poets, writers, and philosophers have frequently used the concepts of spatial proximity, expanding, mingling, and overlap in attempts to define friendship. Consider the following by Edith Wharton: "There is one friend in the life of each of us who seems not a separate person, however dear and beloved, but an expansion, an interpretation, of one's self, the very meaning of one's soul."[22] Or this quotation by Michel de Montaigne: "In the friendship which I am talking about, souls are mingled and confounded in so universal a blending that they efface the seam that joins them together so that it cannot be found."[23]

Some cognitive scientists argue that such concrete metaphors are a way to make sense of otherwise indescribable concepts and feelings, in this case perceptions of oneness and merging, in relatively concrete terms.[24] Researchers have also used these concrete metaphors to quantify and compare such feelings in a number of ways. For example, some researchers have simply asked people to rate how close they feel to a partner or to what extent they would use "we" rather than "I" to describe the merging in their relationship. One of the most successful techniques has not involved words at all. Rather, it asks participants to select from a set of increasingly overlapping circles labeled "self" and "other" the pair of circles that best describes their closeness with a partner (figure 5). These different approaches to measuring closeness and togetherness provide surprisingly similar answers, suggesting that they are tapping into a coherent set of feelings that an individual can have toward another.[25]

Please circle the picture below which best describes your relationship

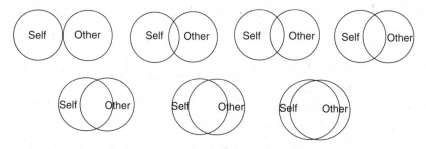

FIGURE 5. Inclusion of other in self scale, from A. Aron, E. N. Aron, and D. Smollan, "Inclusion of Other in the Self Scale and the Structure of Interpersonal Closeness," *Journal of Personality and Social Psychology* 63, no. 4 (1992): 596–612. Copyright © 1992 by The American Psychological Association. Reproduced with permission.

The simple measure for perceived closeness based on interlocking circles has surprisingly strong correlations with both feelings and behaviors associated with friendship. For example, manipulating the closeness felt between strangers can influence how much time one person would devote to helping another, in one experiment increasing that time by 45 percent (two experiments, average d = 0.80).[26] While strangers generally take greater credit for successes on joint tasks and accept less responsibility for failures (three experiments, average d = 0.74), this is not true for partners who feel close to one another (average d = 0.10).[27] And people who feel closer to a partner express a greater willingness to help (six experiments, average d = 1.17) and to sacrifice for a partner's gain (three experiments, average d = 4.07).[28]

These findings have led some researchers to propose that becoming close to someone involves the inclusion of other in self, a mental process by which we treat another's resources and identities as our own.[29] According to this view, we find helping a close friend rewarding because our brain perceives our actions, in some ways, as helping ourselves. We feel distress at losing close partners just as we would feel about losing other aspects of our identity, such as a talent or a prized possession. The closer we are with others, the more our moods depend on their successes and failures.[30] We are even more likely to confuse ourselves with close partners when recalling past events and making judgments about ourselves.

To assess the degree to which people confuse actions they have taken for themselves with actions taken toward close friends, a team of psychologists designed an ingenious experiment involving the recall of past rating decisions. They asked students to rate themselves on a number of personality traits, such as *serious, kind,* and *happy.* Then, using completely different sets of traits, the students also rated their best friend, closest parent, and a familiar media personality. After these ratings were complete, students were then asked to recall which person they had rated for each trait. When students were asked about traits for which they had originally rated themselves, they were much more likely to mistakenly list these as traits for which they had rated their best friend—compared to mistakenly recalling a celebrity (d = 2.17) and even their parent (d = 1.66).[31] Therefore, even in a task as a simple as remembering who one rated on a personality test, people can confuse themselves with close friends.

A number of scholars argue that this psychological merging of close friends with self can be traced to brain organization, as the neural structures dealing with close others share elements with neural structures for the self.[32] To assess this claim, two studies recently replicated the trait-recall experiment while a magnetic resonance imager (MRI) scanned the blood flow in people's brains. In both studies, when people made judgments about a close friend, blood flow increased in the lower (i.e., ventral) part of the medial prefrontal cortex (MPFC), a brain region a few inches back from the middle of one's brow. Notably, this is also a region that is activated when people must make judgments about themselves and when they experience positive emotions. On the other hand, when people made judgments about non-close others, a distinct region lying directly above the ventral MPFC showed activity. Coincidentally, this region also activates in response to negative emotions.[33]

Another study that examined how people's brains react to their friends' names also suggests that confusion of other with self occurs at the level of neural activity. In the study, sixteen women heard their own name, the name of a close other (best friend or sister), and common names that did not refer to someone whom they knew. Using an MRI, researchers took pictures of the blood flow in each woman's brain under each of these conditions. They then assessed the overall similarity of the brain images when hearing one's own name and the name of a close other (adjusting for the effect of simply hearing a common name). This admittedly coarse-grained measure of similarity in brain activity was highly correlated with the degree to which women described their relationship with the friend or sister as close (d = 2.20).[34]

While pointing in exciting directions, these studies are also preliminary and correlative, and further work will be necessary to examine how feelings of closeness actually reflect physiological processes and how they

"I can't remember which one of us is me."

FIGURE 6. "I can't remember which one of us is me." © The New Yorker Collection 2001 Robert Weber from cartoonbank.com. All Rights Reserved.

motivate behavior toward friends. At the same time, as some of the first studies examining the physiological correlates of feelings involved in friendship—in this case closeness—these results are particularly exciting. They suggest, for example, that specific brain networks involved in thinking about oneself may also play a role in thinking about close friends (but not non-close others). And friend- and self-related stimuli (i.e., names) may generate similar activations in the brain. Hopefully, future work will refine these results and provide a clearer understanding of the neural mechanisms underlying the subjective concept of closeness.

Poets have idealized the mutual concern and psychological merging among friends as oneness, but such merging of self with others is rarely complete. Particularly in situations in which there is explicit competition between close individuals, people will often have a first preference for their own interests. For example, several experiments over the last several decades have shown that, when two friends are completing a task they both see as important and there is explicit comparison between their performances, one is less happy when the friend does better. Indeed, such competition may lead one friend to sabotage another's performance.[35] Nonetheless, even the partial confusion of our own identity and interests with those of others is a fascinating phenomenon, and according to experimental and observational studies in the United States and Europe, one that influences behaviors, such as unconditional helping, in friendships and other close relationships. I will explore how closeness plays a role in helping in more detail in chapter 3.

Love

Closeness is a spatial metaphor used to make sense of feelings for a partner. Love does not rely on a metaphor; rather, it is a direct description of feelings that we can have toward others. As an attempt to locate these feelings and make them concrete, people often think of love as residing in a particular place in the body. For people in the United States, for example, the heart is the seat of love, while Trobriand islanders imagine love in the intestines, and members of some West African cultures feel it in the nose. In Nepal, people place love in something translated as the heart-mind (*man*), an organ that is not fixed within the body and can travel with thoughts, remembering, and longing (box 6).[36]

Vast bodies of literature, from the Bible to modern novels, have tried to disentangle the feelings and motivations associated with love. Consider the attempt by Ahdaf Soueif, an Arab novelist, to distinguish the ways that one can express love in Arabic. "'Hubb' is love, 'ishq' is love that entwines two people together, 'shaghaf' is love that nests in the chambers of the heart, 'hayam' is love that wanders the earth, 'teeh' is love in which you lose yourself, 'walah' is love that carries sorrow within it, 'sababah' is love that exudes from your pores, 'hawa' is love that shares its name with 'air' and with 'falling,' 'gharm' is love that is willing to pay the price."[37]

Only in recent decades have scientists tackled the problem of defining the multifaceted feeling of love and developing testable theories about the psychological and biological processes involved in the experience and expression of this complex emotion.[38] The most straightforward way to study love is to ask large numbers of people about their own feelings and motivations toward particular partners. By examining what kinds of statements co-occur more frequently (e.g., "I care about her" and "I like to help her") than others (e.g., "I care about her" and "I hate not being around her"), researchers can then identify clusters of co-occurring statements that may reflect a single, unique dimension of feeling and motivation. Psychologists have productively used this approach to identify key dimensions along which people can feel love toward friends, family, and romantic partners.[39] Common dimensions generally coincide with one or more of the following motivations:

- Benevolence. Wanting to improve a partner's welfare and avoiding behaviors that may hurt a partner.

- Affiliation-positive. Wanting to be with a partner for the rewarding feelings of security, relaxation, and happiness felt in the partner's presence.

- Affiliation-separation distress. Wanting to be with a partner to avoid feelings of discontent, anxiety, or even depression when away from that partner (mediated by fears of social exclusion).

- Sexual attraction. Wanting to have sex or intense physical contact with that partner.

- Poaching avoidance. Wanting to exclude potential competition for that partner (mediated by jealousy and anxiety).

- Wanting reciprocation. Wanting to have these feelings reciprocated by the partner.

Any or all of these motivations may surface in a specific relationship, and whether we define our partner as a lover, close friend, boyfriend, or girlfriend can depend on the exact mix of feelings and motivations that we have toward that partner. Many of these motivations also occur together in predictable ways.[40] For example, passionate love in a romantic relationship may involve high levels of motivation in all of these dimensions (except possibly benevolence), while companionate love that emerges over time in a romantic relationship is characterized by higher levels of benevolence and lower levels of separation distress and, perhaps, sexual attraction.[41] Meanwhile, partners in a close friendship may have high levels of benevolence, a drive to affiliate, and a desire for reciprocation but have no motivation to engage in sexual behavior. Interestingly, only the first and last of the motivational dimensions listed above requires that a person care about what a partner thinks or feels. Therefore, many love-related motivations can be quite selfish.

In recent years, researchers have looked to the physiology of humans and other animals to understand what biological processes might underlie human feelings of love and attachment. Most notably, a growing body of research has revealed how two structurally similar chemicals with common evolutionary origins, oxytocin and vasopressin, promote a behavioral correlate of love (i.e., social bonding) in a range of mammals. Oxytocin is best known for its role in a key maternal activity—producing milk for one's infant. When an infant sucks on its mother's nipple, a signal is sent

---

BOX 6 Love: A Universal Language?

While growing evidence indicates that many of the feelings associated with love arise in a wide range of human cultures (Fisher 2004; Jankowiak and Fischer 1992), there is also striking cultural variability in the appropriate expression of these feelings. For example, in the U.S. there is a relatively strict threshold for what behaviors signal sexual desire or behavior. Consider a letter between Victorian friends, "I hope for you so much, and feel so eager for you . . . that the expectation once more to see your face again, makes me feel hot and feverish." To modern U.S. readers, the letter's reference to hot and feverish feelings likely prime notions of sexual activity, when there is little evidence that it occurred (Coontz 2000, p. 66). This clash of interpretations is strikingly illustrated by the recent and cruel beating of a Somali high school student in Boston who was targeted for holding hands with her friends. For the assailants this was a sure sign of a taboo, a lesbian relationship (Latour 2000). This apparently low bar for identifying sexual relationships might explain the tendency to interpret a wide range of ritualized friendships in other societies as involving a sexual component when there is no evidence for this—and often evidence to the contrary (Brain 1976; Fehr 1996).

This low bar also influences how Americans describe their own friendships. A high school student might say "We're just friends" to refer to a relationship, implying a lack of sexual attraction. Moreover, in the United States there is an aversion to using the term *love* for the feelings that one has toward a friend, focusing instead on the use of more weakly valenced words such as *liking* (Wilkins and Gareis 2006). Contrast this tendency in the United States with the case of Nzema men in southern Ghana who can "fall in love," share their beds, and even marry but never have sex (Brain 1976, p. 55).

Thus, despite underlying similarities in the psychological processes involved in love, there is a great deal of cultural variability in what kinds of gestures and actions are appropriate or inappropriate for expressing particular kinds of love.

---

by spinal nerves to the brain that spurs the release of oxytocin into the bloodstream. Within minutes, the chemical reaches the mammary glands and causes milk to be let down into a collecting chamber from where it can be sucked out by the infant. In short, oxytocin is a key messenger in the primary form of food sharing in mammals.

In addition to this key mammalian function, oxytocin also mediates other kinds of maternal behaviors and facilitates partner recognition in several mammalian species. The observed pattern of oxytocin receptors in particular brain regions among monogamous species suggests that the chemical also plays a role in making long-term mates psychologically pleasing, an effect mediated by the chemical dopamine. However, most of these studies have focused on a small range of non-human animals, particularly mouse-like rodents called voles, and it is important to resist the temptation to generalize these results directly to humans.[42]

Recently, a small body of work has begun to show where this analogy is consistent with the human case. Brain imaging studies have confirmed the direct involvement of oxytocin- and vasopressin-sensitive brain regions during perception of loved ones (i.e., between mother and child and romantic partners). As in other monogamous mammals, these regions are also part of dopamine-producing reward networks, confirming that loved ones really do bring us pleasure. And recently, a novel experiment with British club-goers suggested that the generalized feelings of love often reported by Ecstasy (MDMA) users may also be mediated by oxytocin. The researchers collected blood samples from the clubbers both before and after their evening out. They found that those clubbers who took Ecstasy had substantially increased plasma levels of both oxytocin and vasopressin. More work will need to be done to determine if the increase in oxytocin also correlates with the feelings of love and closeness often reported when using Ecstasy.[43]

Many of the effects observed in humans and other mammals are likely related to the way that oxytocin modulates neural circuitry underlying fear and affiliation in humans, essentially reducing activity of the amygdala and decoupling the amygdala from other brain regions so as to modulate fear responses.[44] Moreover, its activity in reward centers indicates that oxytocin (and its chemical cousin vasopressin) also plays a role in the motivation to approach particular partners. Its action as a moderator of both social inhibition and approach may make oxytocin a particularly important chemical messenger in the cultivation and maintenance of social relationships. However, a number of questions remain in the human case. To what degree does oxytocin play a role in our experienced feelings of love? And how might the system extend beyond mating and kin relationships to mediate social behavior with friends and even strangers?

A great deal of research in the last four decades has focused on the many feelings and motivations associated with love, and recent physiological studies have examined the neural and biological correlates of love's experience. However, unlike studies of closeness, very little is known about how such feelings or chemical pathways are tied to the helping and sharing observed among friends. Psychological scales attempting to measure friend-like love have found strong correlations with measures of subjective closeness, suggesting that the feelings underlying these two concepts are somehow connected.[45] But this is only a starting point, leaving much fertile ground left to cover to really understand how different kinds of love are involved in feelings of closeness and behaviors among friends.

Trust

Trust is generally not defined as a feeling but rather as an expectation. You can trust that the postman will deliver your mail on Monday, which is an expectation of the postman's predictability. You can trust your bank to keep your money safe, and you can trust a close friend to keep a secret or to help when help is needed. The trust most often associated with friendship is not like the expectation of a postman's predictability or a bank's fiduciary responsibility. Rather, it is the trust that one's friend will act in one's interest, if one takes a social risk with potentially negative consequences for oneself, such as divulging a secret or helping at great cost. In the case of a damaging secret, the friend won't pass it on. In the case of costly aid, the friend won't run off with the money. Unlike one's relationship to a bank, which is governed by laws and backed by federal deposit insurance, there is little formal recourse if one's trust is violated by a friend. Rather, trusting a friend involves feeling that the friend is intrinsically motivated to be trustworthy.

In the context of friendship, trust is not only an expectation but also a behavior—a willingness to take social risks with a partner, giving away sensitive information and secrets being a salient example in our own culture. It is often difficult for individuals to articulate clearly why they decide to take social risks. Indeed, a recent set of experiments has shown that what people say about their trust for a particular partner may tell us very little about their willingness to actually take social risks with that partner. These experiments involved manipulating the blood levels of the chemical oxytocin, which has been shown to be crucial for maternal behavior, milk letdown, and pair-bonding and caregiving in many mammals and has receptors in brain regions involved with the formation of social attachments.[46] Based on prior evidence that oxytocin inhibits social defenses and promotes affiliation, a team of researchers bet that artificially increasing levels of oxytocin in a person's blood would also increase that person's willingness to take social risks with an unknown partner.

In these two studies, researchers asked participants to play an economic game called the investment game (box 4).[47] The game involved two players, one called the investor and the other the trustee. The investor can send some money to the trustee with the possibility of a significant return on this investment. However, the trustee decides how much of this return to share with the investor. He could potentially keep it all. So to make such an investment worthwhile, the investor must also expect that the trustee will pass back some of the return on this investment.

In the first investment experiment, half of the investors were given a nasal spray of oxytocin. The other half were given a placebo. Compared to those in the placebo group, investors who received oxytocin were twice as likely to pass all of their money to an anonymous trustee ($d = 0.56$). In the standard interpretation of the game, a simple intranasal injection of oxytocin made people more willing to disregard a social risk and put complete trust in a stranger. Moreover, the oxytocin-treated and placebo-treated investors did not differ in the trust they verbally reported for their partner, suggesting that oxytocin did not change their stated beliefs but rather worked at a more subconscious level in guiding behavior.[48]

A more recent experiment suggests that oxytocin also makes people less sensitive to incoming information about social risks. In this experiment, investors (some given oxytocin and some a placebo) played six trust games with anonymous individuals and then were informed that trustees had abused their trust about half the time. After hearing about the regular abuse of trust, investors played six more games. Not surprisingly, those individuals who had received placebo nasal sprays reduced their transfers after hearing news of the betrayals. However, investors who received oxytocin made no significant change to their transfers, suggesting that they were insensitive to the news about social risk (difference in investments by oxytocin treated and non-treated investors after feedback, $d = 0.43$).

The potential implications of these two experiments are exciting, specifically that oxytocin may play an important part in regulating our willingness to take social risks with a partner and also our willingness to trust blindly without regard for past behaviors. As I described earlier, some researchers have proposed that oxytocin influences trusting behaviors by inhibiting defenses and making people more willing to take a social risk.[49] These trust experiments also suggest that what people say about trust may not have much bearing on behaviors. Specifically, oxytocin-treated investors verbally reported no greater trust for their partners than did placebo-treated investors. However, they took much greater social risks with the same partners. In such situations, psychological and physiological processes that occur under the radar of everyday awareness may be much more important than what people say. More research is needed to replicate these tantalizing findings and to understand why some self-reported concepts, such as psychological closeness or attributions of friendship, have shown measurable behavioral consequences, whereas others, such as trust, have not.

HOW WE COMMUNICATE FRIENDSHIP

So far, I have described some of the important feelings, thoughts, and motivations that one can have regarding friends. But in what ways do friends communicate their feelings and intentions toward one another? Recall the Wandeki man at the beginning of this chapter who expressed feelings of friendship by stating that he would like to eat his partner's intestines, or high school students who communicate their friendship through what appears to be malicious teasing or practical jokes.

One might imagine numerous ways to signal feelings of friendship. Consider saying with a stone-cold face to a stranger on the street, "I am your friend and want to help you whenever you are in need." You have said something, but you have told the stranger very little about your actual intent or feelings toward him. Philosophers of friendship are adamant on this point. Friendship is best expressed through actions, not verbal promises. The use of language may be an important part of expressing friendship, but it is through the quantity and quality of communication (i.e., through letters, hanging out, retelling familiar stories, and even body posture and facial expressions), rather than through simple promises, that friendship is most credibly expressed. Here, I describe three common ways of credibly communicating one's intentions to a relational partner. These include exclusive behavior, honest expressions of emotion, and accepting vulnerability in one's partner.

Exclusive Behavior

Exclusive behaviors that cannot be scaled up to a large number of potential partners are particularly hard-to-fake signs of interest in a friend. Spending time with a friend, writing personalized letters, sharing meals together, and talking on the phone are all activities that take time and effort and thus limit one's ability to do the same with other people.[50] These actions honestly signal a unique interest and investment of time in a specific person, precisely because it would be very difficult to extend the same behaviors to a large number of partners. Contrast these gestures with writing a mass email, addressing a crowd, or sending computer-printed

Christmas cards to thousands of people, as is often done by political candidates seeking to prime the sympathies of potential donors. Because these modes of communication can be broadcast to a large number of people, they are unable to signal exclusive interest in a particular partner.[51]

Giving gifts is a common signal of exclusivity, and one that has been studied extensively by anthropologists and sociologists.[52] In the U.S., good gifts are uniquely valuable or symbolic to the receiver, incur a large (but not necessarily monetary) cost to the giver, require time to find, and cannot be exchanged.[53] Money is the antithesis of a good gift—it has no extrinsically greater value to any one person than to another, does not require a long search, and can be easily exchanged. Indeed, the ways that people modify money in attempts to make it an appropriate gift provide a window into the symbolic importance of exclusivity in gift giving. For example, when people do give money as a gift, they frequently make it distinctive, by giving a particularly crisp bill personally acquired from the bank, by enclosing it in a personalized card or decorative wrapper, or by writing notes or decorations on a check. Retailers have also tried to find ways to make money an appropriate gift by creating gift cards with sufficient variety in design, color, wrapping, and purchasable item (e.g., bookstore, clothing outlet, home improvement store) to make consumers feel that it satisfies at least some of the criteria for a good gift.[54]

These attempts to transform money into an acceptable gift illustrate the importance placed on gifts being personalized and exclusive. In chapter 6, I discuss in more detail why exclusive behaviors, and gift giving in particular, are likely an essential element in the long-term regulation of friendships.

Honest Expressions of Emotion

People also signal their feelings toward partners with hard-to-fake expressions. One of the best examples of such expressions is a Duchenne smile, which, unlike many ways of smiling, appears to be an involuntary response to emotion. A Duchenne smile uses both a muscle near the lips (zygomatic major) *and* the muscle that surrounds the eyes (orbicularis oculi) that, when contracted, creates crow's feet at the corners of the eyes. Contrast this with a polite smile, which is easier to fake but that uses only the mouth muscles. As the Duchenne smile's namesake claimed, inertia of the orbicularis oculi while smiling "unmasks a false friend."[55]

Other expressions of affection that people use as cues of liking by a friend involve body postures and gestures, such as leaning toward a partner and moving one's arms, hands, and fingers in various ways.[56]

Moreover, behavioral studies have shown that friends are more likely to talk, smile, and laugh together than are non-friends.[57] However, it is not clear how such behaviors are especially hard-to-fake signals of affection, raising important questions about why such behaviors might be credible signals of one's feelings for and intentions toward a partner.

Accepting Vulnerability

Putting oneself in a vulnerable position vis-à-vis a partner is also a credible sign of commitment. For example, by sharing potentially damaging personal secrets, we demonstrate that we trust a friend and that we expect the relationship will last a long time. Thibaut and Kelley refer to this as giving a partner control over our fate by conceding some control over our social standing, identity, and reputation. For example, in the Cretan mountain village of Hatzi, the secrets that women share exclusively with their best friends could potentially threaten their reputation in the village if divulged. This places them in a very vulnerable position vis-à-vis their closest friends.[58] Another context for accepting vulnerability is by reacting to teasing, practical joking, and playful insults in an equally playful manner. This shows that one believes a friend's intentions are good, and when teasing is reciprocal it communicates that each partner is certain about the other's intentions.[59] One of the most honest ways of accepting vulnerability is getting close enough to a person to touch him or her and, perhaps even more important, to permit oneself to be touched. Not surprisingly, these are powerful signals of affection and trust and also appear to reinforce such feelings.[60]

The most direct sign of commitment is to forgo immediate self-interest and to act to maintain a relationship or fulfill a partner's wishes. Such acts include staying in a relationship given a more attractive alternative, making a sacrifice to help a partner, sharing, and avoiding any indication that one is making conscious calculations about the benefits and costs of helping. Helping and sharing with little concern for the balance of favors is a potentially costly signal of one's commitment to a partner and a friendship. This shows that one cares about the partner, and not simply about what one is getting out of the relationship. It also shows that one thinks one's partner feels the same way. Otherwise, such unconditional support would be socially risky, exposing one to potential exploitation.

In this chapter, I have sketched a rough outline of friendship based on behaviors and feelings commonly observed among self-described friends. Friends share with and help one another when needed, and this cannot be explained in terms of a norm of reciprocity, an attempt to maintain balance, or a concern about the shadow of the future. The psychological system underwriting these behaviors is frequently described in laymen's terms, with concepts such as closeness, love, and trust; and behavioral and neurobiological studies are beginning to identify how these feelings are related to behavior. Finally, an important part of friendship is making sure that each partner knows that the feeling is mutual, which requires signals that are difficult to fake on the part of either partner.

This outline is a starting point but raises more questions than it answers (see box 3). Do the behaviors and feelings gleaned from studies in the United States and Europe arise in other cultural settings? And how do these differ from the behaviors and feelings observed in other kinds of relationships—such as those with kin or romantic partners? Moreover, if friends pay so little attention to past behaviors and the future consequences of their behaviors, how do they avoid exploitation?

The next chapter will tackle the problem of cross-cultural similarities and differences in friendship. At a superficial level, the examples from the beginning of this chapter suggest that there are great differences in how friends behave toward one another across cultures. However, examining underlying processes can often reveal deeper similarities. The greeting of a Wandeki man to his friend is only superficially aggressive. At a deeper level, it expresses closeness, much like the potentially gruesome statement one might hear in English, "I want your heart to be next to mine." On the other side of the world, in a U.S. high school, teasing and practical jokes are acceptable because friends know that the intentions behind such behaviors and statements are well meaning (as long as they don't go too far). Each of these cases represents a more general principle: that in the everyday working of friendships, it is not the particular behaviors that matter most, but rather the meanings they convey and the intentions presumed to underlie them.

2 Friendships across Cultures

In 1848, the Russian priest Andrej Argentov moved with his wife and a maidservant to start a small church on Siberia's arctic coast—to win the souls of nomadic Chukchee reindeer herders who lived in the region. During one of Argentov's first forays into Chukchee territory, a herder named Ata'to asked him to establish a ritual friendship. At that time, Argentov probably did not know that a key obligation of such ritual bonds involved common sexual access to spouses in the union. Nevertheless, after a short time the Chukchee friend paid a visit to Argentov's new home and claimed his right in the relationship. When telling this story, the Russian ethnographer Bogoraz-Tan likely understates the emotional intensity of the ensuing confrontation: "[Ata'to] had some companions with him, and so refusal was of no avail. At the critical moment, however, the maid-servant consented to take the place of the mistress."[1] Soon after this incident, Argentov, his wife, and his maidservant left their lonely Siberian church, never to return.

Based on this skeletal account, it is difficult to accept that the ill-fated relationship between Ata'to and Argentov was what we would call a friendship. Group marriage between two couples was a common Chukchee practice. It hinged on feelings of affection and goodwill among all four participants, transcending the simple "sharing" of spouses. However, in this particular case, it is unclear whether Ata'to was exploiting a cultural norm for his own aims, targeting helpless Russian missionaries in a foreign land, or truly hoping that a long-term relationship might develop from the consummation of the group marriage. Whatever the reason for Ata'to's demand, Argentov's unfortunate attempt at making friends clearly illustrates the difficulty of exporting our own expectations of friendship to other cultures and societies.

In numerous cases similar to the Chukchee one, anthropologists have shown that the ways of living deemed "natural" by many Westerners—monogamous marriages, households composed largely of nuclear (rather than extended) families, and the necessity of love before marriage—would appear strange in other human communities.[2] And upon closer inspection, relationships described as friendships in other societies, such as Chukchee group marriages, often violate certain Western social norms. Notably, people in other places and times have sanctified friendships through public rituals, inherited them from parents, and entered them based on the dictates of family and community elders, practices that would be difficult to reconcile with Western notions of friendship.[3] Therefore, we cannot simply rely on observations about friendship in our society for a general understanding of it as it may have existed in other human groups in the present and past.

Argentov's arctic misadventure is a striking reminder of how friendships can differ across cultures, but there may also be behaviors, feelings, and expectations among friends that are highly consistent, perhaps even universal, in human communities around the world. The only way to find out what is unique to specific societies and what recurs across human friendships is to review in a systematic manner the ways people have cultivated friendships in other places and times. This chapter therefore surveys the reports of anthropologists, missionaries, and travelers who have found themselves living in a wide range of cultures. Sifting through these reports, I ask several questions about the universality of friendship. Is something like friendship even practiced in most societies? And if so, are expectations of friends in Western contexts, such as honesty, secret sharing, emotional support, and material assistance, equally important in these communities? Moreover, does the behavioral and psychological system described in chapter 1—with unconditional help and sharing, feelings of love and closeness, and communication of these feelings through gift giving and other signals—operate in other cultural contexts?

EXPLORING THE ETHNOGRAPHIC RECORD

Over the last two centuries, anthropologists and other travelers have documented thousands of cultural groups around the world, leaving behind a wealth of material for answering the questions posed above. Particularly relevant to these goals are the Human Relations Area Files (HRAF) at Yale University, which houses a database of the world's best-documented cultures, sorted and filed by geographic location and cultural characteristics.

Assembling the HRAF has been no small feat. It has involved numerous researchers carefully reading thousands of pages of text and indexing each paragraph for specific topics of interest. The coding system, known as the Outline of World Cultures, includes over 700 topics that capture major elements of life in human communities, such as diet, kinship, geography, and warfare. In this way, the HRAF has cataloged and indexed more than 350,000 pages, making these 400 cultural groups among the best documented in the world.[4]

With its 400 coded societies, the HRAF is the ideal place to start a systematic cross-cultural exploration of friendship. However, there are several questions to address before using it to study friendship in diverse cultures. First, how do we know when we actually have found a case of friendship in the ethnographic record? Second, when does an absence of friendship in the record mean that friendship doesn't exist? And third, which cultures among the vast number in the record should we target in an analysis?

Knowing What to Look For

On the problem of defining *virtue*, Plato's Meno writes: "And how will you enquire, Socrates, into that which you do not know? What will you put forth as the subject of enquiry? And if you find what you want, how will you ever know that this is the thing which you did not know?"[5] The same problem posed by Meno in defining virtue confronts anyone hoping to understand friendship, especially as it occurs in diverse cultural settings. The crux of the problem is this: To define friendship-like relationships, we must identify a number of concrete examples upon which to base the definition. However, without a prior definition, how can we identify examples for deriving it? This conundrum intensifies when we travel to lands with other languages, where we are not even certain what the appropriate word for *friend* is. Known as Meno's paradox, this catch-22 of conceptual development is one of several theoretical and methodological challenges that confront any study of a Western concept in comparative perspective.

In describing informal modes of interaction, ethnographers have frequently deployed terms such as *friend* and *friendship* as conceptual primitives (i.e., you know it when you see it). Until this point, I have followed their lead, but how do we really know a friendship when we see it? What characteristics of relationships in other cultures have led researchers to infer that the best translation is *friendship*? Here, I begin simply by analyzing observers' descriptions of friendships in diverse cultures. While observers' use of *friendship* is a good place to start, ethnographers have also used other idioms to describe relationships that upon closer inspection are quite similar to friendship as we know it. For example, ethnographers frequently described such relationships as "artificial" or "fictive" forms of kinship, because members of the culture themselves referred to their friends with kin terms, such as *brother, sister, father,* or *aunt.* Identifying common elements within these multiple descriptions of friendship, fictive kinship, and artificial kinship provides a starting point for determining whether certain obligations, expectations, or behaviors reliably occur together in a wide range of cultures.[6]

What Does It Mean When We Don't Find Friendship?

In many cases, even reading a book-length description of a culture will reveal no mention of something like friendship. The most obvious explanation for such an absence is that there is nothing like friendship to be found. However, there are several reasons why a single description of a culture might fail to make any mention of friendship, despite the presence of an equivalent relationship in that culture. These reasons include lack of documentation, lack of searching, and the theoretical bias of an ethnographer.

The cross-cultural record is a mixed bag in terms of the quality of data that are available for any given culture. Some groups have been thoroughly documented by generations of anthropologists, with thousands of pages devoted to their lifeways, customs, and social organization. For most groups, however, we know little more than the culture's name. Even within the HRAF, which only describes cultures with substantial documentation, the total pages devoted to a particular culture numbers anywhere from five to twenty thousand! This variability in the depth of description devoted to each culture introduces problems when studying friendship. For example, does the quantity of text devoted to friendship in a society reflect the importance of the relationship or rather that ethnographers have spilled more ink describing that society? Conversely, the absence of any mention of friendship in a culture may mean nothing more than incomplete documentation.

A second, more pernicious problem is what one anthropologist has referred to as "anthropology's love affair with kinship," and the related neglect of friendship in describing small-scale societies.[7] Early in the twentieth century, anthropologists discovered that it was a straightforward task to collect detailed genealogies of people in small villages and to construct relatively simple, tree-like maps charting the relationships between every individual in a society (figure 7).[8] In many societies, such kin relations ostensibly guided behaviors, obligations, and feelings between two individuals. Thus, the concept of kinship gave anthropologists what appeared to be a clear-cut system for analyzing behavior that depended on nothing more than who gave birth to whom, who was married to whom, how old one was, and who was male or female.[9] A prominent anthropologist's description of Tale society in Ghana provides one striking example of the view that kinship, in this case patrilineal ties, provides the underlying "skeleton" for society: "The whole of Tale society is built up round the lineage system. It is the skeleton of their social structure, the bony framework which shapes their body politic; it guides their economic life and moulds their religious ideas and values."[10]

Whether for these or other reasons, documenting the kinship system of a culture became routine in anthropological fieldwork in the twentieth century. Meticulous genealogical charts were a key part of an ethnographer's toolkit for describing a society. Indeed, in one of the discipline's most influential field manuals, *Notes and Queries on Anthropology*, two-thirds of the section on social structure is devoted to family, kinship, and lineage, with no mention of extrakin relationships.[11] This methodological focus certainly reinforced perceptions of kinship's importance in daily life. For example, according to the genealogical method endorsed in *Notes and Queries*, the ethnographer first identified a list of terms for people related by either birth or marriage (e.g., *father, mother, cousin, aunt, father-in-law*) and then identified through questioning and observation the traditional behaviors toward such individuals. Such a method is very good at discerning how kin relationships affect behavior, but it ignores other kinds of relationships that might exist between people.[12]

In a discipline so focused on kinship, there has often been little mention of extrakin relationships, such as friendship.[13] For example, in the HRAF, for every twelve paragraphs devoted to kin ties, there is only one discussing friendship. Despite this theoretical bias, I will show that with sufficient diligence one finds numerous descriptions of friendship hidden away in the ethnographic record, and these descriptions reveal patterns of social organization much more complex than those based solely on marriage and biological descent.

Which Cultures Should Be Considered?

Finally, we arrive at the problem of deciding which cultures to include in a cross-cultural survey. Ideally, such a survey would focus on cultures that have been extensively documented by numerous observers, so that one

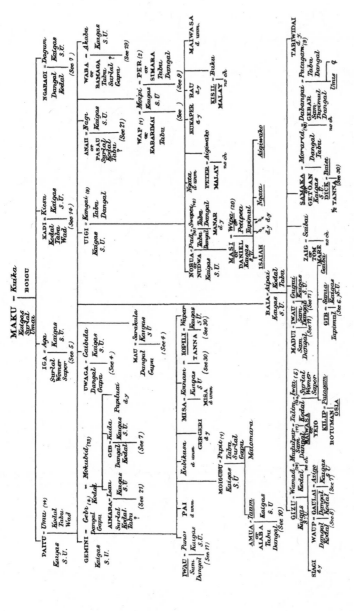

FIGURE 7. The bony skeleton of society. Genealogy of Maku from Murray Island, Torres Straits, W. H. R. Rivers 1900. Collected and drawn by W. H. R. Rivers during the expedition to Torres Straits between Australia and Papua New Guinea. The patriarch of the genealogy, Maku, is at top center. Following standard kinship notation, marriage is indicated by equals signs and progeny by vertical lines downward (Rivers 1900).

FIGURE 8. The sixty societies in the Probability Sample File

can see when observations are corroborated (and contradicted). The sample should capture a wide range of human societies, with equal representation of different parts of the world and different ways of making a living. Moreover, it should counterbalance the focus in chapter 1 on large, industrial nations by turning to small-scale societies—farming villages, bands of hunters and gatherers, and nomadic herders. Therefore, I will focus on a sub-sample of four hundred HRAF cultures specifically chosen by anthropologists in the 1960s to ensure representative coverage of peasant and small-scale societies around the world. This sub-sample of the HRAF is known as the Probability Sample File (PSF).[14] In this way, I have whittled the sample from thousands of cultures in the ethnographic record to four hundred of the best-documented and finally to a subset of sixty that fairly represent a world sample of small-scale societies. The sixty cultures in the PSF span the six inhabited continents and include a wide range of subsistence styles, ranging among nomadic herders, hunter-gatherers, shifting horticulturalists, and settled farmers. Although this sample is useful for counting purposes and will be the basis of the quantitative analyses in this chapter, many of the most thorough descriptions of friendship lie outside of it. Therefore, I will also draw from beyond the sample of sixty societies in providing illustrative examples (figure 8).

ARE THERE SOCIETIES WITHOUT FRIENDSHIP?

For many Westerners, Atevi society may seem like an alien culture. It lacks any concept of friendship, liking and loving are unknown feelings, and in contrast to the paucity of terms for *trust* in the Atevi language, there are fourteen words for *betrayal*. In her many descriptions of the Atevi, C. J. Cherryh describes how the absence of emotional bonds affects not only the nature of Atevi society but also how the Atevi interact with neighboring cultures. With their striking divergence from our expectations of human relationships, the Atevi are prime candidates for understanding how human communities might exist without friendship. The one problem is that the Atevi are not human. They are creations of science fiction. C. J. Cherryh, over the course of ten novels, has developed a fascinating thought experiment as to what life would be like in an alien society without friendship, love, or trust.[15]

Cherryh's exercise in anthropological science fiction shows that it is possible to imagine a society without friendship. But are there real-world instances of human communities where something like friendship does not exist? Such examples would definitely challenge assumptions that friendship is a universal aspect of human life.

A systematic review of the HRAF database reveals numerous descriptions of friendship, with more or less detailed accounts of the feelings, obligations, and purposes of the relationship. Among Lepcha farmers in eastern Nepal, special friends, or *ingzong*, look after one another in emergencies, help with farm work, and promise to care for one another's children in the event of one's death. Special friends who live far away fulfill a special purpose—offering a place to stay during expeditions for much-needed goods, such as cloth and salt. Lepcha say that the god Komsithing first thought of the institution when he was drunk. He subsequently forged *ingzong* with all the foreigners who possessed things he did not have himself: with Sikkimese for their oxen, Tibetans for rugs, Bhutanese for fine cloth, and Indians for their copper vessels. While *ingzong* provide clear material benefits, the best *ingzong* also love one another.[16]

Among women living in the small Cretan mountain village of Hatzi, close friends provide a safe haven for speaking openly, sharing their problems, and disclosing secret affairs. Worries lessen when confiding in friends. Friends make mundane tasks, such as baking, fun and pleasurable. Most important, a woman's friends understand and appreciate her in a male-dominated environment, where how she feels and how she must behave often come into conflict. Anthropologists have often claimed that in male-dominated peasant villages, such as Hatzi, married women are sequestered in homes, surrounded by kin, and not in a position to make or have friends. However, far more women in Hatzi feel closer to friends than to their family members.[17]

Ju/'hoansi foragers in Namibia and Botswana cultivate long-term *hxaro* ties with partners based on gift giving and mutual assistance over vast geographic expanses. *Hxaro* partners regularly exchange gifts—beads, arrows, tools, clothing—as tokens of their affection and as signs that they "hold each other in their hearts." These far-reaching networks allow people to live and forage in other areas when food or water in their home territory has disappeared, or when social relations have become strained at home. If they were to infringe on the land of others without such ties, the owners would try to expel them. People draw on *hxaro* ties when searching for a spouse, and they share meat with *hxaro* partners when they have a large kill. While Ju/'hoansi cultivate many *hxaro* from genetic kin, they have equal numbers of *hxaro* non-relatives as well.[18]

· These are only a few examples among many, including labor-sharing relationships among farmers and gardeners, "stock associates" who take care of each other's cattle, and war companions who protect one's downed body during battle. Indeed, ethnographers have described friendship-like relationships on all inhabited continents and among groups at all levels of social complexity.

Despite the prevalence of friendship in the ethnographic record, there are also a number of societies where the HRAF database contains no mention of friendship or friendship-like relationships. As discussed earlier, the absence of friendship might be due to many factors, and a careful analysis suggests that whether friendship is reported in a culture depends heavily on how many ethnographies one reads. For example, if one focuses on cultures described by the equivalent of only one or two academic books (fewer than 500 text pages), friendship is reported in less than a third of all groups. However, as the number of pages devoted to a culture in the HRAF increases, there is also a clear increase in the probability of finding a reference to friendship. With the equivalent of three or four academic books per culture (500–1,000 pages), one finds a description of friendship in about half of all cultures. And not until one examines cultures described by more than 2,000 pages (more than eight hefty books!) is one certain to find references to friendship in a culture (see appendix A). Contrast this with the fact that kinship figures prominently in ethnographic descriptions of nearly all cultures.

These results suggest that finding friendship in the ethnographic record is a direct function of how hard one looks. And by extending this search to literature outside of the HRAF database (and by reading the HRAF material more carefully), friendship is described in all sixty cultures in the PSF.

Despite the high prevalence of something like friendship in documented human societies, there are a small number of cases in which ethnographers, like the novelist C. J. Cherryh, have stated explicitly that a society lacks any term or concept closely analogous to friendship. This is the case for seven of the four hundred cultures described in the HRAF.

In three of these societies, the ethnographer's statement is clearly contradicted by later observers. Consider the confident statement by one anthropologist, Roy Barton, about Ifugao farmers of the Philippines: "Friendship is not a tie. In our sense of the word it does not exist. Persons between whom there is no tie may become chummy, but that, the Ifugao's nearest approach to friendship, carries no such implications of loyalty, mutual admonition and assistance as our word connotes." The statement clearly dismisses the possibility of finding something like friendship among the Ifugao. However, in other descriptions of the Ifugao (some by Barton), there are references to friends who invite one another to feasts, who mediate disputes, who serve as hosts in foreign lands, and who avenge the murders of their friends. Moreover, there are several descriptions of individuals feigning friendship to obtain material goods or to lure an individual into an ambush. Without a concept of friendship, it would be impossible for the Ifugao to feign it. For these reasons, Barton's initial dismissal of Ifugao friendship is hard to accept without serious qualifications.[19]

When we remove this and similar cases where claims are convincingly contradicted by other authors or later descriptions, we are left with five cases of highly collective cultures in which exclusive friendships are strongly discouraged as a threat to the broader community. In such cases, friendships are viewed as seeds for disruptive factions and conflicting loyalties. For example, among Kogi farmers of Colombia, the priests, or *Máma*, discouraged friendships, describing how a Kogi man should behave:

> He has no friends, because friendship does not exist among the Kogi. There are men of his family and men of other families; there are people of the same *Túxe* [neighborhood] and there are people of other *Túxe*, but friends do not exist. The *Máma* himself asks that the men not be friends with one another since this would lead to adultery. If two men are together frequently, they will become enamored of each other's wives and all would end in a fight. . . . There is no reason to call a man a friend if perhaps tomorrow you would have to call him enemy!

The same reality prevails on the other side of the world among the Muria Gond in northern India. Here, exclusive friendships run counter to the strongly communal nature of Muria society. In this society, children

---

BOX 7 Greedy Institutions

In one episode of the cartoon sitcom *The Simpsons*, the miserly C. Montgomery Burns, owner of Springfield's nuclear plant, shares his years of business wisdom with Lisa Simpson's third grade class. According to Burns, the key to a successful business is to root out three demons: family, religion, and friendship.

Burns's view of a successful business is what sociologists have referred to as a "greedy institution," or an institution that demands total commitment from its members, making them abandon all other kinds of loyalty—in Burns's case, to family, religion, and friends themselves. Religious cults are a prime example of greedy institutions, often requiring members to cut off ties with outside family members and friends (Coser 1974). Another particularly good example of a greedy institution is the Christian monastery of medieval Europe. According to Christian theology, Christ's command to love one's neighbor implied universal good deeds to all, whether friend or foe. Friendships threatened a general love for all human beings by focusing on exclusive partners and taking one's eye away from the way of God. As one influential monastic thinker, Basil, asserted: "The brothers should maintain mutual love for each other, but not so that two or three at one time conspire to form cliques, for this is not charity but sedition and division. It is a sign of the evil behavior of those who join together in such a way." This concern also framed regulations, such as how a novice training to be a Dominican in Rome was supervised even on walks so that he and his companions did not always choose the same partners and avoided getting to know each other too well (McGuire 1988, p. 30).

---

and youth grow up together in common dormitories, and from childhood they are trained to do things together, to move, to work, to mourn, and to rejoice as a group. The "exclusive, passionate devotion" of friendship between two people threatens this social unity.[20]

In each of these cases, communities, neighborhoods, or families are what sociologists call "greedy institutions" because they demand undivided commitment from their members (box 7). In such communities, friendships pose a threat (at least a perceived one) by potentially dividing one's loyalties and are therefore discouraged by the heads of these communities.[21] Of interest here is that friendships must be explicitly discouraged, suggesting that they might arise naturally if no prohibitions, exhortations, or penalties were in place. Moreover, even in these cases, further reading suggests that friendships, in one form or another, do indeed arise, though perhaps more rarely and secretly than in most societies.

The overarching conclusion from this initial foray into the ethnographic record is that despite documentary and theoretical biases against finding friendship, a suitably thorough search regularly reveals descriptions of something like friendship in a wide range of human societies. Indeed, it is very difficult to find societies like C. J. Cherryh's fictional Atevi where something like friendship is definitively nonexistent.

HOW UNIVERSAL ARE WESTERN NOTIONS OF FRIENDSHIP?

That ethnographers have described friendship or something like it in a broad range of cultures tells us very little about how such relationships might work in each of these cultural settings and how each instance differs from our own cultural stereotypes. The Chukchee practice of group marriage described at the beginning of the chapter is striking, but it might exaggerate the true extent to which friendships differ across cultures. To understand how widely friendships vary and how closely they approximate Western notions requires a more systematic review of friendship-like relationships in the cross-cultural literature.

Here, I begin by examining behaviors, feelings, and expectations that are often assumed to define friendship in the United States and Western Europe. Using these qualities as a basis for comparison, I show how they map (or fail to map) onto social relations in non-Western settings. Although many Western notions of friendship will travel poorly to other societies, there will also be several that are common in a wide range of societies, permitting a working definition of friendship that is meaningful in many human contexts.

Westerners routinely describe a common list of qualities, rules, and behaviors that are expected of friends. Friends (at least good ones) like one another, enjoy one another's company, and maintain mutual goodwill. They help one another in times of need, listen to one another's problems, make sacrifices, and provide emotional support when necessary. They share confidences and can be trusted not to divulge important secrets. Their relationship is personal and private, and it does not answer to a higher authority. They engage in constructive conflict management, and they try to resolve differences among themselves. Friends should not go to court to resolve a dispute. Ideally, friends do not care what they get out of the relationship but value the friendship for its own sake. They are honest

TABLE 1. Characteristics of friendships in sixty societies

Characteristic	No. Described	No. Disconfirmed	Percent Described	Percent Disconfirmed[a]	Ideal in U.S.?
Behaviors					
Mutual aid	56	0	93	0	Y
Gift giving	36	0	60	0	Y
Ritual initiation[b]	24	1	40	4	N
Self-disclosure	20	2	33	10	Y
Informality	17	0	28	0	Y
Frequent socializing	11	6	18	55	Y
Touching[b]	11	0	18	0	N
Feelings					
Positive affect	47	0	78	0	Y
Jealousy	0	0	0	0	N
Accounting					
Tit-for-tat[c]	7	5	12	71	N
Need	32	0	53	0	Y
Formation and Maintenance					
Equality	18	14	30	78	Y
Voluntariness	11	7	18	64	Y
Privateness	3	2	5	66	Y

[a] Percent of those described.
[b] Included because of mention in other cultures.
[c] Included because of theoretical arguments claiming friendship is based on tit-for-tat.

with one another, feel free to express themselves to one another, but do not pass judgment. Finally, unlike partners in kin or work relations, one can choose one's friends.[22]

A review of the cross-cultural material shows important consistencies and discrepancies between the U.S. ideal of friendship and friendships in the sixty PSF cultures (table 1). In the next sections, I describe in more detail how the salience of these behaviors, feelings, and qualities varies across cultures, why some appear more regularly, and what this means for our understanding of friendship as a human relationship. I start with eleven aspects of friendship commonly described in the U.S. and discuss the others later.

Behaviors

In the ethnographic record, numerous descriptions emphasize how behaviors among friends signal goodwill and also underlie assistance in real emergencies. Here, I examine how five behaviors commonly described among friends in the U.S. (i.e., mutual aid, gift giving, self-disclosure, informality, and frequent socializing) are described in the cross-cultural record.

Mutual Aid.   Studies of friendship in Western contexts consistently find that people see friends as people they can trust to offer help, to care for them, to look out for their interests, and to make sacrifices in times of need.[23] Of all the qualities of friendship considered here, mutual aid is also the most frequently cited behavior in cross-cultural descriptions of friendship (described in 93 percent of societies and never disconfirmed).

The words and phrases used to describe friendship in other languages frequently reflect the kinds of mutual aid and support expected of friends. For example, the Western Tibetan word for friend translates to "happiness-grief-identical" (*skyidug-chik-pa*), and the Northwest Coast Salish term means "giving partner." Mortlockese speakers living on small atolls and islands in Micronesia describe friends as *pwiipwin le sopwone wa,* "my sibling from the same canoe." The phrase stems from a story of two men who found themselves adrift at sea for many days, sharing dwindling food supplies and lifting each other's spirits until rescued. Based on this experience of mutual aid in adversity, the men swore to treat each other like brothers—to care for each other, to cooperate, to be of one mind, and to share land and resources.[24]

Among other kinds of help, friends in different cultures loan one another money, help with tilling and harvesting fields, assist one another in disputes, give honest advice, keep secrets, act as go-betweens during courtship, offer lodging and hospitality in foreign lands, support a friend's reputation, and help bear the costs of weddings and funerals. In many cases, partners are expected to provide support in a number of these domains, and in chapter 7, I will discuss the kinds of aid most often provided in more detail. The enormous variety of help friends provide one another might lead us to dismiss a unified definition of what friends do for one another. However, as discussed in the section on accounting (later in

---

BOX 8 "Round in Circles"

Among the far-flung Massim islands of Papua New Guinea, seafaring voyagers would travel hundreds of miles annually across dangerous seas to engage in a remarkable system of gift giving and trade. First described in detail by the anthropologist Bronislaw Malinowski (1922), the Kula ring involved reciprocal visits between residents of different island communities, often speaking mutually unintelligible languages, in which partners exchanged two kinds of local jewelry—shell armbands in one direction and shell necklaces in the other. Any Kula trader had partners in at least two directions, and so a shell armband received from one partner would soon be passed to a different partner, but always in the same direction. The global effect of this practice was the perpetual (and puzzling) circulation of Kula valuables over vast ocean expanses, with necklaces moving clockwise and armbands moving counterclockwise.

At the local level, the exchange of these valuables occurred between Kula partners who maintained strong mutual obligations of hospitality, protection, and assistance. A good Kula relationship was "like a marriage," and Kula partnerships often lasted a lifetime. Ideally, Kula valuables were given generously and accepted unconditionally. Bartering among partners was bad form (Damon 1980).

Since the exchange of Kula valuables appeared to serve no economic purpose, Malinowski used the system as an example of trade that was motivated by cultural rather than economic concerns. However, at a deeper level, Kula gift giving played an important role in the Massim economy. By

(continued)

---

this chapter), what unites these cases is not the kind of help that is given but rather *how* such help is given.

Gift Giving.   Gift giving is probably a universal element of life in human communities and is a hallmark of friendship in Western society.[25] The HRAF texts also suggest that giving gifts is an important signal of friendship in a wide range of human societies (in 60 percent of all societies, no disconfirmations). Viewed by a cynical outsider, the transfer of gifts may look like the mere movement of non-usable trifles among people. For example, in one of the most thorough descriptions of gift giving in a small-scale society, anthropologist Bronislaw Malinowski showed that shell jewelry literally traveled in circles among island traders off the coast of New

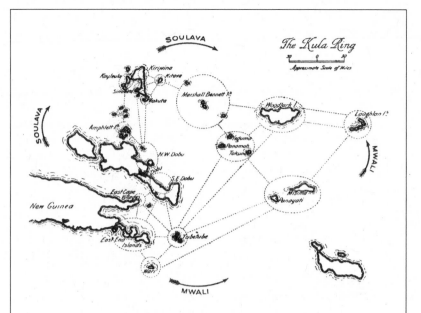

FIGURE 9. Map of Kula ring, from Malinowski 1922

providing a way to forge and reinforce far-flung trust relationships, the system permitted safe travel and commerce between otherwise hostile island communities. Indeed, under the ritual veneer of Kula exchange and the protection of their Kula partners, traders could enter diverse communities to trade commodities such as pigs, yams, pottery, obsidian, and betel nuts (Leach and Leach 1983; Landa 1994).

Guinea (box 8). Most Westerners would consider this movement of gifts, called the Kula ring, a program in "re-gifting" taken to extremes. However, in most cases, these and other gifts are not valued for their direct economic uses. Rather, gifts are bestowed as an expression of the giver's feelings and goodwill for his or her partner.[26] Indeed, I will show in chapter 8 that gift giving plays an important role in the ability to avoid exploitative partners and maintain the evolutionary viability of friendship-like relationships.

Self-disclosure.   Many Americans think that sharing secrets and talking about personal issues is an essential part of friendship. Indeed, some psychological theories (developed in the U.S.) suggest that divulging sensitive information plays a crucial role in fostering commitment in a relationship by both signaling trust in a partner and providing a partner with one form of "collateral."[27] In a number of other societies, sharing secrets is also an important part of friendship. Among inmates at Bomana Prison in Papua New Guinea, close friends confide the details of their court cases, reveal the true story of their crimes, and discuss the prison rules they break. And among Kanuri farmers and fishers in Nigeria, youth often have one particularly special friend, called an *ashirmanze* ("secrets man"), with whom they can share their most intimate thoughts, hopes, and fears.[28]

However, despite these examples, self-disclosure is only occasionally mentioned as a behavior among friends in the cross-cultural record (mentioned in 33 percent of societies, and disconfirmed in 10 percent of those cases). Indeed, in many societies, talking and sharing secrets is much less important among friends than is providing material support. (These societies are discussed in more detail in chapter 7.)[29] More important, there are some cases where self-disclosure is actually avoided. In her study of farmers in central France, for example, Deborah Reed-Danahy describes how individuals avoid disclosing personal information to friends (even close friends), as such sensitive information might be deployed in the case of interfamily disputes. In western Congo-Brazzaville, Kunyi cultivators avoid disclosing too much to friends because such disclosures may lead one to lose *ngolo* (power) and to become more vulnerable to witchcraft. In such cases, one may have complete faith in a friend for the time being but must still be careful of the consequences if the friendship were to break up in the future.[30] Therefore, while it may be difficult in the United States to imagine close friendships without self-disclosure, in other parts of the world, discussing personal matters and secrets is not a common element of friendships. This provokes the question, why is self-disclosure so common among friends in the United States?—a topic explored in more detail in chapter 7.

Informality.   In the United States, friendships are relationships in which one can violate rules of formal conduct, and this is also a common expectation in many other cultures. In societies where one must behave in very formal and respectful ways toward family members and elders in the community, friendship is a context in which one can let one's guard down (mentioned in 28 percent of societies and no disconfirmations). As one Muria Gond boy of India stated, "When you are friends you can fart together."

Depending on the society, friends can use nicknames for one another rather than formal terms of address based on kinship or status. In China, for example, where there are strong norms for using honorifics with elder acquaintances, among close friends (and only among close friends), one can use the term *younger brother* regardless of whether the friend is older or younger. Close friends can tease and mock one another and make rude comments about one another's appearance in ways that would normally lead to quarrels and fights among acquaintances. For example, among Shona farmers in Zimbabwe, where charges of sorcery are serious matters, friends can call one another witches. Friends can also leave formality at the door and enter into open discussion in ways that are not possible within other types of relationships. Among Thai peasants, who generally avoid unpleasant interactions, friends are expected to give frank and critical advice, a behavior not even observed among kin.[31]

In several cultures, however, certain kinds of friendship permit only limited degrees of informality. Among Andaman Islanders, for example, one class of friends was reportedly prohibited from talking but was nonetheless responsible for helping one another if one member of the group was in trouble. In some cultures, rituals were a turning point where relations between friends transformed from ribald joking and informality to more tempered relations of formal respect and exchange.[32] Nonetheless, friendship more often than not serves as an outlet where one can engage in otherwise unacceptable behavior.

Frequent Socializing. In the U.S., good friends frequently spend time together.[33] However, frequent socialization was described as an important element of friendship in only a small proportion of the sixty PSF societies. Moreover, in an equivalent number of societies, friends rarely interacted on a regular basis (mentioned in 18 percent of societies and disconfirmed in 54 percent of those). Infrequent meeting generally stems from the practice of cultivating friendships with people precisely because they live at a distance. For example, Turkana cattle herders in East Africa purposely forge bond friendships (*lopai*) with other herders who live outside their yearly orbit of migration. Such friends provide links to far-off regions where the Turkana might travel and trade, where they can beg for cattle in times of need, and where they can gain entry to grazeable land in bad dry seasons.[34]

Similar kinds of far-off friends in other societies rarely meet but serve as valued assets who can offer safe haven during journeys, back one in disputes in foreign lands, pool the risk of cattle herding, and provide information and rare goods. Therefore, the advantages of having friends in far-off places can outweigh the immediate benefits of frequent socializing. Even in the U.S., the emphasis on frequent socializing may be more important in defining casual friends, whereas closer friends can withstand and even benefit from decreased contact.[35]

Feelings

Although behaviors are publicly observable, feelings and thoughts are more difficult to extract from the ethnographic record. For example, anthropologist Robbins Burling spent three years living among Garo villagers of northeastern India, who generally avoided publicly expressing their feelings. For this reason, it is not surprising that despite hundreds of references to friends and friendship in his detailed 377-page ethnography, the only time that Burling mentioned affection or love in the context of friendship was when describing his own feelings of affection and admiration toward his friends in the field.[36]

Despite the obvious difficulties in observing and recording psychological states, feelings and thoughts are important in understanding friendship. For example, among U.S. high school students there are no clearly marked public rituals or events by which individuals become friends or affirm their friendship. In public life, students often interact agreeably with others whom they might not consider close friends and might even think of as "jerks" or "assholes." Conversely, friends frequently are permitted (even expected) to treat each other in a manner that to an outside observer would seem decidedly unfriendly. Therefore, friendly behavior alone, even if we have a deep understanding of local norms and behavior, can tell us very little about the status of the relationship between two people.[37]

More generally, friendship is not just based on behaviors but also represents the *potential* for certain kinds of behaviors in specific contexts (e.g., help when one needs it). Such behavioral potential is mediated in large part by psychological states, such as intentions and motivations toward an individual.[38] The next section explores two kinds of feelings observed among friends in the U.S. (positive affect and jealousy) and how they map to cross-cultural descriptions of friendship.

Positive Affect.   Despite the challenges in observing feeling states, ethnographers have frequently noted feelings of warmth, affection, intimacy, and closeness among friends in diverse societies (78 percent confirmed, no disconfirmations). Such expressions range from simple statements, such as "I like her" or "I fancy him," to deeper expressions and feelings of caring, love, and affection. Among Santal farmers of India, friends' hearts are "bound fast," and among Bena farmers of Tanzania, friends are "overcome with emotion" for one another. Ashaninka gardeners of Peru even compose songs for their friends to express their mutual fondness.[39]

Unlike the distinction often made in the United States between liking (which can occur among friends) and loving (which can occur among romantic partners and kin), in some societies it is common to use the same term to describe one's love for close friends, family, and romantic partners. Among Pashtun herders in northern Pakistan, the love felt for close friends is extolled in poetry and compared with the feelings of lovers. And the word for the love of spouses among Lepcha farmers in eastern Nepal is the same one used for trading friends.[40]

Jealousy.   When asked to list the characteristics of a good friendship, people don't generally say friends should be jealous. Indeed, jealousy among friends was not mentioned once in the PSF. However, ethnographers have described possessiveness toward friends in a number of other societies, from Tarascan farmers in Mexico to Pashtun herders in Pakistan and Copper Inuit youth in Canada. In such cases, people jealously guard their friends, worried that others may replace them. And such feelings can often destroy a relationship. Consider the case of Juan and Pedro, two youth in a Guatemalan village who were the best of *camaradas*. Juan and Pedro hung out day and night, attended festivals together, and protected each other during fights with other youth. They had been doing so for several years. However, during one of the village's fiestas, Juan began to dance with another boy who had been courting Juan as his *camarada*. Pedro was drunk and angry. He hit Juan in the face, and the two started a scuffle that led to a fight. After that night, the two never spoke directly again.[41] The neglect of jealousy as a feeling among friends may be due in part to the fact that it is not an ideal but rather an unwelcome by-product of relational regulation (see chapter 6). Nevertheless, it deserves further attention as an element of regulation that may be ignored in many academic or popular treatments of friendship precisely because it is not ideal.

Accounting

Ziyou said, as recorded in the Confucius analects, "Too close accounting in service to a lord will lead to disgrace, while in relating to friends and companions it will lead to estrangement."[42] In evolutionary accounts of cooperation, help among non-kin is frequently modeled with some form of accounting over outcomes, whether one's behavior is contingent on a partner's past behaviors, the balance of accounts, or the prospects of gain in the future. Such accounting practices have their place in social exchange, but they also neglect many other ways that people can make decisions in their relationships. People care about more than just behaviors and outcomes. They act according to what they think the *other* person is thinking. People reciprocate gestures they perceive as favors more than those perceived as bribes. They are more likely to forgive an unintentional slight than an intentional one. And people bear substantial costs to help people they perceive as close friends, often with little thought to past behaviors or the consequences of their actions.[43] In these cases, people do not simply make decisions based on past behaviors and the costs and benefits of current options. They also rely on judgments about a partner's intentions and make knee-jerk, stimulus-response decisions based on how close they feel to a partner.

Need.   In the U.S., for example, friends don't follow a norm of reciprocity by which one gives only because one has received, they don't regularly try to balance accounts, and they don't appear to give based on the future consequences of their actions (chapter 1). Rather, as C. S. Lewis writes, "The mark of perfect friendship is not that help will be given when the pinch comes . . . but that having been given, it makes no difference at all."[44]

Reports of friendship in the ethnographic record are consistent with this view. In the vast majority of cultures, help was contingent on the need of a partner, and most of the rare mentions of something like tit-for-tat contingency specifically stated that it was not acceptable among friends. In Thailand, for example, tit-for-tat exchange characterized a substandard kind of friends, "play friends," who "feast with you when you can feast them, betray you if it is their profit to do so, and certainly disappear if you become a non-entity."[45] Indeed, how friends keep (or don't keep) accounts can tell us more about their relationship than do the kinds of help provided. Among Tzeltal villagers in Central America, for example, ritual friends and distant acquaintances provide help in the same array of tasks, including plowing, cultivating, harvesting, firing pottery, and house building. What distinguishes ritual friends from acquaintances is the timing and expectation of return. Acquaintances expect either immediate compensation or an explicitly stated deadline for repayment. Ritual friends do not.[46]

The distinction between close friends' loose accounting and the kinds of balanced reciprocity observed among acquaintances maps onto a distinction made nearly a half-century ago by anthropologist Marshall Sahlins in describing how people help one another in small-scale societies. Sahlins proposed that people share goods and render assistance according to three general principles. First, people may engage in generalized reciprocity, with no clear expectation of return. Sahlins proposed that this is most common among close kin and members of the same household, where goods are shared and help is rendered with little calculation. Sahlins's definition of close kinship was quite loose, focusing more on residential proximity, subjective closeness, and a history of mutual support, and therefore could encompass friendship as well. As people interact with more-distant individuals—whether defined geographically, subjectively, or genealogically—they engage in balanced reciprocity, which rests on what I described in chapter 1 as a norm of reciprocity. Although people may deny that calculations are made, they expect some form of return over the short or medium term. Sahlins's final form of reciprocity, negative reciprocity, may not seem like reciprocity at all. It involves haggling and bartering to get the best out of a deal and usually involves the immediate exchange of goods and services. It requires very little trust and is the usual mode of exchange among strangers. Close friends clearly follow a form of generalized reciprocity and are expected to avoid the latter two kinds of accounting.[47]

Although friends avoid strictly balanced reciprocity in providing aid, there is one domain where it can be expected—gift giving and other methods of signaling one's feelings about the relationship. For example, Ju/'hoansi hunter-gatherers in the Kalahari desert cultivate friendships, called *hxaro*, based on the regular exchange of gifts and mutual aid. Gifts among *hxaro* partners must be reciprocated adequately and within a reasonable period of time. At the same time, the provision of aid within *hxaro* relationships may become quite imbalanced as partners meet one another's indefinite and uncertain needs.[48] Therefore imbalances in expressing friendship through gifts may be inappropriate, whereas imbalances in assistance because of varying fortunes may be completely acceptable. This distinction between balanced reciprocity in gift giving and need-based criteria for helping a friend will play an important role in evolutionary models of friendship outlined in chapter 8.

Formation and Maintenance

In the U.S., friendships are frequently cited as being maintained among social equals, voluntarily entered into (and ended), and privately regulated by the two friends. Here, I examine how these emerge as requirements of friendship in other cultural settings in the ethnographic record.

Equality. Social equality between partners is often claimed to be an essential element of friendships. And friends in Western contexts often try to downplay the personal attributes and styles of interaction that make them appear unequal.[49] There are some cases in the cross-cultural record where friendship is also limited to social equals, often because wealthy individuals will not entertain friendship with poorer peers—because they see no benefit from it, because they are concerned that the person would then have the right to all their property, or because they fear the social stigma that may come from befriending people of a lower social status.[50] Nonetheless, the cross-cultural record describes numerous cases of friendships among individuals who are clearly not social equals. For example, Claudia Rezende portrays the friendships that arise among Brazilian mistresses and their maids, a situation in which authority is impossible to erase. Based on this, she argues that friendship constitutes an "idiom of togetherness and affinity" rather than one of equality. Similarly, friendships often develop between masters and their apprentices, ritual parents and their children, employers and employees, and chiefs and their subjects. These are all instances of what scholars have referred to as "lop-sided friendships."[51] In several societies, especially in South Asia, friendships explicitly unite individuals across castes, where many of the authority relations and asymmetric behavior norms continue to apply. And in societies where all relationships are viewed in terms of hierarchy, it is nearly impossible for a friendship to avoid being placed in a hierarchical idiom, such as equivalents to older brother, maternal uncle, or father.[52]

Voluntariness.   A common assumption in the U.S. is that people are free to choose their friends and to end those friendships.[53] However, there are many examples in the cross-cultural record where individuals' choices in beginning and ending friendships are highly limited. The limits on choice include: outside parties such as families playing a role in arranging or permitting friendships, the passing of friendships through family lines for many generations (box 9), prior participation in pre-arranged ceremonies or relationships, or serious public or supernatural sanctions associated with ending the relationship. In Cameroon, among Bangwa farmers, for example, friends (as well as a future spouse) are assigned to a baby at birth, a practice that changed only with the coming of colonial influence. Among Iroquois farmers in the U.S. and Canada, a healer may "assign" a gravely ill person to a new friend as part of the healing process. Meanwhile, blood brotherhood, a common form of friendship in some societies, can be as binding as marriage, requiring elaborate ceremonies to annul the union and to avoid the supernatural sanctions that might result from the break.

BOX 9 When Friendships Bind

In many societies, mutual aid relations are of sufficient importance that they are inherited through generations, arranged by parents, and maintained for life. For example, twentieth-century Serbian peasant families would pass godparent relationships (often formalized friendships) to successive generations. In his extensive study of Serbian godparenthood, the demographer Eugene Hammel documented that in 90 percent of male baptisms, the godparent of the father is related to the godparent of the son. The transmission of godparenthood relationships is sometimes so reliable that there are several anecdotes of individuals determining that they are biologically related not by directly tracing their kin ties, but rather because they maintain godparenthood relationships with the same lineage (Filipovic 1982; Hammel 1968). Similar patterns of inheritance exist in societies around the world, including the Kwoma of Papua New Guinea (Whiting 1941), the Shluh of Morocco (Hatt 1974), the Tonga of Zambia (Colson 1971), and the Rundi of Burundi (Czekanowski 1924).

Although these examples violate Western notions that one freely chooses (and rejects) the objects of one's affection, they also suggest that something like friendship can exist without the kind of choice expected of friendship in the U.S. Interestingly, the sociologist Graham Allan challenges the notion that friendship is largely a matter of choice even in the West. Although the limits on choice are not as extreme as those found among groups such as the Bangwa, Allan argues that broad social factors such as class, race, and gender as well as the settings in which we participate heavily constrain our set of potential friends. For example, our friends are limited by the activities that we pursue, and the centers of these activities, whether church, school, work, or play, can segregate us by such factors as race, class, and gender in sometimes subtle ways.[54]

Privateness. In the U.S., the two members of a friendship are most responsible for negotiation and enforcement of rules in the relationship. For example, an obligation to a friend does not necessarily imply an obligation to his family (although it can), friendships are not recognized under the law (as are marital and kin ties), and friends are supposed to sort out problems among themselves.[55] It is true that in some cases, friendships may be extremely private. For example, Gurage farmers of Ethiopia kept their bond-friendships in the utmost secrecy and as "guarded as personal amulets and secret ritual formulae."[56] And bond friends among Tikopia gardeners on this Pacific island would seal their friendship privately by chewing betel leaves together, and their friendship could not be enforced by any higher authority.

There are also numerous examples in the cross-cultural record in which outside parties play a role in defining and enforcing relationship norms.[57] In many other cases, friendships are publicly recognized, with community members interested in their maintenance and exerting external influence through shaming, gossip, and even heavy sanctions against ending a relationship. Often, one's poor treatment of a friend becomes public knowledge, making it much more difficult to find a new partner. As the French doctor Victor Tixier described of the Dhegiha-Osage of North America, if someone who has broken a friendship "wants to secure a new companion, he has difficulty finding a brave who will be attached to him. New friends demand that horses be given to them as a token of future faithfulness."[58]

A Working Definition of Friendship

The eleven notions of friendship common in the U.S. differ a great deal in the degree to which they are described across cultures (figure 10). In many places, for example, self-disclosure is not only unnecessary but also avoided among friends. Friends may see one another only rarely precisely because individuals strategically cultivate friends in far-off places. Individuals may have very little choice in either beginning or ending their friendships. And there are many ways that friendships can be unequal.

Based on this initial reading of the cross-cultural data, only four of the eleven elements—mutual aid based on need, positive affect, and gift giving between partners—represent recurring parts of friendship-like relationships as they arise in numerous groups.[59] This finding fits well with experimental work in Western settings (chapter 1) that demonstrates a change in feelings toward close friends that motivates one's willingness to sacrifice for them. Based on the cross-cultural consistency and functional coherence of these four elements of friendship, I propose a working definition of friendship-like relationships that represents their minimal regulatory requirements: A friendship-like relationship is a social relationship in which partners provide support according to their abilities in times of need, and in which this behavior is motivated in part by positive affect between partners. A common way of signaling this positive affect is to give gifts on a regular basis.

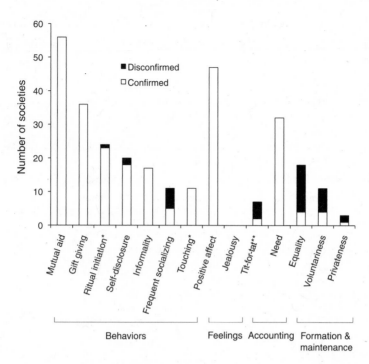

FIGURE 10.   Characteristics of friendships in sixty societies. *Included because of mention in other cultures. **Included because of theoretical arguments claiming friendship is based on tit-for-tat.

Here, I use the term *friendship-like relationship* to differentiate it from the colloquial usage of *friendship* in the U.S., which carries culturally specific expectations and connotations. The term also leaves open the possibility that individuals may participate in such a relationship even if they do not refer to it regularly as a friendship. For example, partners who participate in other formally defined relationships, such as sibling ties, marriages, or boss-employee relationships may enact mutual aid mediated by positive affect (and thus belong to a friendship-like relationship), while not specifically referring to one another as friends. Similarly, although I derived this definition by examining friendships inferred by cross-cultural observers, the definition can also apply to relationships that are linguistically classified as existing between genealogical kin or clan members. Indeed, chapter 3 will explore how friendship and genetic kinship can overlap in the same relationship. The term *friendship-like relationship* clarifies how such ties can arise with biological kin and other formally defined relationships and how they can exist even without a word for friend. Nonetheless, the term is quite unwieldy, so for the rest of the book, I will use it interchangeably with the simpler term *friendship*, hoping that the reader will keep these issues in mind.

FRIENDSHIPS UNIMAGINED BY WESTERNERS

The previous analysis started with Western notions of friendship and examined how well they fit with friendship-like relationships described in other cultures. While this helped identify the practices, feelings, and expectations that can travel across borders to friendship-like relationships in a wide range of other cultures, it ignored those aspects of friendships in other societies that are absent from the Western stereotype. Here, I describe two elements of friendship-like relationships in other cultures—commitment rituals and physical touching—that are lacking in Western notions of the relationship (table 1).

Commitment Rituals

In the modern U.S., the commitment ritual par excellence is a wedding, in which two individuals vow to spend their lives together in mutual protection and support. The details of these rituals can vary dramatically by religion and personal preference, but they also share key features (e.g., formality, public witnesses, and memory devices) that play a role in maintaining the promises of commitment made in the ceremony. In most of the U.S., weddings and the state of union they imply (i.e., marriage) are steeped in expectations of procreation in a monogamous and heterosexual union.[60]

By contrast, in many parts of the world, commitment rituals have also been used to confirm friendships grounded in a promise of mutual support and protection. Consider the *kasendi* ceremony practiced among nineteenth-century Lunda farmers in Zambia, by which two men became blood brothers and were expected to assist each other when needed. In the presence of friends and family members, the two men in the ceremony sat opposite each other, holding hands, each with a vessel of beer by his side. They made cuts on their clasped hands, stomachs, right cheeks, and foreheads and picked up blood from each of these cuts with a blade of grass. Each man washed the blood-soaked grass in his own beer vessel. Then they exchanged vessels and imbibed the other's beer. While the men drank the beer, their friends beat on the ground with clubs and cried out, ratifying the treaty. The friends of the two men then drank some of the beer. To end

FIGURE 11.   Blood brotherhood commitment ritual among Lunda speakers in Zambia, from Wood 1868, p. 419

the ceremony, the men exchanged gifts, generally drawn from their most precious possessions (figure 11).[61]

In fifteenth-century France, friends could even enter a legally binding contract to cement their relationship. Specifically, two friends who wanted to combine their households could register their relationship by signing a contract, or *affrerement*, in the presence of witnesses and a notary. In such contracts, partners declared that they would combine their goods, possess them in common, and live together, sharing "one bread, one wine, and one purse." Partners usually testified that they entered the contract because of their mutual affection for each other. For example, in December 1443, Jacob Elziari and Jacob Martin, two farmers and friends, concluded an *affrerement* in Aix, France. The contract included all their goods, with the obligation to live a common life with their wives and families. They agreed not to end the *affrerement* except by mutual agreement; the party that asked for a dissolution would have to pay a penalty.[62]

Although the details of such ceremonies vary from culture to culture, the formal and elaborate cementing of a close friendship before an audience (and sometimes with signed contracts) is documented in hundreds of cultures around the world, including many European societies in past centuries. In numerous cases, the ritual marks the culmination of many years of friendship between two people who now want to take the next step and declare publicly that they will support each other for the rest of their lives. In other cases, the ritual brings together two relative strangers who are expected to grow to love and support each other in the relationship over the long term. In other instances, the ritual is part of a much longer, multi-generational process, whereby adults pass on their formal friendships to their respective children (box 9). Indeed, it appears that the modern United States and Europe may be outliers in this respect, since for the vast majority of Westerners there are no formal rituals to show commitment to a friend or friendship.

The most commonly cited examples of friendship ceremonies involve the sharing of blood by which the two participants become blood brothers or blood sisters. While this custom was apparently widespread in sub-Saharan Africa and to a lesser extent in Asia, Europe, and North America, there are many other kinds of ceremonies with no exchange of bodily substances. And far from being an exotic custom restricted to "other" people, rituals to consecrate friendship were, until recent centuries, standard practice in Christian churches in some parts of Europe. In medieval Greece, for example, liturgies for "brother-making" in the Eastern Orthodox Church suggest that friendships could be sanctioned as spiritual bonds under the aegis of the church.[63]

For many cultures, the only evidence we have that something like friendship existed lies in outsiders' detailed descriptions of such rites. Like kinship and marriage, these rites and the relationships that they entailed were publicly observable and verifiable, and thus much easier to record than the psychological motivations involved in informally cultivated friendships. Therefore, the vast majority of reports on rituals of friendship focus on the details of the rite—how blood is drawn, who officiates, how people sit, who is present—rather than psychological aspects, such as the reasons for having it, the motivational changes it entails, and the degree to which sanctions are believed to hold.

Nonetheless, a number of detailed accounts of such rituals give insights into the psychological role played by rituals in regulating a friendship. Rituals appear to promote commitment to a friendship in two ways. First and foremost, participating in a ritual signals one's exclusive commitment to that relationship. Such rituals also signal this commitment to the broader community, as friends, families, and neighbors generally witness the ceremony. Moreover, as a reminder of one's commitment, these rituals often produce portable symbols of the relationship, much like wedding rings in modern marriages. For example, among Azande farmers of north-central Africa, men would carry a blood-smeared lock of their blood brother's hair in a neatly woven pouch or wooden cylinder. By requiring unambiguous, exclusive, public, and material signaling, these rituals make it very difficult for a person to deny having an obligation to a ritual friend.[64]

Second, rituals add a threat of punishment for failure to help a friend. In the case of Azande blood brotherhood, participants reportedly believed that a friend's blood stays in one's stomach and will become poisonous if the friend is betrayed. Other rituals create less supernatural and more social forms of sanction. Betraying a friend to whom one has made a public commitment is more likely to lead to shame and reprisal in the public arena. If anything, betrayal of one friend will make it more difficult to find other friends in the future. Whether bolstered by supernatural or social sanctions, this class of rituals adds another emotion to the regulatory system operating between friends. Rather than simply helping a friend because one feels good about it, fear of reprisal can also become a potent motivator in the context of ritual and public sanctions.[65]

Touching.   Touching at appropriate moments and in acceptable ways is an important signal of friendship. Upon meeting after a long hiatus, male friends in the U.S. may hug rather than shake hands. However, there is no universal rule as to what is acceptable touching among friends (box 5). In the right place, it can be perfectly normal for (same-sex) friends to hold hands or intertwine a few fingers while walking down the street, caress each other, grasp the leg or arm of the other while sitting, or sleep in the same bed. Close friends in a Taiwanese village might display more physical contact in public than do married couples. Figure 12 shows two Fore friends from Papua New Guinea nestled together as they take a nap. And anthropologist John Honigmann describes the way that close friends among Kaska foragers and traders in northern British Columbia would express their affection for one another: "The emotions are rarely expressed verbally, but are abundantly demonstrated tactually. Such expression is most strongly developed in girls, who are often seen holding hands, sitting close together, hugging and wrestling. Boys, too, often sit resting against each other's bodies."[66]

Meanwhile, in other places, touching between friends may only be

FIGURE 12. Two Fore friends resting together; photograph provided by Shirley Lindenbaum

appropriate at certain turning points—a double-handshake after a long separation or walking with arms around each other, but only when going home drunk.[67]

DISCUSSION

This foray into the ethnographic record has shown that relationships similar to friendship arise in a wide range of societies, raising important questions about why the same suite of behaviors and feelings recurs with such regularity. While friendship-like relationships appear quite commonly in the ethnographic record, many of the Western ideals of friendship often fail to apply. In other cultures, friends are occasionally chosen by parents and relatives, and once a friendship is established, ending it might be as difficult as getting divorced. In a number of societies, sharing personal matters and secrets with friends is not only unimportant, but carefully avoided. And friends can be of unequal status, so much so that an outside observer might judge that one friend is exploiting the other.

As one peels away those aspects of friendship that travel poorly to other societies, a core pattern is revealed that consistently appears in descriptions of friendship across a range of societies. Friends help one another in times of need, and they do so predominantly because they feel positive emotions toward one another. Moreover, a frequent way of communicating feelings for a friend is through the exchange of gifts. This skeletal description of friendship fits remarkably well with the system revealed by experimental studies in Western countries (described in chapter 1).

Notably, friendship is defined as much by feelings and psychological states as it is by behaviors among friends. In many cases, friendship represents the *potential* for particular kinds of behavior that is underwritten by positive feelings toward a partner (e.g., help when one needs it). To focus exclusively on behaviors in defining friendship can lead to numerous problems in interpretation: mistaking practical joking as unfriendly conflict, confusing courteous behaviors among acquaintances for friendship, and failing to differentiate various mechanisms underlying the provision of aid.

This chapter has derived a working definition for friendship that unifies what is known from cross-cultural research about behaviors and feelings among friends that recur in a wide range of cultural settings. The next two chapters examine in more detail the similarities and differences between friendship and two other kinds of close relationships—those based on kinship and sexual attachment.

3 Friendship and Kinship

Och, he's no friend of mine; he's more a friend, if you know
what I mean.

Irish villager in the 1960s

In different places and times, the word *friend* and its linguistic relatives
have displayed remarkable flexibility in meaning. In the quote above, an
Irish villager uses *friend* in two very different ways to explain his relationship with a biologically unrelated associate. By stating, "he's more
a friend," the villager follows the standard, English sense of a non-kin
relationship. Yet, in a feat of verbal acrobatics, he first uses *friend* to mean
a blood relative by declaring, "he's no friend of mine." This ambiguous
usage of *friend* is not an isolated incident. More than eight hundred years
ago, speakers of Old English would use *freond* for both kin and non-kin
loved ones. And modern speakers of Norwegian, Danish, and Icelandic
continue to use *freund* for blood relatives, preferring cognates of another
word, *vinr*, for friends.[1]

The fact that a word like *friend* can rove so nimbly across apparent relationship categories is a warning to those who would seek clear distinctions
between kin and friends. Despite this ambiguity, scholars have frequently
drawn such divisions. In China twenty-five hundred years ago, Confucius
classified friendship as a relationship apart from family ties, requiring
unique rights and obligations. In the *Oxford English Dictionary*, the first
definition for friendship states that it is "not ordinarily applied to lovers
or relatives." And social scientists frequently contrast friendship and kinship as if they are mutually exclusive relationships. For example, kin are
biologically related while friends are not.[2] Kin relationships last a lifetime,
whereas friendships can end at any moment.[3] And, you can choose your
friends, but you can't choose your family.[4]

Are friends, as such distinctions suggest, fundamentally different
from kin? Or, as the Old English *freond* implies, do they spring from a
common source? This question is at the center of current debates about

the evolutionary origins of not just friendship, but all close relationships among humans. Some theorists in evolutionary psychology have proposed that kinship and friendship are fundamentally different relationships and rely on independent modes of thinking and feeling. According to this view, the cognitive underpinnings of kinship, friendship, and other kinds of relationships, such as romantic ties, evolved from different selection pressures. Helping among kin reflects selection for conferring benefits on biological relatives. Long-term bonding with sexual partners results from selection for extended biparental care. And friendship reflects selection for the capacity to cultivate cooperative ties beyond those involving biparental care or biological relatedness. According to such accounts, these are distinct kinds of relationship and are the outcome of independent adaptations deploying different psychological machinery.[5]

An opposing camp of researchers has argued that such clean distinctions between kinship and friendship are unfounded. Rather, they claim that all close relationships—kin-based, romantic, or otherwise—are manifestations of a unified suite of psychological processes, including love, forgiveness, jealousy, and commitment. The core of the argument is that close friends, immediate kin, and close romantic partners frequently share feelings described in terms of closeness, warmth, and love. Partners in these various kinds of relationships also tend to help each other from a genuine concern for the other's well-being. Combining these observations, the close relationship perspective proposes that friends who feel closer are more likely to help one another for the same reason that kin or lovers who feel closer are more likely to help one another. In short, friendship is a direct application of the feelings and behaviors that underlie kinship or sexual attachment to non-kin, non-sexual partners (figure 13).[6]

In the next two chapters, I argue that each of these perspectives possesses a kernel of truth, but neither accounts for the full complexity of feelings and behaviors in friendship-like relationships. Close relationships of all kinds share similar psychological and behavioral components. To varying degrees they are based on feelings of closeness as well as motivations to help one another, to be together, and to jealously exclude relationship competitors. Despite such commonalities, friendship does differ from kinship (and romantic relationships) in several key ways. These differences raise difficult questions for those who would propose that friendship is simply an extension of kin or romantic ties.

In this chapter, I compare kinship and friendship using ethnographic accounts, observational studies, and behavioral experiments. By reviewing studies that use a range of research methods rather than just one, we can

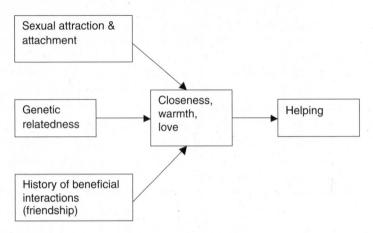

FIGURE 13. Close relationships hypothesis: helping among genetic kin and among close friends is mediated by similar feelings, such as subjective closeness.

get a richer picture of how friends differ from kin in helping and sharing. A key finding from these diverse orders of data is that close friendship and kinship, though similar in the feelings and behaviors they inspire, guide behavior in subtly different ways. Key differences include the ways that feelings of closeness and the perceived costs of aid affect helping behavior and how people provide aid to friends and kin during large-scale, life-threatening emergencies.

While differences exist, the common view of friendship and kinship as fundamentally different does not mean that they cannot occur within a single relationship. I finish by discussing how relationships are often hybrids, simultaneously recruiting the psychological machinery for relating with kin and with friends.

KINDS OF KINSHIP

Before we can use kinship as a point of comparison, it is important to clarify what we mean by the term. In most organisms, the task is simple. Kinship is defined in terms of genetic relatedness, or the probability that two individuals will share a gene from a common ancestor. According to this definition, I am more related to my brother than to my mother's sister and more related to my aunt than to my cousin. This definition plays a key role in the evolutionary theory of inclusive fitness, which proposes that genes, and the behaviors they entail, can increase in a population by promoting their replication not only in their host organism, but also in other organisms. Since closely related individuals (let's say full sisters) are more likely to share a particular genetic variant, a genetic variant that makes one sister selectively help the other is more likely to replicate itself when the sister receives help. Given this argument, under a wide range of conditions, we would expect closely related individuals to help one another, to share food, and to protect one another even when such aid bears a significant cost to the giver.[7]

In 1964, W. D. Hamilton formalized this intuition and derived this elegant inequality predicting when one would expect genetic kin to help one another:

$$RELATEDNESS > \frac{COST}{BENEFIT}$$

The *cost* in the inequality is the reproductive cost incurred by the helper, and the *benefit* is that accrued by the recipient of aid. *Relatedness* is the percentage of genetic variants that the two individuals should share by common descent. For example, my coefficient of relatedness (R) is one-half with my parents, children, and full siblings (whom I will call *immediate kin*), one-quarter with grandparents, aunts, uncles, nieces, and nephews (*extended kin*), and one-eighth with full first cousins (*distant kin*). As the closest common ancestor connecting me with another person goes back deeper in history, my coefficient of relatedness with that person gets very close to zero.

Hamilton's inequality nicely sums up an earlier claim by theoretical biologist J. B. S. Haldane that he would willingly lay down his life to save ten cousins. Haldane's assertion fits the above inequality as follows. Each cousin has a one in eight chance of sharing a genetic variant with Haldane by common descent. The benefit to his cousins as a whole is ten times the cost to himself (ten lives saved to one life lost), and one-eighth is greater than one-tenth. If similar life-or-death situations were to arise repeatedly over generations, we would expect the cousin-saving genetic variant and the behaviors it entails to spread throughout a population. And therefore Haldane expected that he would be biased toward saving his ten cousins.[8]

Hamilton's rule is derived from a simplified set of assumptions, and many factors not accounted for by those assumptions can change the theory's predictions. For example, the reproductive potential of relatives matters, so that one should be more inclined to help a relative who promises to have a long reproductive career ahead. And competition with relatives matters. If one is in fierce competition for resources with one's own close relatives, then one will be less likely to help them. Later in the chapter, I will describe how inheritance rules can set up this kind of competition. Despite these elaborations of the model, extensive studies with non-humans, and some with humans, have since confirmed Hamilton's general prediction about altruism toward genetic kin.[9]

This theory has nonetheless spurred considerable debate in the social sciences, with scientists arguing over whether helping behavior is the result of the "power" of genetic variants or the decision to help those who are socially defined as people whom you should help.[10] Of course, not all helping and bonding among humans or other animals can be explained solely in terms of kin selection. Consider male chimpanzees living in Kibale National Forest, Uganda. These males often develop strong social bonds with other males that involve grooming, meat sharing, help in disputes, and patrolling together. However, in some cases, these buddies are no more related to one another in a biological sense than are any two randomly chosen males. One explanation for these buddies' lack of relatedness is that due to the reproductive cycle of chimpanzees, there are few biological brothers of equal age and rank with whom to form alliances. In this case, age mates, who may not be genetically related, become the most attractive allies.[11] However, such observations are not challenges to kin selection per se since the theory does not aim to explain *all* altruistic behavior. In this case, a theory of reciprocal altruism—whether based on conditional aid or a more friendship-like mechanism—may account for the chimpanzees' relationships instead.

Most direct challenges to kin selection hinge on the fact that humans possess a unique capacity for language and symbolic thought. With the ability to manipulate symbols in novel ways, humans can create kin ties with genetically unrelated partners. For example, all over the world, parents frequently adopt and care for children who are not their genetic progeny, calling them the equivalent of "son" or "daughter." People can treat unrelated friends as brothers, sisters, aunts, or uncles. They can render existing kin relationships closer (e.g., by calling a cousin a sister) or more distant (e.g., by renouncing a child or parent). And individuals creatively reconfigure genealogies for a wide range of goals, including gaining access to material resources, circumventing incest taboos, and claiming a common ancestor for political purposes.[12] Such manipulations can extend to very large groups. For example, political factions defined by sharing a common ancestor in the father's line (i.e., patrilineal clans) have been known to include more than a million members, many effectively unrelated in a biological sense (with coefficients of genetic relatedness approaching

BOX 10 "I don't call her 'Mother' any more!"

Contrary to the popular idiom that you can't choose your relatives, in many societies it is quite appropriate to create and dissolve social kin ties over the course of one's life. In some cases, partners who have cultivated a close friendship begin to think of and refer to each other as kin. For example, Carol Stack (1974) observed that poor blacks in the U.S. often redefined their friendships as "going for kin," which implied a new level of obligation and helping. In other societies, a kin relation is assigned (often at a ritual), and a deeper relationship may or may not flourish among the partners (Titiev 1972). Kin relations can also be renounced. Among Chuuk islanders in the Pacific, a biological sibling who consistently fails to help may be "divorced" and the relationship broken off (Marshall 1977). And among Hopi agriculturalists in the U.S. Southwest, the ethnographer Mischa Titiev described the heated proclamations renouncing the use of "mother," "brother," or other kin terms when they were angry with a relative. In particular, he described how he would infuriate his Hopi housekeeper by calling a now-renounced ceremonial mother "her mother," leading to furious exclamations: "I told you that I don't call her 'Mother' any more!" What remains to be shown in such reclassifications of kinship is whether they correlate with actual helping behavior (Titiev 1967).

zero).[13] Such extensions of kinship beyond biological relatedness are often referred to as *social kinship* (box 10).

Societies differ a great deal in their use of social kinship. In the mainstream U.S., people maintain numerous bonds with partners they would not describe as either social or genetic kin. They have friends, acquaintances, coworkers, girlfriends and boyfriends, drinking buddies, and business associates. Of course, each of these partners has the possibility of becoming social kin. One can marry a friend's brother and thus become a special kind of sister—a sister-in-law. One can become the godfather of a friend's child, or one can adopt a child. Yet with all of these ways to extend kinship, social and genetic kin encompass only one part of the universe of social relationships in the mainstream U.S.[14]

A very different approach to kinship was taken by early-twentieth-century Hopi farmers living in the American Southwest. Nearly all relationships among the Hopi, even those among genetically unrelated individuals, were expressed using kinship terms and concepts. A Hopi might have fourteen "mothers" in addition to his biological parent and six "brothers," only one of whom was a biological sibling. This was the outcome of a conscious strategy by parents to connect their child with a wide network of contacts and resources. In short, parents ordained ceremonial parents from outside the circle of clan relatives to provide children with "another group of relatives." In this system, every person with whom one interacted was classified in terms of kinship. Indeed, newcomers, including some early anthropologists, were often incorporated into the system through such ceremonial adoptions.[15]

Societies also differ in how strictly social kinship maps onto genetic kinship. In some segments of U.S. society, kinship designations closely fit the biological facts of reproduction, with two important exceptions—in-laws and adoption.[16] In many societies, however, factors other than genetic kinship, such as co-residence, a history of mutual support, and feelings of goodwill, play a greater role in defining social kinship.[17] Among the South Fore in Papua New Guinea, for example, one can call a close friend an uncle, or *anagu*, as a sign of closeness, regardless of one's genetic relation. In fact, a man may refer to such an unrelated *anagu* as a "true" *anagu*, while disparaging a genealogically related uncle who is less dependable and less helpful by calling him a "small" *anagu*.[18]

The fact that people can manipulate kin concepts so easily to create social kinship beyond genetic kinship raises interesting questions. Is a social brother treated same as a biological brother? Do parents invest equally in adopted and biological children? Is social kinship simply a vacuous act of naming and classification that has little material consequence?

Among the varieties of social kinship, adoption has been the most thoroughly studied in a quantitative sense, and the results are mixed. In many island societies in Pacific Oceania, for example, nearly 25 percent of children are raised in non-natal homes, and adoptive parents bear considerable costs to provide food, shelter, and clothing. These adoptions appear to be inexplicable in terms of kin selection. However, several features of these adoptions suggest that kin-selected behaviors still play a role. In these societies, most adoptions occur between close or extended genetic kin rather than strictly unrelated individuals, and in many cases adopted individuals do not enjoy the same inheritance rights or leadership opportunities as their non-adopted siblings. For example, in the Northern Gilbert Islands adopted children cannot become leaders of the kinship groups into which they are adopted.[19] Therefore, in Oceania, social kinship does not stray far from genetic kinship, and even then, genetic and social kin are treated differently.

The Oceania case differs markedly from some modern industrialized

societies in which the level of investment in children can be quite independent of genetic relatedness. For example, a recent study of U.S. first
graders shows that those in adoptive families were as or more likely to
enjoy parental investment as were those living with both genetic parents.
And investment was measured in many ways, including the probability
of being in a private school, the number of reading, math, cultural, and
extracurricular activities, assistance with schoolwork, number of meals
spent with the child, and parental involvement in school. And these results
cannot be explained in terms of different parental education, income,
wealth, or age.[20]

To complicate matters further, stepparents in the U.S. (who may or may
not be adoptive) exhibit strong tendencies in the opposite direction. Many
studies show, for example, that stepfathers on average provide less direct
care, monetary support, financial aid for continued education, playtime,
and homework help to their stepchildren than do biological fathers.[21] In
addition, studies connect living with a stepparent to greater risk of abuse,
neglect, and suboptimal growth.[22]

The disparate findings from adoptive and surrogate parents suggest that
motivational factors other than those related to kin selection are at work
in parent-child relationships. In the case of adoptive parents and stepparents, there may be a fundamental difference in motivations, with adoptive
parents primarily interested in having a child and stepparents primarily
interested in having a new partner, with children simply a necessary part
of the package. It is also important to note that these are general trends
and that each person—whether an adoptive parent or stepparent—may
have unique motivations that do not fit the overall pattern.

A limited set of studies dealing with other kinds of social kinship con-
firms that kin-selected motivations are not the only factors underlying
the provision of help and support. For example, among Binumarien horticulturalists in Papua New Guinea, social kinship is nearly as important
as biological kinship in determining who helps whom in their garden (box
11).[23] Moreover, a recent study of cooperative whale hunting among the
Lamalera of Indonesia suggests that lineage membership can be a better
predictor of working together than raw genetic relatedness.[24] These examples illustrate how humans often systematically channel their help and
cooperation in ways unpredicted by kin selection. Nonetheless, it would be
too great a leap, as some anthropologists have proposed, to assume that the
theory of kin selection is irrelevant to people's behavior. Indeed, in many
cases, social kin tend to be biologically related, suggesting that, despite
the striking human capacity to manipulate the names and trappings of

---

BOX 11 Helping in the Gardens

In the early 1970s, Binumarien was a community of sweet potato gardeners and pig keepers in the Eastern Highlands of Papua New Guinea (population 172). Binumariens defined social kinship in ways very similar to Americans and Europeans—through marriage, adoption, and step-relations—and these social kin ties were important sources of help in caring for pigs and the constant maintenance of their sweet potato gardens. Distantly related social kin (R < 0.125) dramatically increased the number of potential helpers in these tasks, tripling the number of "parent-child" relations and quadrupling the number of "siblings." Moreover, such social kin were nearly as likely to help one another in the gardens as were strictly genetic kin. Of genetic parent-child relations, 78 percent helped in the gardens, compared to 79 percent of socially defined parent-child relations and 60 percent of children-in-law or parents-in-law. For siblings, the respective percentages of helpers were 57 percent, 35 percent, and 46 percent. Compare these to the low probability of helping (19 percent) among individuals unrelated either socially or genetically. In the Binumarien case, social kinship permitted a large expansion of the universe of reliable helpers in gardening and likely in other activities of daily life (Hawkes 1977, 1983).

---

kinship, human behaviors toward social kin are still constrained by kin-selected behavior to some extent.

COMPARING AND CONTRASTING
CLOSE FRIENDS AND IMMEDIATE KIN

The many ways that humans define kinship pose a challenge to any comparison one might make between friends and kin. Here, I focus on a particular kind of genetic kin—immediate kin, such as parents, children, and siblings, who have a very high coefficient of relatedness (R = 0.50). The reason for this is that there are many striking similarities between close friends and immediate kin. In both cases, people appear to help for the sake of helping, rather than from a fear of punishment or out of some expectation of return.[25] When people talk about immediate family and close friends, they use similar vocabularies, revolving around concepts of love, loyalty, and goodwill. Indeed, people often explicitly incorporate close non-kin friends into their families through the creative use of kin terms, such as brother, sister, aunt, and uncle.[26]

Insight into the cross-cultural extent of this similarity between immediate kin and close friends comes from a study of the rules of relationships in the United Kingdom, Hong Kong, Japan, and Italy. The study examined people's rankings of rules for nearly two dozen common relationships, from family members to coworkers, roommates, and even repairmen. The rules included norms for eye contact; touching; discussing finances, sex, and religion; swearing; joking; and repaying debts. The authors found that each culture had fundamentally different conceptions of the rules that apply to their close relationships. However, in each culture, the rules for close friends were most close to those for immediate kin—specifically siblings and parents.[27]

With such apparent similarities in the feelings, expectations, and rules found among close friends and immediate kin, it is unclear how we might differentiate between these two kinds of relationships. Are the psychological mechanisms that guide behavior toward friends the same as those that influence behavior toward biological kin? Or do these two kinds of relating draw from different psychological machinery? If so, to what degree can these systems overlap in any given relationship?

In this section, I compare close friends and immediate kin in terms of their feelings of closeness, their helping behavior, and their modes of accounting. Then I contrast close friends and immediate kin, showing two important differences. First, immediate kin depend less on feelings of closeness in decisions to provide help. Second, immediate kin and close friends differ in their response to the costs of helping, with immediate kin increasing and close friends decreasing aid when the costs increase.

Feelings of Closeness and Love

If we are to believe that people around the world accurately describe their private thoughts, close friendship and immediate kinship feel very similar. Yahgan foragers in South America say that close friends are like "loving brothers and sisters," a sentiment echoed among native Hawaiians, Pacific Island Chuukese, Ghanaian Akan Ashanti, and Thai farmers. And people from small-scale societies in all six inhabited continents say that they feel the same closeness, nearness, warmth, or strength for close friends and immediate kin.[28] This is in contrast with the kinds of feelings generally expressed for distant kin, acquaintances, or strangers.[29]

As described in chapters 1 and 2, researchers have tried to quantify and compare these feelings in a number of ways, with greatest success focusing on a spatial metaphor of closeness and merging with a partner. Though focused almost exclusively on young adults and adolescents in North America and Europe, the findings of these studies are consistent

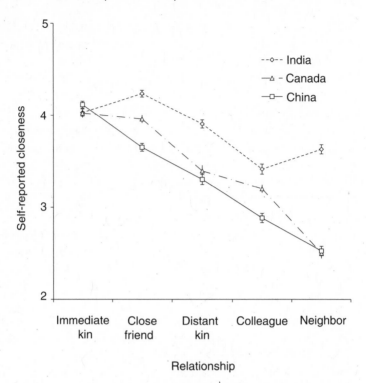

FIGURE 14.   Subjective closeness for kin and friends in India, Canada, and China, from Li et al., "Rethinking Culture and Self-Construal: China as a Middle Land," *Asian Journal of Social Psychology* 9, no. 3 (2006): 245. Copyright © 2006 by Wiley-Blackwell. Reproduced with permission.

with what people said in one of the few cross-cultural studies of subjective closeness. In the study, participants from China, India, and Canada consistently rated feeling most close to close friends and immediate kin, but less close to distant kin, colleagues, and neighbors (figure 14).[30]

We know very little about the degree to which feelings for close friends and close kin reflect similar physiological processes. Several recent studies have shown that people have greater activation in one brain region, the ventromedial prefrontal cortex, when viewing one kind of biological kin—their mother. Interestingly, this brain region also activates when people make judgments about themselves and close friends—but not non-close others—and when people are thinking about positive emotions (chapter 1).[31] This suggests that there are at least some common mechanisms underlying the similar feelings shared by immediate kin and close friends. However, as of 2008 only one published study explicitly compared the neural correlates of close friendship *and* kinship. It examined the brain activation of women when viewing a picture of their own child and when viewing other individuals, including a best friend, an acquaintance, a child with whom they were well acquainted, and an unknown child. Unfortunately, the researchers used best friends as a control condition, and thus it is difficult to determine what if any similarities or differences existed between viewing a best friend and one's own child in the study. Studies following similar protocols have provided insights into the differences and similarities in brain activation when perceiving offspring and romantic partners at different stages of a relationship.[32] Although such studies are limited in what they can tell us about how our brains operate in interactions with friends and kin, they could provide an initial snapshot of similarity and difference if they explicitly compared close friends and other kinds of immediate kin (e.g., parents, siblings, children) and also examined correlations with reported feelings of closeness for specific partners. However, until that time, we can say very little about the biological substrate for these similar feelings of closeness.

Despite the relative lack of physiological data, whether we look to broad-brush cross-cultural observations or more fine-grained quantification of feelings, similar findings emerge. Feelings about close friends and immediate kin are similar. But do these feelings influence how people actually behave toward one another?

Helping Behavior. In the 1600s, the author and philosopher Thomas Browne professed a love and sympathy for his close friends that rivaled feelings for the nearest of his "bloud": "I confesse that I doe not observe that order that the Schooles ordaine our affections, to love our Parents, Wifes, Children, and then our Friends, for excepting the injunctions of Religion, I doe not find in my selfe such a necessary and indissoluble Sympathy to all those of my bloud. I hope I doe not breake the fifth Commandement, if I conceive I may love my friend before the nearest of my bloud, even those to whom I owe the principles of life."[33] Browne's words are like a love poem. They also reflect previous findings that close friends and immediate kin have much in common in terms of the feelings they inspire. But do these comparable feelings of closeness and love translate into actions?

A glance at the cross-cultural record suggests that many people expect the same number or even more good deeds from close friends as they do from immediate kin. Among Zande farmers in early-nineteenth-century north-central Africa, a murderer could allegedly count on finding refuge with a close friend, but not with his biological brother. At the same historical time in Papua New Guinea, Kwoma gardeners of New Guinea expected that one should side with a friend over any but closest kin in disputes. In many other settings, friends are expected to help one another as much as they would help their immediate kin. In the words of one Lozi farmer living in Zambia, friends "treat friends like kinsmen."[34]

Unfortunately, ethnographic observations often deeply confound norms of behavior (i.e., what should be done) with actual acts. And for every claim that friends and kin are equivalent in action, there is often an opposing view, such as the following Malay proverb: "Friends and acquaintances are as the leaves that the wind of misfortune blows away; blood-relations are as the sap of the tree which always falls near the parent stem."[35] What we need in this case are systematic studies of the relative degrees to which people actually help friends and kin. Luckily, researchers have taken several approaches to measure helping behavior in the context of these relationships.

The most straightforward method involves asking people to report the aid given to and received from their significant others. In studies in numerous countries, close friends and immediate kin are quite comparable in terms of the aid they provide. Whether one focuses on suburban Canadians, working-class Brits, low-income Australians, poor urban Chileans, or middle-class Californians, friends provide a comparable share of support for small services, such as giving a ride, helping with childcare, lending household items, and providing emotional support and companionship, to that provided by parents, siblings, or children. Friends also provide more substantial kinds of help, such as financial assistance, sick care, and housekeeping, at comparable levels to siblings (though less than parents).[36]

Another line of research has asked people how they would behave in hypothetical situations, for example if a friend needed to be rescued from a burning building. One such vignette study asked U.S. college students what they would do if a good friend, family member, acquaintance, or stranger was just evicted from his or her apartment. Students could choose one of seven kinds of help, which had been rated in terms of cost by other students in an earlier study (box 12). These included:

1. Do nothing (rated cost = 0.0).

2. Give him or her an apartment guide (0.6).

3. Help him or her find a new place to live by driving him or her around for a few hours (2.9).

4. Offer to have him or her come stay with you for a couple of days (provided you had space) (3.6).

BOX 12 Vignette Experiments

Suppose a researcher wants to know whether people are more likely to risk their lives to save a close friend from a life-threatening situation than to rescue a stranger. It would be ethically and practically impossible to set up such an experiment with real-life consequences. In such situations, researchers often turn to vignettes, or hypothetical scenarios, to examine how different incidents and situations elicit different thoughts, feelings, and behaviors.

Vignettes possess several advantages over traditional experimental and observational designs. They can expose individuals to multiple, comparable situations, permitting very fine-grained manipulation of context while presenting realistic and ecologically valid scenarios. They also permit the study of scenarios that would not normally be possible in laboratory settings (Alexander and Becker 1978; Rossi and Anderson 1982).

By manipulating the details of such vignettes, one can see how certain kinds of contexts provoke different responses and hypothetical behaviors. For example, one recent study of helping among friends and siblings asked participants to rate how willing they would be to rush back into a burning building to save the life of a friend or a sibling (Kruger 2003).

However, vignette studies also possess an important shortcoming that should be considered when interpreting results. It is not clear that what people say they would do in a hypothetical situation bears any relation to what they would actually do if the situation arose in real life. As Stanley Kubrick stated in an interview in the 1960s, "I once saw a woman hit by a car . . . and she was lying in the middle of the road. I knew that at that moment I would have risked my life if necessary to help her, whereas if I had merely read about the accident or heard about it, it could not have meant too much" (Southern 1962). In recent decades, experimental economists and psychologists have begun to examine how much responses to vignettes reflect behaviors in the real world. In many situations, it appears that the two match quite closely. In others, there are notable differences. Until more is known about the nature of these differences, we might assume that vignette studies tell us something about how context affects behavior, but not about how much (Ajzen, Brown, and Carvajal 2004; Bertrand and Mullainathan 2001; Wiseman and Levin 1996; Southern 1962; Sharpley and Rodd 1985; Johnson and Bickel 2002; Madden et al. 2004; Gillis and Hettler 2007).

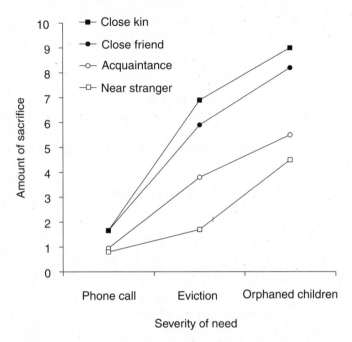

FIGURE 15. Acceptable cost to help by type of relationship and severity of need, from Cialdini et al. 1997. Copyright © 1992 by The American Psychological Association. Reproduced with permission.

5. Offer to have him or her come stay with you for a week (provided you had space) (4.2).

6. Offer to have him or her come stay with you until he or she found a new place (provided you had space) (5.9).

7. Offer to let him or her come live with you rent-free (provided you had space) (8.3).

The researchers also asked students about other hypothetical situations, one in which a partner died in an accident leaving his or her two children without a home, and another in which a partner required aid in making a phone call. In all three cases, students said they would bear more cost to help friends and family members than to help acquaintances and strangers, with slightly higher levels of cost borne for family members than for close friends (figure 15).

While consistent with the cross-cultural data, these vignettes do pose a problem as methodologies. They rely on what people say they do or would do, which may bear very little resemblance to what they actually do in

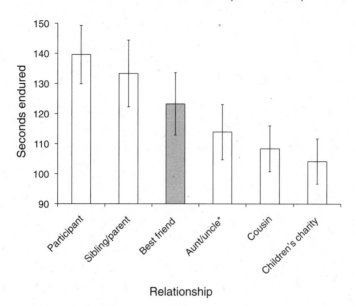

FIGURE 16. Amount of time spent enduring pain to send money to kin and friends, adapted from Madsen et al. 2007. *Aunt/uncle could also include other relatives with relatedness = 0.25, such as grandparents, nieces, or nephews. Error bars indicate standard errors. Adapted with permission from the *British Journal of Psychology,* © The British Psychological Society.

situations with real costs and consequences. More recently, a group of researchers at the University of London tried to resolve this problem with a novel experiment. They asked university students to endure pain from a physical task in return for a proportional reward given to one of the following individuals—the student herself, a sibling or parent, a same-sex best friend, an extended kin member (aunt or uncle), a distant kin member (cousin), or a national children's charity.

The task was a standard ski-training exercise—sitting on thin air supported by one's back against a wall, with calves and thighs at right angles to each other. Any normal person who has tried this knows that it leads to discomfort and pain within seconds. For every 20 seconds that a participant held this position, a reward of .70 British pounds was mailed to the appropriate recipient. As expected by theories of kin selection, individuals endured more pain for themselves (approximately 140 seconds) than for immediate kin (approximately 132.5 seconds), more for immediate than for extended kin (approximately 113 seconds), and more for extended than distant kin (approximately 107 seconds) (figure 16).

Where did friends fit in? The time endured for a best friend (approximately 123 seconds) was somewhere between immediate and extended kin, and also much greater than that endured for a charity (approximately 103 seconds) or distant kin (approximately 107 seconds).[37]

Therefore, people not only say that they help friends at comparable levels to immediate kin, they actually do. In this tightly controlled experiment, they endured longer periods of pain for close friends than for extended kin, distant kin, and strangers. The only people for whom participants sacrificed more were immediate kin and themselves, suggesting at least in this case a slightly stronger bias toward helping immediate kin than close friends.

Modes of Accounting

Friends and immediate kin not only provide similar kinds of help. They also make decisions to help in similar ways. As I described in the previous chapter, friends in a wide range of cultures are expected to help when their partner is in need, a norm that is also common among close family members.[38] To assess whether such norms translate into real behavior, Lenahan O'Connell conducted a study of 108 U.S. families who had built their own homes. She asked the families what kinds of help they received from kin and friends in building their home, and also how they ultimately returned the favor. The builders received help from both friends and family, on average three helpers per builder, and the share of help was divided equally between the two kinds of relationships. Help often included physically demanding and time-consuming tasks, such as making cabinets, putting up the house frame, installing plumbing, and laying the foundation. The interviews happened well after the actual home building, and thus any unreciprocated help described at the time of the interviews had endured for a considerable length of time.

When asked about repayment, many participants said that they had not given it much thought. Others described "paying" friends with services of highly unequal value in the broader U.S. economy. One man, whose friends were responsible for making the cabinets and installing plumbing, stated that he paid his friends back with "fried chicken and garden vegetables." Quantitative data also showed high degrees of non-reciprocation. Participants reported they had reciprocated the favor with less than a third of their helpers, low for both friends (31 percent) and relatives (26 percent). Moreover, they described no intention to repay nearly half of their helpers (42 percent of friends and 52 percent of family). Although these results suggest a slight skew toward unbalanced aid among family members, the more remarkable result is the low level of reciprocation within either category of relationship. Compare this to hired contractors, who are reimbursed nearly 100 percent of the time.[39]

Closeness and Helping

Researchers have proposed that the similar levels of help observed among close friends and immediate kin are causally related to the similar feelings of closeness felt among partners. According to these accounts, feelings of closeness are a proximate psychological mechanism by which one's interests merge with those of another. Partners who feel closer to each other are less likely to think in terms of "my stuff" and more likely to think in terms of "our stuff," less likely to think in terms of "my well-being" and more likely to think in terms of "our well-being."[40] This explanation also fits nicely with the lack of accounting and strict reciprocity observed among immediate kin and close friends. With kin, we may learn these feelings very early, or we are perhaps endowed with them at birth. With close non-kin friends, according to the argument, we build these feelings up over time as a relationship develops. Some psychologists have described this familial treatment of unrelated others as "psychological kinship" (figure 13).[41]

The main evidence for this proposal comes from four vignette experiments conducted over the last decade in which individuals were asked about the kind of help they would provide for a partner in a number of situations: if the partner needed to make a phone call, had just been evicted from an apartment, had died in an accident leaving his or her two children without a home, needed rescue from a burning house, needed help with an everyday errand, or needed an organ or blood donation.[42]

The experiments revealed three interesting findings. First, individuals who felt greater subjective closeness said they were more willing to help each other (average d-statistics = 1.22). Second, and consistent with kin selection theory, individuals said they would bear a greater cost to help more closely related kin. Finally, subjective closeness and genetic relatedness were correlated so that more closely related kin felt greater closeness. Thus, the experiments showed a complex interweaving of biological relatedness, subjective closeness, and helping behavior.

Further analyses also showed that subjective closeness mediates at least some of the effect of genetic relatedness on helping. Specifically, the effect of genetic relatedness on willingness to help was substantially reduced when one simultaneously considered the effect of subjective closeness on helping. These findings have led some researchers to conclude that the same underlying feelings of subjective closeness are involved in helping among non-kin friendship and kinship. Such subjective closeness may be a cue to kinship, much like facial or attitudinal similarity, but one that is built up over years of living, working, and eating together.[43] According to this argument, friendship is simply an application of a psychological system originally evolved to detect kin (figure 13).

There are several problems with the close relationships argument, three of which I will describe here. First, subjective closeness never completely accounts for the greater tendency of kin to help one another. That is, given the same level of subjective closeness, one is on average more likely to help kin than non-kin.[44] Second, the willingness to help immediate kin and to help friends is affected in different ways by subjective closeness. This is illustrated nicely by a vignette study conducted by two psychologists, Howard Rachlin and Bryan Jones, at the State University of New York, Stony Brook. Rachlin and Jones asked college students to think about friends and relatives at differing levels of social closeness. Then, for each partner, they asked students how much money they would be willing to sacrifice to give their partner $75. A completely selfless individual would sacrifice $75 (or more) to give his or her partner $75. A completely selfish individual would give nothing to send his or her partner $75. As expected, people were willing to give up far more for individuals to whom they felt closer, whether these were kin or non-kin. Indeed, some individuals were willing to sacrifice *more* than $75 to send very close partners $75. And, in line with theories of kin selection, students were more likely to help kin over friends at the same level of subjective closeness (figure 17).[45]

But there was also a more subtle difference between kin and non-kin in how subjective closeness predicted helping behavior. At very high levels of closeness, the amount people would forgo for non-kin and kin was practically the same. However, at low levels of closeness, students were willing to forgo nearly twice as much for kin as they were for non-kin. In short, kinship modified the effect of subjective closeness on students' willingness to sacrifice for another person. Sacrifice among kin was much less sensitive to subjective closeness than it was among non-kin, suggesting that feelings of closeness influenced the behavior of non-kin more strongly. This observed interaction between biological kinship and subjective closeness in determining altruistic behavior raises important questions. What are the psychological and physiological processes underlying this interaction? Do the effects of subjective closeness and biological kinship reflect two differ-

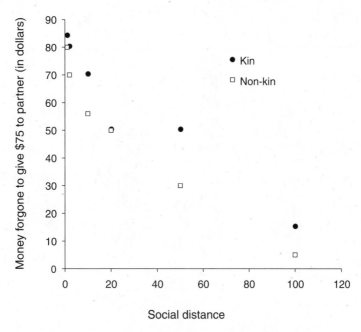

FIGURE 17. Money forgone by subjective closeness and biological relatedness. Reprinted from *Behavioural Processes* 79, no. 2, H. Rachlin and B. A. Jones, "Altruism among Relatives and Non-relatives," pp. 120–123, copyright © 2008, with permission from Elsevier.

ent systems that can overlap in a single relationship? I will discuss these issues in more detail later in the chapter.

How the Costs of Helping Change the Willingness to Help

Kin and friends also respond very differently to changes in the cost of helping. One would expect willingness to help to decrease as the cost of helping increases. However, in several observational studies, kin are paradoxically more likely to provide support when the cost of helping goes up. For example, in one study of suburban Canadians, kin provided a greater proportion of large services (e.g., major repairs, children's daycare, long-term healthcare) than small services, whereas friends showed the opposite trend. A similar reversal in the effect of cost was observed among middle-class women in Los Angeles (figure 18).[46]

The different effect of cost on helping among immediate kin and close friends is also illustrated nicely by a recent study. Steve Stewart-Williams asked Canadian college students how much help they had given

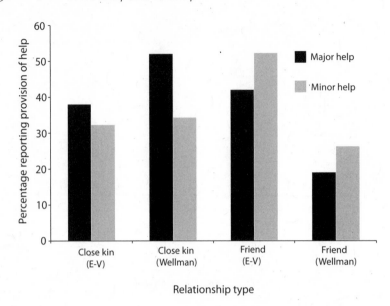

FIGURE 18. Changing probability of providing aid based on cost of help; data from Essock-Vitale and McGuire 1985 and Wellman and Wortley 1990

and received from close kin (siblings), friends, and cousins in the last two months (figure 19). As kin selection theory predicts, students were more likely to provide aid to siblings than to cousins and to cousins more than to acquaintances. However, the story for close friends was more complex. When help involved little material cost (e.g., emotional support), students reported giving more help to close friends than any other relationship category. When help was moderately costly (e.g., help during an illness, help with chores and errands, help with housing or finances), help given to close friends dropped to the level given to siblings. Finally, when asked about hypothetical high-cost help (willingness to donate a kidney or to risk injury or death to save a person's life), students' willingness to help close friends dropped to the same level as that for cousins.

At a coarse approximation, close friends and immediate kin enjoyed comparable levels of support. However, at a finer level of analysis, people behaved differently toward close friends and immediate kin. Increasing the cost of help reduced the tendency to help close friends but actually increased it among close biological kin. This suggests that benefits to one's partner and costs to oneself are weighed differently in kin and non-kin relationships.[47]

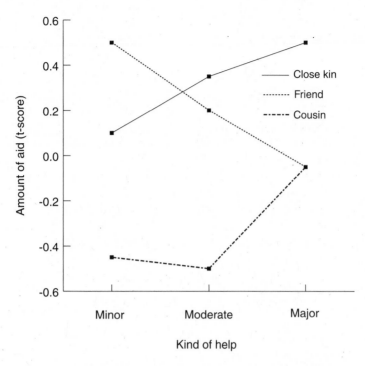

FIGURE 19. Amount of aid by kind of help and type of relationship. Reprinted from *Evolution and Human Behavior* 28, no. 3, S. Stewart-Williams, "Altruism among Kin vs. Nonkin," pp. 193–198, copyright © 2008, with permission from Elsevier.

The difference between friends and kin is most apparent in the way that kinship generally takes precedence over friendship in the case of life-or-death emergencies. In one study of survivors of a fire, for example, individuals reported that they were much more likely to search for family members than for friends.[48] Other research has found that immediately following man-made and natural disasters, family members are the first to be helped, with friends, neighbors, and strangers receiving progressively less attention. For example, following the onset of the Gulf War in January 1991, Iraq launched forty Scud missiles against coastal Israeli cities for five weeks. Schools and universities were closed, people stayed home from work, and many individuals left the coast to live with relatives. During this time, they turned to immediate kin more often to discuss war-related concerns and to check in after missile attacks than they did to discuss personal matters in their daily lives, whereas they turned less to friends to discuss such wartime issues (d-statistic between kin and friends for discussing personal matters = 0.06, d-statistic for calling after an attack = 0.98).[49]

Despite the apparent similarity between immediate kin and close friends, both in terms of feelings and behaviors, these two kinds of relating differ subtly in how feelings are translated into behavior. First, feelings of closeness cannot account entirely for the willingness to help among kin, since we treat friends and immediate kin of equal perceived closeness unequally—on average favoring immediate kin. Second, kin are less influenced by perceived closeness when choosing to help. In short, we help very close friends and kin at near equal levels, but if asked to help a brother with whom we felt only as close as a distant acquaintance, we would be much more likely to help the brother than the acquaintance. Finally, friends and immediate kin respond differently to the costs of helping. When the costs of helping go up, friends are less likely to provide aid, but among kin we observe precisely the opposite effect.[50] These differences indicate that helping among close friends cannot be explained solely as an application of psychological processes that motivate support among immediate kin.

WHY MAKE FRIENDS?

People have developed numerous ways of socially extending their kin ties and cultivating friendships. In some societies a great deal of effort—in terms of travel, gift giving, and socialization—is spent on such relationships. Amidst this flurry of social activity, a practical question arises: Why bother making social kin and cultivating friendships at all? Why not eschew all of this social work and rely on one's existing set of biological kin for aid and support? In addition to avoiding the cost of creating new ties, it would also reduce conflicts of loyalty and the danger of exploitation.[51] The fact that we put so much time into these endeavors raises questions about their payoffs.

In the 1960s, anthropologist Sandra Wallman asked this very question of Basuto subsistence farmers in southern Africa. A single Basuto household rarely commanded all of the four resources—oxen, land, seed, and food for workers—necessary for farming. To remedy this situation, households frequently entered into cooperative partnerships called *seahlolo*, whereby each household provided some combination of these resources. The two *seahlolo* partners were expected to work together and harvest together, dividing the crop equally. While classical views of rural, non-industrial societies suggest that Basuto farmers would make every effort to enter into *seahlolo* with kin, in the majority of cases they chose non-related partner households.

Wallman found two reasons why individuals did not choose kin. First, kin often did not have the right resources to be good *seahlolo* partners. For example, two brothers may both own an ox, but neither owns seed or land. Second, a common concern that "there was always trouble with family" led to the perception that close friends were better partners. As one man said, "I know [my friend] and we shall not quarrel."[52] These two accounts reflect more general reasons that arise frequently in the ethnographic record: immediate or extended kin are not enough when it comes to eking out a living, and kin are one's most immediate competitors.

Kin Are Not Enough

In the ethnographic record, immediate or extended kin are not sufficient to meet the needs of individuals for three primary reasons. First, one's kin often have access to the same resources as oneself, and non-kin may be the only way to reach novel, non-overlapping goods. Second, in large-scale enterprises, where success depends on the number of hands one can muster, there may not be a sufficiently large pool of immediate or extended kin to succeed. Third, one is occasionally thrust into an environment where there are simply no kin to draw from.

The first potential benefit of friendships is that they permit access to resources well outside one's sphere of immediate kin. No matter how supportive immediate or extended kin may be, family, by virtue of their proximity and common material inheritance, are usually limited in the kinds of resources and aid they can provide. As in the case of Basuto farmers, one may have no brothers who can provide seed. In a small village in Andalusia, Spain, a farmer may have no kinsmen with sufficient leverage to plead his case in a water dispute.[53] In Melanesia, a gardener may need to travel to other islands to acquire obsidian and high-quality pots.[54] In each of these cases there are gaps in the kin network, and friendships provide a means to fill these gaps. Among Orokaiva gardeners in Papua New Guinea, for example, there may be no kin in safe areas when warfare breaks out near home.[55]

Second, friendships can be useful because they can extend one's ability to draw on large quantities of help at critical moments. In some especially conflict-prone societies, having more allies than one's enemy is an important prerequisite for success in combat. Relying on kin alone places inherent limits on the size of factions, and so to be successful one must depend

BOX 13   Kin and Allies in an Ax Fight

On February 28, 1971, in southern Venezuela, a fight broke out in a village of Yanomamo gardeners. Beginning as a dispute over plantains between a village woman and a visiting man, the fight soon drew in combatants on both sides brandishing long clubs, machetes, and—in the case of one fighter—an ax. After an extended skirmish and the felling of a young man with the blunt side of the ax, the fight subsided, de-escalating with insults and hostile stares. The next day, some members of the visitors' coalition packed up and left the village.

We know about the ax fight because anthropologist Napoleon Chagnon and filmmaker Tim Asch happened to be in the village when it broke out, and they documented the event as it unfolded on 16mm motion picture film and in 35mm still photographs. Eight years later, Chagnon and graduate student Paul Bugos analyzed the fight to test kin selection theory, examining the genetic relatedness of people who fought in the same coalition. Kin selection theory would predict increased relatedness among allies, and this is what Chagnon and Bugos found. The first male antagonist's supporters were eight times more related to him than they were to the second antagonist, and the second male antagonist's supporters were two times more related to him than they were to the first (Chagnon and Bugos 1979; Biella, Chagnon, and Seaman 1997).

For this reason, the ax fight is usually (and correctly) cited as support for kin selection theory. However, Chagnon and Bugos also showed that another kind of tie, cemented through marriage, influenced who fought with whom. Nearly all of the participants in the fight were genetically related to some degree. Indeed, the original disputants were the equivalent of cousins. Against this background of low-level relatedness, marriage alliances provided another rationale for choosing sides. For example, such ties-by-marriage help predict how members of the largest set of extended kin ($R \geq 0.25$)

(continued)

on augmenting kin with other allies and thus surpassing the competition. For example, in a well-documented dispute within a Yanomamo village in Venezuela, both factions reached sizes that would never have been possible if both relied solely on ties of close genetic kinship. Rather, allies based on marital and other ties augmented each of the coalitions to include dozens of individuals (box 13).[56] In addition to large-scale disputes, cooperative

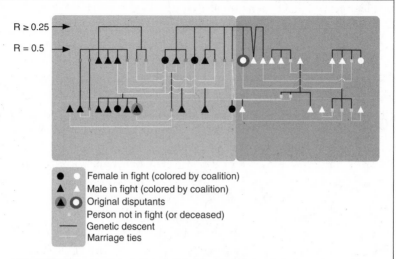

R ≥ 0.25 →

R = 0.5 →

● ○ Female in fight (colored by coalition)
▲ ▲ Male in fight (colored by coalition)
▲ ○ Original disputants
      Person not in fight (or deceased)
——   Genetic descent
      Marriage ties

FIGURE 20. Genetic and marital relationships of participants in the ax fight. White lines indicate marital ties. The height of black lines connecting members of the elder generation (the top tier in the chart) indicates degrees of genetic relatedness. Short lines indicate genetic relatedness at the level of immediate kin (R = 0.50). Tall lines indicate genetic relatedness at the level of extended kin (0.25 ≤ R < 0.50). Note that there are eleven relationships at the level of distant kin (0.125 ≤ R < 0.25) that are not reflected directly from the figure. Interestingly, these distant ties cross-cut the coalitions more often (eight times) than they lie within them (three times). These ties occur between the following ID numbers in the Yanomamo Interactive database (cross-cutting: 259-1109, 1335-1109, 723-950, 2505-910, 2505-2248, 2505-2209, 1744-517, 1897-517; within coalition: 1062-910, 1062-2248, 1062-2209). Data from Biella, Chagnon and Seaman 1997 and Chagnon and Bugos 1979.

ultimately chose sides in the fight (top center of figure 20). And such ties also wove together groups of extended kin into larger, competitive coalitions. Thus, the ax fight provides one particularly well-documented example of how both close genetic kin and less closely related allies can play a role in forming coalitions in the human case.

enterprises, such as fishing, hunting, or whaling expeditions, can also make non-kin friends a useful way to augment one's group to necessary sizes.[57]

The third way that non-kin friends can become useful is when kin are absent. This can occur after migrations, after marriage and the move to a new village, or simply if one's kin have died.[58] In each of these cases, non-kin friendships can be important conduits for support and aid that would otherwise have been provided by kin. I'll discuss this last reason for making friends in more detail in chapter 7.

Competition and Conflict

In a recent book on the declining state of friendship in the modern world, Digby Anderson deplores the fact that friendship lacks the same legal status as other kinds of relationships, and his argument focuses on laws of inheritance among kin.[59] However, the lack of formal inheritance among friends is not unique to modern society. Indeed, in only a handful of societies, including the Fore of Papua New Guinea and Zuni farmers in the U.S. Southwest, is there a norm whereby non-kin friends can inherit property. And this general inability to inherit property may make friends especially valuable.

In many human groups, kin-biased rules of inheritance also have the unfortunate consequence of pitting sibling against sibling as jealous heirs for the same resources.[60] In such cases, friends are often above the fray, precisely because they are excluded from inheritance. Among Bangwa subsistence farmers in Cameroon, for example, kin are wrapped up in petty disputes and witchcraft accusations, while close friends, who are removed from such intrafamily feuds, become an important source of support. If a farmer becomes sick, he calls a friend to identify (or "divine") who might be responsible. Most of the time the party identified is a kinsman. And as friends have no rights to one's property, a dying man will call his friends, not his relatives, to administer the will and police the scramble for inheritance. Similarly, among Chuuk gardeners in Oceania, friendships are unfettered by sibling rivalry or squabbles over inheritance. These friends, or *pwiipwi*, offer an informality, intimacy, and confidentiality normally lacking in relations among siblings, and they are sought because they provide emotional support in a sea of tenuous emotional relationships.[61]

IS FRIENDSHIP JUST CONFUSED KINSHIP?

Kin and friends are often pitted against each other in popular thought. More than two millennia ago, Euripides opined, "One loyal friend is worth ten thousand relatives." And there is no dearth of proverbs making the opposite claim. Consider the earlier Malay proverb that paints friends as being as fickle as leaves in the wind, whereas kin are as solid as sticky sap. The famed ethnographer Bronislaw Malinowski was often rebuked by the Trobriand Islanders with whom he lived because he would use the phrase "They are friends" (*lubaym*) to refer to two individuals from the same clan.[62] "No!" they would answer. "This man is my kinsman; we're kins-men—the clan is the same!"

Despite the common tendency to differentiate friendship and kinship, people also have difficulty making clear-cut distinctions. Individuals in the United States and Britain frequently claim that their closest friend is a relative.[63] And as one sociologist experienced when he tried to interview Montrealers about their kin and friends: "It was difficult for Montrealers to relate friendship to kinship and vice versa in large part because the languages, thought, and emotions of friendship and kinship are experienced as intersecting and complementary, often at the same time."[64]

In this chapter, I argue that both the intuition that friends and kin are different and the difficulty in differentiating them stem from two basic facts. Friendship and kinship reflect different psychological processes and have subtly different consequences for behavior. However, these may also arise simultaneously in a given relationship. For example, biological kin can augment their relationship with feelings of friendship, an observation that led writers of the Icelandic sagas to clarify the relationship between two cousins, "There was great friendship between them besides kinship."[65] In short, people can be both kin and friends.

The psychological studies described in this chapter provide greater insight into this idea by showing that two factors—genetic relatedness and subjective closeness—interact in any given relationship to influence one's willingness to help. Although subjective closeness is a characteristic of friendship, one can also cultivate such feelings among biological kin. In short, friends and kin are not mutually exclusive categories of relationship. Rather, friendship and kinship are two psychological systems that can exist in hybrid form in the same relationship but have different consequences for behavior.

The fine-grained, quantitative findings from experiments and observational studies have provided new insights into how kinship and friendship interact in guiding helping behavior. However, they have two important limitations. First, they have depended almost exclusively on judgments about hypothetical vignettes. Moreover, the one study in this genre that focuses on a real sacrifice uses a behavior—undergoing the ski-training squat to send money to friends—that is only loosely connected with the kinds of choices we face in the real world. Therefore, it is not clear how the findings from these experiments generalize to behaviors that reflect more real-life decisions. Second, the studies are limited to the United States and Europe. Similar fine-grained experiments that are conducted in diverse cultural settings and experiments that examine real behavior will extend these insights and ultimately provide a better picture of how helping and sacrifice depend on both friendship and biological kinship.

By comparing and contrasting kinship and friendship, this chapter introduced the notion that some relationships can actually be a hybrid of the two. In chapter 4, I make a brief excursion into two other kinds of love—romantic and sexual—and how these relate to friendship. I call this quartet of relationships "Four Kinds of Love" and offer some evidence as to how the four ways can be connected to one another or kept apart—both in real life and in the minds of those who study these matters.

4 Sex, Romance, and Friendship

Shall I compare thee to a summer's day?
Thou art more lovely and more temperate.
WILLIAM SHAKESPEARE, Sonnet 18

Intensely close relationships often inspire speculation about sex.[1] Consider the more than one hundred sonnets that Shakespeare devoted to a young man he named "fair youth." In these sonnets, Shakespeare writes romantically of sweet love and how the youth's beauty awakes and delights his heart. Although these verses lack the explicit sexual references of some of Shakespeare's other poems, their passionate romantic language has spurred long-standing controversy about whether the poet was involved in a homosexual affair. Similar debates surround intensely emotional letters written by the poet Emily Dickinson to her friend and eventual sister-in-law, Sue Gilbert. Between a man and a woman, a friendship requires much less than sonnets and emotional letters to arouse suspicion or comment. Indeed, the entire premise of the Oscar-nominated film *When Harry Met Sally . . .* rests on the question "Can men and women be friends or does sex always get in the way?"[2]

Neither Shakespeare nor Dickinson is in a position to clarify the sexual content of their relationships. But more recent speculation about famed talk-show host Oprah Winfrey and her best friend of thirty years, Gayle King, provides such an opportunity. In many respects, the two are an item. They frequently vacation together and are often seen together in public. Winfrey confesses that they phone each other four times a day, and that she builds a "Gayle wing" in each of her houses. Gayle has admitted, "If Oprah were a man, I would marry her." The intimacy and closeness of this relationship has fueled frequent intrigue in tabloids and the Internet about the sexual content of their relationship.

In an interview in her magazine *O*, Oprah Winfrey discusses her friendship: "I understand why people think we're gay. There isn't a definition in our culture for this kind of bond between women. So I get why people have to label it—how can you be this close without it being sexual? How else can you explain a level of intimacy where someone *always* loves you, *always* respects you, admires you? . . . Something about this relationship feels otherworldly to me, like it was designed by a power and a hand greater than my own. Whatever this friendship is, it's been a very fun ride—and we've taken it together."[3]

Winfrey's description of her friendship raises important questions about how sex, romance, and friendship fit together. Is it possible to have a "romantic" friendship that feels like our stereotype of romance, with near obsessive affiliation and preoccupation with a partner, but that removes sexual desire? Or are partners in such relationships simply deluding themselves and suppressing hidden sexual urges? How does sexual desire and behavior enter into friendships, and to what degree does it change a relationship? In this chapter, I explore these questions, differentiating the biosocial motivational systems underlying different kinds of love that partners can feel for each other and examining how these different faces of love can interact in a single relationship.

THREE FACES OF LOVE

Buddhist teachings distinguish between two kinds of love. The first, *kama* (found in the title *Kama Sutra*), is sensual love focused on self-gratification that throws up obstacles to enlightenment. The second, *metta*, is unconditional, benevolent, and most related to feelings of friendship. Mangaia islanders in Polynesia draw another distinction. They use *inangaro kino*, literally "terribly in love," to define a state of romantic passion in which a lover is single-mindedly fixated on the beloved and jealously wants the beloved to reciprocate the obsession. If separated from the beloved, a potentially fatal feeling of *atingakau*, or heartbreak, can take hold. *Inangaro kino* can, but does not always, involve deep caring and benevolence for a partner and is a feeling distinct from sexual desire.[4]

The anthropologist Helen Fisher, who has spent more than a decade studying the nature of love, proposes that these recurring folk distinctions reflect real differences in how our brains and bodies function in different relationships. Fisher argues that three distinct but intertwined drives—lust, romantic love, and attachment—play unique roles in the drama of human mating. *Lust* is simply the craving for sexual gratification and can occur quite independently of passionate love or long-term attachment. *Passionate* or *romantic love* involves heightened preoccupation with a particular partner and includes signature symptoms like sleeplessness, increased energy, mood swings, possessiveness, separation distress, emotional dependency, and obsessive thoughts about the beloved.[5] Finally, *companionate love, partner attachment,* or *friend-like love* is a calmer kind of love, involving feelings of benevolence, security, and deep affection for a long-term partner. Behaviorally, it is characterized by mutual feeding, grooming and aid, gift giving, maintenance of close proximity, separation distress, cooperation, and gestures of affiliation. Defined thus, companionate love shares many of the feelings and behaviors associated with friendship.[6]

Fisher also argues that these three kinds of love address different tasks that our ancestors repeatedly faced in the quest to reproduce. *Lust* is an indiscriminate drive that motivates one to have sex with a range of appropriate members of one's species. *Romantic love* leads people to disrupt existing habits and routines and to invest in building a new component of one's social niche—a relationship with the beloved. In this state, lovers become intensely focused on attracting the other's interest, engaging in courting and relationship-building behavior, and reconfiguring their existing social networks to include the beloved. Passionate love also spurs protective feelings for the relationship, most notably jealousy toward would-be competitors.[7] Whereas passionate love involves expanding one's social niche, *companionate love* motivates one to maintain an existing niche, through mutual help and affiliation. According to this evolutionary account, romantic attraction is relatively short-lived and encourages people to focus their energy on mating with a particular partner. Meanwhile, partner attachment remains after the honeymoon, promoting cooperation with a partner long enough to rear a child into infancy.[8]

In many cases, these three kinds of love—lust, romantic love, and companionate love—follow a dependable temporal trajectory, captured eloquently by a Ju/'hoansi forager of the Kalahari Desert of Botswana: "When two people are first together, their hearts are on fire and their passion is very great. After a while, the fire cools and that's how it stays. They continue to love each other but it's in a different way—warm and dependable. . . . Look, after you marry, you sit together by your hut, cooking food and giving it to each other—just as you did when you were growing up in your parents' home. Your wife becomes like your mother and you, her father."[9]

In the U.S., researchers documented a similar trajectory in an extensive study of over a thousand heterosexual relationships at different stages of development. In the study, women were asked to rate their feelings of passionate love, defined as a wildly emotional state, with associated tender

FIGURE 21.   Feelings of companionate and passionate love by relationship length; data from Traupmann and Hatfield 1981

and sexual feelings, elation and pain, anxiety and relief. Then they were asked to rate their feelings of companionate love, described as low-keyed emotion, with feelings of friendly and deep affection.[10] Figure 21 displays the trajectories of these two ratings, using years of marriage as a proxy for the length of the relationship. Although companionate love is stronger among newlyweds than among daters, it slowly declines to pre-marriage levels. Passionate love, on the contrary, stays steady into marriage, and then declines to much lower levels than existed while dating. This mirrors findings from other studies of romantic couples that show that when the romantic-sexual aspect of a relationship ends, it may be redefined as a friendship rather than completely terminated.[11]

Recent studies have begun to show that such perceptual shifts also reflect biological changes. For example, one team of Italian researchers compared the blood of students who reported being "truly, deeply, madly in love" early in a relationship (i.e., less than six months) with students in longer-term romantic relationships (i.e., two years). Participants in early stages of romantic love had slightly higher levels of cortisol, perhaps reflecting increased arousal and stress, as well as substantially higher levels of nerve growth factor (NGF), a hormone that regulates neuronal maturation and survival but that may also modulate the expression of vasopressin—a neuropeptide linked with the formation of social bonds (NGF d-statistic = 0.79, cortisol d-statistic = 0.41). Moreover, the high levels of cortisol and NGF in partners who were madly in love were only temporary. When tested at least one year later in their relationship, the students' cortisol and NGF levels had dropped to levels observed among singles.[12]

Brain imaging studies have also shown that viewing a romantic partner activates different areas of the brain, depending on whether the relationship is at its early stages or has lasted a long time. Interestingly, several of the regions activated only by a long-term partner overlap with those activated when a mother views a picture of her child, suggesting some similarities with the neural activation underlying the experiences of maternal and longer-term companionate love. Supporting this view is the observation that one of these activated brain regions—the pallidum—and an associated hormone that we first encountered in chapter 1 (oxytocin) both play a role in pair-bonding in other monogamous mammals as far-flung as prairie voles, California mice, and marmosets. Some scholars speculate that they play the same role in humans.[13]

These converging lines of evidence from sociology, psychology, and neurobiology suggest that feelings of love often change over the course of a romantic relationship, moving from a focus on intense passionate obsession to a companionate form of attachment. Many couples follow this trajectory, but not all do. This is a very important qualification. Sexual desire may never lead to romantic love, and romantic love may fizzle before partner attachment sets in. These are obvious and well-accepted deviations from love's trajectory.

People can also encounter love in other combinations—attachment with neither sex nor romantic love, and romantic love without attachment or sex. Indeed, in the following sections, I systematically review how people can feel these three kinds of love—sexual desire, romantic attraction, and companionate love—in almost any combination, sometimes in ways that violate common scientific and popular stereotypes of how love progresses.

Is Sexual Desire Necessary for Romantic Love?

If you ask a random U.S. college student, he or she will contend that sexual attraction is necessary for romantic love.[14] And many scholars of love have echoed this point. Consider the claim of two prominent relationship researchers: "It is apparent to us that trying to separate love from sexuality is like trying to separate fraternal twins: they are certainly not identical, but, nevertheless, they are strongly bonded. . . . For romantic personal relationships, sexual love and loving sexuality may well represent intimacy at its best."[15]

Similar reasoning underlies claims that the loving relationships of Shakespeare and his "fair youth," Emily and her eventual sister-in-law, Oprah and Gayle, and Harry and Sally must also involve sexual desire. But is it impossible to love with such focus and passion and yet not have sexual desire for a partner? Is sexual desire necessary for romantic love?

In many situations sexual desire is an important component of romantic love.[16] However, several lines of evidence indicate that sexual desire is not always necessary. For example, in one of the early landmark studies of love involving the self-reports of over a thousand individuals, psychologist Dorothy Tennov found that 61 percent of U.S. women and 35 percent of U.S. men reported feeling romantic love, including intense need for affiliation, separation distress, jealousy, and intrusive thoughts, without feeling "any need for sex" with their partner.[17] Moreover, research conducted across different cultures and historical periods has found that many individuals develop passionate infatuations with partners without any clear presence of sexual desire.[18] Recent in-depth studies of so-called passionate friendships among adolescent girls in the U.S. show a similar pattern. Such relationships involve inseparability, jealousy, possessiveness, preoccupation, intense separation anxiety, and fascination with one another.[19] They also involve behaviors, including stroking, holding, and cuddling, generally reserved in the U.S. for lovers and parents and children. However, there is little evidence that these relationships involve sexual desire or behavior.[20]

Young children who have not yet begun adrenal puberty provide further evidence that one can have intense, romantic attachments without sexual desire. The adrenal and gonadal hormones that elevate day-to-day sexual desire increase dramatically between the ages of four and eighteen.[21] Therefore, if sexual desire were a prerequisite for romantic attraction, one would expect a general increase in romantic attraction over childhood and adolescence. However, this is not the case. In one study, two hundred youths aged four to eighteen were asked to think about an other-gender boyfriend or girlfriend for whom they had intense feelings and to rate their agreement with statements such as "I am always thinking of so-and-so" or "When so-and-so hugs me, my body feels warm all over." Children of all ages reported intense, obsessive preoccupation with their partner, and the intensity of infatuation was not associated with age. Ultimately, it is not possible to determine whether the subjective experience of infatuation was fundamentally the same for four-year-olds as for eighteen-year-olds. Indeed, the subjective qualities of infatuation likely change as sexual desire enters the situation more prominently. Nonetheless, the fact that very young children report signature feelings of romantic attraction at levels comparable with much older adolescents suggests that sexual desire is not a necessary condition for the intensifying preoccupation, separation distress, and heightened need to affiliate observed in romantic attraction.[22]

Is Romantic Love Necessary for Partner Attachment?

In the mid-1960s, sociologist William Kephart asked more than a thousand college students, "If a boy [or girl] had all the other qualities you desired, would you marry this person if you were not in love with him [or her]?"[23] Being in love is a defining expression for romantic love, and by today's standards, many of the students were unromantically practical. Based on their answers to these questions, one-third of men and *three-quarters* of women did not feel that being in love was necessary for marriage. Compare this to the situation in the 1990s, where most American men (86 percent) and women (91 percent) answered the same question with a relatively unanimous "no." Indeed, by the 1980s, the majority of American men and women claimed that romantic love was so important that if they fell out of love, they would not consider staying married.[24]

The fact that American attitudes could change so dramatically within three decades suggests that we should see a great deal of cross-cultural variation in the importance of romantic love in entering a long-term marital commitment. In the 1990s, students in relatively affluent nations (Brazil, Australia, Japan, and the United Kingdom) rivaled their United States contemporaries in romanticism, also answering "no" when asked the question. However, nearly four out of ten Russian students were willing to accept marriage without being in love. And in traditional, developing nations such as the Philippines, Thailand, India, and Pakistan, most students were willing to marry someone with whom they were not in love.[25]

Of course, marriage does not imply that partners experience or exhibit attachment or companionate love. For example, in a study of seventy-three societies, only 56 percent of marriages were classified as "intimate," in which partners ate, slept, worked, and spent their leisure time together. Marriages among Trobriand Islanders off the coast of Papua New Guinea exemplified such intimate ties. Wives and husbands lived in the same house along with their children. They spent most of their work and leisure hours together, eating, sleeping, talking, joking, and sharing household tasks, including childcare. Spouses signaled their mutual love by giving gifts and calling each other *lubaygu*, literally, "my friend."

In the other 44 percent of societies, however, marriages were distant and aloof, as exemplified by unions among Rajput farmers of Khalapur, India. The custom of *purdah* secluded Rajput women: Wives lived in an enclosed courtyard, tending young children, performing chores, eating, sleeping, and cooking. Meanwhile, men spent their leisure time talking and smoking in the men's quarters, where the husbands also slept. The household arrangement reinforced the husband's bond with his mother but also estranged husband from wife. Husbands would view the role of the wife as sexual and reproductive. Only after the mother-in-law died might the wife and husband become something like companions. Thus, a lack of romantic love in marriage may also be accompanied by a lack of friendship, and this pattern seems to be widespread in many societies.[26] Nonetheless, in many other societies, particularly in which arranged marriages are relatively common, there is an explicit expectation that partners come to love one another.[27] Indeed, it seems that in many contexts, companionate love, like friendship, can arise without the initial impetus of romantic attraction or sexual desire.

How Do These Systems Interact?

So far I have questioned the common assumption that love must follow a particular trajectory that begins with sexual desire, moves to romantic love, and ends with the friend-like feelings of companionate love. Ample evidence suggests that it is possible to have romantic love without sexual desire and furthermore to bypass both of these and go directly to companionate love.

Most evidence for the distinctions among sexual desire, romantic attraction, and companionate love rests on what people say and do. Reviewing what is known about the neurobiology of love, Helen Fisher makes a persuasive case that these three drives also recruit different neural structures and chemicals in the brain (table 2). For example, sexual desire and behavior in humans both rely on the hormone testosterone. Men and women with higher testosterone levels have sex more often. Male athletes who inject testosterone have more sexual thoughts, more morning erections, more sexual encounters, and more orgasms. And using a testosterone patch increases sexual interest and sexual activity in menopausal women with low sexual desire. However, these hormones and brain regions do not play central roles in romantic attraction or partner attachment.[28] Moreover, the

TABLE 2. Three kinds of love and how they differ

Kind of Love	Experience	Behavior	Chemical Mediators	Primary Brain Regions
Lust/sexual desire	Craving for sexual gratification	Seek sex	Testosterone	Hypothalamus, amygdala
Passionate love/ romantic attraction	Exhilaration, intrusive thoughts of beloved, jealousy	Exclusiveness, separation anxiety, proximity maintenance	Dopamine, norepinephrine, serotonin	Ventral tegmental area, dorsal caudate nucleus
Companionate love/ friend-like love	Calmness, security, closeness, social comfort, benevolence	Proximity maintenance, mutual grooming and feeding, separation anxiety, helping and sharing	Oxytocin, vasopressin	Pallidum, nucleus accumbens

SOURCE: Fisher 2006

brain regions activated when viewing the picture of a partner with whom one is "truly, deeply, and madly in love" only minimally overlap with those activated when people are sexually aroused.[29]

The second kind of love—passionate love—includes feelings of ecstasy, intense energy, sleeplessness, craving, and mood swings. Fisher argues that these feelings depend on the elevated expression of dopamine, a chemical connected to feelings of enjoyment and motivations to perform rewarding activities. Simply viewing the photo of a romantic partner increases blood flow in the brain's ventral tegmental area—the starting point for cells that send dopamine to other parts of the brain. Viewing a partner also activates the tail of a C-shaped neural structure called the caudate nucleus, a region that receives dopamine from the ventral tegmental area and plays a role in learning about rewards.

For an understanding of the third kind of love—companionate love or friend-like love—Fisher turns to studies in non-human mammals, especially the mouse-like prairie vole. She argues that feelings and behaviors associated with companionate love are mediated by the activity of two neuropeptides that we first met in chapter 1—oxytocin and vasopressin. Moreover, Fisher suggests that two brain regions essential for pair-bonding in prairie voles—the ventral pallidum and nucleus accumbens—are also important in the cultivation of long-term attachments in humans.

Many of the brain processes and networks involved in these three kinds of love likely interact and overlap. Dopamine expressed when falling in love leads to a chemical cascade, including the release of testosterone, which may in turn increase sexual desire. By similar mechanisms, sexual activity may also increase romantic attraction. And the central role of reward systems in all three kinds of love may also mean that commonly used drugs that suppress dopamine pathways—such as some antidepressants—can have unintended consequences for all three kinds of love, an argument made recently by Fisher and psychiatrist J. Anderson Thomson.[30]

A growing body of evidence suggests that these three faces of love (i.e., sexual desire, romantic attraction, and companionate love) reflect different motivational systems, with each depending on different but interconnected neural and hormonal systems.[31] This functional independence allows humans to have sex without bonding, but, more important, to bond without sex or sexual desire.[32] Although these systems can operate independently, there is a great deal of evidence that they also have the capacity to interact, reinforce, and counteract each other.

Sexual desire or romantic love, for example, does not imply friendship. Nonetheless, these faces of love often coincide. The norm for many

Americans and Brits is to think of one's romantic partner as one's best friend.[33] And among adults, romantic partners and spouses report higher levels of friend-like motivations than do best friends.[34] Indeed, it seems that romantic partners are often more friend-like, in terms of feelings and behavior, than are one's closest friends.

What causes many romantic, sexually involved couples to become such close friends? Thus far, most insights into the process have arisen from research on a diminutive, mouse-like rodent that lives on and under the plains of North America—the prairie vole. Prairie voles engage in life-long monogamous relationships involving grooming, nest sharing, and pup raising. These cooperative bonds develop quite rapidly. An initial exchange of pheromones and then a remarkable twenty-four-hour marathon of sex seem sufficient to start the relationship on its long course. Detailed experiments with these rodents have identified chemicals and neural pathways initiated by this prolonged bout of mating that are responsible for the long-term preference for one's partner. Interestingly, this process involves the two hormones that continue to arise in our discussions of bonding, trust, and friendship—oxytocin and vasopressin.[35]

In humans, for example, oxytocin is released in women during sexual activity, is involved in mother-infant bonding, and acts as a link between infant suckling and the expression of milk. Moreover, administration of oxytocin increases trusting behavior among humans in economic games, so much so that they appear to disregard acts of betrayal (chapter 1).[36] The implications of these combined findings are tantalizing. If sexual behavior and physical contact alter neurohormonal networks in the human brain that involve the release and reception of oxytocin, as they do in prairie voles, then this might explain how romantic couples can become such "super-friends." But before you attempt a twenty-four-hour sex marathon with a new partner in hopes of solidifying a permanent pair bond, don't rule out other non-sexual ways of altering the oxytocin system. It seems that humans can become very committed partners even when delaying sex for periods of time that might seem intolerable to prairie voles. Moreover, close friendships may also recruit the same systems without ever requiring sexual desire or activity.

The three kinds of love—sexual desire, romantic love, and companionate attachment—often arise in a predictable trajectory. However, they also involve relatively independent neurological and psychological systems, and it is possible to have sexual desire without romantic love, romantic love without companionate attachment, and companionate attachment without sexual desire. In the next section, I consider in more detail how friendship, which has many core similarities with companionate attachment, overlaps with sexual desire and behavior.

SEX AND FRIENDSHIP

Despite the complex relationship between friendship and sexual behavior, scholars have generally taken one of two opposing approaches to the problem. One camp focuses on the non-sexual aspects of close friendship, such as mutual aid and feelings of goodwill, treating innuendos of sexual behavior with skepticism.[37] The other camp focuses on the potential sexual content of close friendships, using anecdotes and signs of physical contact to infer that such close friendships are primarily sexual.[38]

For example, among Naman herders in twentieth-century Namibia, there was a close relationship known as *soregus,* into which members of either the same or opposite sex could enter. Implying deep friendship and mutual assistance, *soregus* was initiated formally by one of the parties through drinking from a bowl of water and then handing the rest to the other to drink.[39] Some authors have emphasized reports where *soregus* partners have engaged in sexual behavior, such as mutual masturbation. Other authors have expressed skepticism about the frequency with which *soregus* involved sexual behavior, focusing rather on the mutual assistance among partners, especially in economic terms.

In this case, as in many others, there is insufficient data from such accounts to determine how frequently sexual behavior occurred in these close friendships. However, this does not stop speculation on either side. In *GLBTQ: An Encyclopedia of Gay, Lesbian, Bisexual, Transgender and Queer Culture, soregus* is described as an "especially intimate bond of association . . . that included sex between men and between women."[40] Indeed, in a recent submission to the South African parliament regarding whether to broaden the institution of marriage to include lesbian and gay people, *soregus* was used as a native precedent for same-sex unions. Interestingly, the document translates *soregus* as "homosexual." This is quite a stretch, since it was known that women could be *soregus* with men, and its literal meaning is "sharing of water," an important form of mutual aid in the dry deserts of southern Africa.[41]

One of the few cases where we have good quantitative data on frequency of sexual desire and behavior among friends is from U.S. college students and adults. Depending on the age group, one- to two-thirds of individuals report feeling sexual attraction toward one of their same- or opposite-sex friends.[42] And sex with friends is common. In one study of

U.S. college students, one-half had had sex with a platonic friend (51 percent), and one-third (34 percent) on more than one occasion.[43] A consistent finding from such studies is that men more than women harbor romantic or sexual interests for their friends.[44]

It is also possible to have sex with a friend without any feelings of romantic interest, an observation clearly shown by two studies of love in several kinds of relationships. In the first study, done in the 1990s, eighty-four New Haven adults answered a series of questions about how they felt toward their same-sex best friend and immediate kin (mother, father, and sibling) as well as their lover or spouse. Some questions captured the kinds of companionate feelings associated with friendship-like relationships, such as feelings of warmth and closeness and motivation to help. Other questions captured feelings characteristic of romantic relationships, such as physical attraction, preoccupation with one's partner, and the exclusive importance of that partner in one's life. Among the New Haven adults who participated in this study, best friends had "love profiles" that were relatively similar to kin. And not surprisingly, they felt less passionate love toward best friends than toward romantic partners.

Ten years later, researchers at Michigan State University asked students the same questions about a particular kind of sexual relationship, called a "friend with benefits." A "friend with benefits" is a euphemism for a friend with *sexual* benefits; it refers to someone with whom one has sex, without the exclusivity of a romantic relationship. They are defined as "just friends," but they are attracted enough to each other to have sex.[45]

What the researchers found suggests that not all sexual relationships require passion (figure 22). Students felt no more passionate love for "friends with benefits" than they did for their best friends or family members. Indeed, their love profiles looked very similar to those with best friends, rather than those with romantic partners or spouses. This fits closely with data from a study of college students at Arizona State University, where fewer than half who were sexually attracted to a friend were also romantically attracted to that same friend.[46] Therefore, it appears that many people can have sex with a friend, while avoiding the feelings associated with romantic attraction.

In most societies where quantitative estimates exist, the vast majority of sexual behavior occurs among individuals of the opposite sex.[47] For this reason, there is often greater suspicion that opposite-sex friendships will lead to sex or that sex will disrupt the relationship. Writing in sixteenth-century France, Michel de Montaigne, for example, claimed that friendships between men and women could not achieve the intensity of male-male

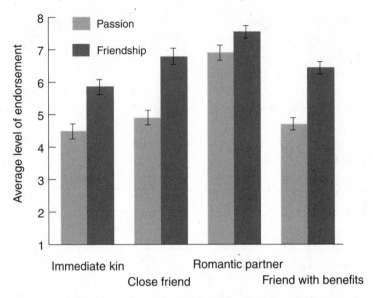

FIGURE 22.   Feelings of passion and friendship by relationship type; data from Sternberg 1996, Bisson and Levine 2009. Note the similarity of the "close friend" and "friend with benefits" profiles.

friendships, because sex or marriage always gets in the way. This claim has become a recurrent theme in Western pop culture, showing up in film, television, fiction, and nonfiction. Of the fifteen societies of the Probability Sample File where ethnographers have noted whether opposite-sex friendships were possible, in 47 percent they were—but they were often less common. However, in the other 53 percent of societies, there were specific concerns about women and men being friends. These were mostly related to the suspicion about the sexual content of that relationship. For example, in one ethnography of Greek village life, if a woman was treated to a drink by an unrelated man, generally a sign of friendship, people took this as evidence of a sexual liaison. Among Iroquois farmers, ritual friendships among men and women were not forbidden, but neither were they "exactly appropriate," since they might involve sexual attraction. Often these prohibitions were most severe during marriage or reproductive life. Thus, for example, among the Igbo of Nigeria in the early twentieth century, opposite-sex friendships were entirely possible, but only for women after menopause.[48]

Even in the United States, where opposite-sex friends are more accepted, opposite-sex friendships are much less common than same-sex friendships and are often problematic.[49] Opposite-sex friends frequently avoid flirting as an explicit relational strategy.[50] And marriage often leads to a much greater loss of opposite-sex friends than same-sex friends—a trend generally attributed to social norms, jealousy, or perceived sexual competition.[51] Sexual attraction, especially when non-reciprocated, can also disrupt existing relationships. In a recent study of U.S. college students who had non-coerced sex with partners they considered to be "just friends," one in four said it caused the friendship to end. The overwhelming reason was that one of the two partners was sexually attracted to the other, while the other was not so attracted.[52] Such differences of opinion about a relationship are common. Sociologists and psychologists have found that the level of romantic interest of one partner in an opposite-sex friendship is generally uncorrelated with the interest of the other.[53] Many people place categorical prohibitions on themselves when it comes to potential sex with a friend; 39 percent of women and 20 percent of men in a 1981 study said they abstain from sex with friends because they fear it could ruin a friendship.[54] Therefore, while it is possible for friends to have a sexual relationship, there are many factors, including social norms, jealousy, and differing expectations, that make it particularly difficult.

FOUR KINDS OF LOVE

This chapter and the previous one examined how friendship fits with two other kinds of close relationships—those among kin and those involving sex. At the surface, all close relationships share feelings of warmth, love, and goodwill, suggesting that they stem from a single psychological or psychosocial system. However, there are also crucial differences in these ways of relating.

When compared to kin, friendship differs in three important ways. First, helping among friends depends strongly on feelings of closeness, whereas helping among kin is less sensitive to such psychological states. Second, immediate kin are less sensitive to the costs of helping than are friends, even increasing their help as the costs increase. Third, in life-or-death emergencies, people overwhelmingly turn their attention to kin, and only secondarily to friends. Despite these differences, kin and friends are not mutually exclusive categories, because it is still possible to augment a kin relation with feelings of closeness characteristic of friendship, and it may also be possible to convince oneself that a biologically unrelated friend is truly like a brother or sister.

In sexual relationships, there are several kinds of love, of which companionate, or friend-like, love is only one. For example, one can have a romantic relationship with no friendship, in which one is obsessed with a partner but lacks any degree of goodwill toward him or her. It is also possible to have a purely sexual relationship without romantic interest or the goodwill of friendship. At the same time, sexual and romantic relationships can come to include friend-like feelings and behaviors, and friendship can unfold into sexual or romantic interests. More research is necessary on the topic, but it may even be possible to have "romantic friendships," which involve intense devotion to (and jealousy over) a partner but lack sexual desire or activity.

Many questions about how these kinds of love interact and influence behavior remain. For example, the fact that helping among kin depends less on feelings of closeness begs the question: Do we feel the psychological processes involved in kin-directed altruism the same way that we feel the subjective closeness that can increase helping among friends? If kin-biased helping is more phylogenetically ancient than friend-biased helping, is it possible that such psychological processes pass under our conscious radar and no longer come into awareness?

More important, where did the psychological mechanisms involved in friendship come from? Are they direct applications of the psychological system underlying kin-biased support? Are they derived more proximally from the systems involved in long-term bonding between sexual partners (without the sex)? Or do they stem from a completely different set of processes? The fact that it is very difficult to distinguish between friendship and companionate love in long-term sexual relationships points to the middle alternative. However, more work is necessary to assess such claims.

Researchers are only beginning to tease apart how these different kinds of love arise and interact in people's bodies and brains, and only when we have a better map of such internal processes will it be possible to disentangle the feelings and behaviors associated with kinship, close friendship, and different kinds of sexual relationships. While many of the mechanisms underlying behavior in close relationships are still to be mapped out, one recurring observation is emerging from psychological and behavioral research. Specifically, there are at least four ways of loving that can co-occur in any relationship: sexual desire, romantic attraction, friendship, and kinship. Each of these kinds of loving appears to involve different psychological states and motivations, different linkages between these psychological states and behavior, and different neurohormonal systems underlying feelings and behavior. Yet these systems are deeply entangled in a way that makes love the complex system of feelings and motivations that has for so long challenged poets and scientists alike.

5 Friendship

Childhood to Adulthood

The road from first meeting to friendship is not always easy. Along the way, one must often deal with conflicting loyalties, cope with rejections and breakups, avoid exploitation, and figure out just what one's friends expect of the friendship.[1] Overcoming these challenges requires a number of truly remarkable social skills. Among other things, one must be able to read others' wishes, needs, and intentions, forgo immediate self-interest at appropriate times, negotiate interpersonal boundaries, and know when (and when not) to forgive. Given all of these requirements, it is not surprising that some adults never find making friends easy or natural.[2]

In Japan, it is commonly believed that children are too immature to handle these issues. Instead of having close friendships (*shinyuu*), they are expected to have only playmates (*tomodachi*).[3] Prominent theories of child development in the United States and Europe are less categorical about the limitations of youth, but they also propose that learning how to cultivate friendships is one of the central challenges faced by adolescents as they mature to adulthood.[4]

How do children learn the suite of skills, expectations, and behaviors necessary for the cultivation and maintenance of friendships? In this chapter, I document what is known about the development of friendship as people progress from childhood to old age. Limited studies in different societies suggest that children learn the skills and expectations of friendship in strikingly similar ways. However, along with learning the broadly shared norm of providing mutual aid, there is a great deal of room for variation between individuals in how they learn the art of making and having friends. Therefore, I finish the chapter by reviewing what we know about how individuals within the same society come to differ in how they make and keep friends, focusing specifically on gender and personality differences.

FRIENDSHIP ACROSS THE LIFE COURSE

Obviously, infants are not born with all of the skills and motivations necessary to make friends that an adult has. Rather, over the first two decades of life, these capacities emerge, with earlier skills providing the foundation for more complex skills and motivations. In the United States and Europe, toddlers develop attachments to specific friends with whom they prefer to play and interact. Over time, children learn that some of these attachments are different, in the sense that they involve expectations of turn taking, sharing, mutual aid, and support. By their teens, adolescents differentiate among their friends, describing some friends as "best," others as "close," and others as just acquaintances. By adulthood, notions and practices of friendship become relatively stable. However, even at later ages, subtle changes can occur, as people face questions such as whether heterosexual men and women can be just friends or whether a spouse, brother, or parent counts as a friend. The following sections review what is known about the development of friendship from toddlerhood to old age, focusing on studies conducted in the United States and Europe and drawing when possible from studies conducted in other societies.

Playmates and Friends

Even before they can say "friend," many human toddlers cultivate bonds showing the first signs of friendship. Observational studies in group care settings have shown that as early as one year of age, toddlers differentiate their peers, directing a disproportionate amount of their social effort and positive expressions, such as smiles and laughter, toward one or two children within a group. Such dyadic bonds are generally mutual and can be quite stable, occasionally lasting years (box 14).[5]

It is very tempting to call such nascent dyadic bonds friendships. Consider the description of a relationship between two children in Hampstead Nursery, a residential nursery for English children orphaned and homeless during World War II: "Reggie (18–20 months) and Jeffrey (15–17 months) had become great friends. They always played with each other and hardly ever took notice of another child. This friendship had lasted for about two months when Reggie went home. Jeffrey missed him very much; he hardly played during the following days and sucked his thumb more than usual."[6]

But do these early playmate preferences really count as friendships? Two key elements of such attachments, the preference for interacting with a particular partner and the communication of positive affect through smiling and laughter, both map squarely onto common behaviors of

---

BOX 14 Behavioral Observation

How do we know about friendship among infants if infants can't talk? Most of our knowledge about toddler friendships comes from painstaking observational studies of infants in playgroups, usually in nursery schools and other group settings. Video recording is a very useful tool in such studies, allowing observers to rewind footage and catch observations that might have been missed the first time around. For measures of proximity, researchers may use a specific perimeter, such as one child being within three feet of another during at least 30 percent of the combined observations of the two children. For measures of who drives affiliation in the pair, it is possible to observe how many times one child approaches the other relative to how often the child leaves. For measures of shared positive affect, they might use observations of coordinated laughter or simultaneous smiling. With such simple observational measures, researchers are able to examine the degree to which the same two children are more likely than by chance to be near each other, and whether longer-term companions are more likely to share positive affect. Once children can talk, most studies of friendship focus on what they say rather than what they do. However, behavioral observation studies provide the best link with studies of relationships (and perhaps friendship) among non-human animals, where researchers must rely on behaviors to infer the existence of a social relationship (Howes 1983, 1988; Howes, Hamilton, and Philipsen 1998; Hinde and Atkinson 1970).

---

friends. Yet while such friendly behaviors and presumably the feelings underlying them represent important precursors to friendship, they are not unique to friendship.

Mutual attraction and affiliation are necessary elements of friendship, but these form the basis for at least two kinds of relationships that begin to emerge in childhood. The first kind fits closely with lay notions of a playmate, a relatively temporary relationship based on enjoying shared interests and common activities. The second kind of relationship is longer lived, and, most relevant to our comparison with friendship, demands mutual help and support.[7]

It is not clear at what age mutual support becomes a defining feature of friendship, especially since children may know this before they can articulate it verbally.[8] However, behavioral studies suggest that there is a crucial transition that begins by about five years of age. For example, in studies that have examined children younger than five years of age, children make almost no distinction between friends and acquaintances when deciding to share (five studies, average d = 0.22). Although these children preferred to interact with certain partners over others, they did not treat these play-mates differently in terms of sharing. This appears to change soon after four or five years of age. Among older children, adolescents, and college students, there is a much stronger effect of friendship on sharing behaviors (ten studies, average d = 0.81). The underlying cause of this change is far from clear. Is it a natural development that occurs in most children? Does it spring from the increasing practice with friendship that children receive in preschool and kindergarten? Or is it simply because older children have more things to share?[9]

Regardless of its developmental origins, the importance of mutual sharing and aid as a feature that differentiates friends from playmates is also observed in many cultures, although the developmental timing has not been systematically studied. Timbira horticulturalists and foragers of Brazil characterize the friendship of youth as associated with mere liking and hanging out, while the adult form of friendship implies the difficult responsibility of loyalty and mutual aid. And the contrast between friends and playmates continues into adulthood in many cultures. For example, Russians make a similar distinction between *drug* and *priiatel'*. One Mus-covite living in Israel expressed the difference between these terms this way: "*Priiatel'* is someone with whom you share some interests, you can sit with, drink with, talk with, maybe exchange some favors. But a *drug*, a real friend, is one you can go to and say, 'I just killed a man. Help me get rid of the body . . . ' I have two friends I know I can count on to do this."[10]

Similar (though perhaps less demanding) distinctions between close and casual friends are frequently made in a diverse range of societies.[11] As I will describe in the next section, an important part of childhood and adolescence is learning about the importance of mutual support, trust, and loyalty in friendships, and how this distinguishes some good friends from casual friends and playmates.

The Language of Friendship

The ability to speak entirely changes the landscape of friendship. With language, friends are able to communicate their precise needs to one another, are able to negotiate norms more flexibly, and can coordinate activity in much clearer ways. Children also begin to use the word *friend* strategically to claim temporary relationship status and gain entry into groups, with such phrases as "If you play over here I'll be your friend" and "You're my friend, right?"[12]

BOX 15 Imaginary and Supernatural Friends

Friendships are apparently so necessary in young childhood that children often make them up. In the United States and Europe, nearly half of all children will have an imaginary friend at least once during childhood, with the earliest imaginary friends appearing at age two or three (Gleason 2002; Taylor 1999; Gleason and Hohmann 2006). Some companions are kept for months and others for years, with some children maintaining imaginary companions until age ten and even age eighteen (Taylor 1999). Although children are often aware that their friends are imaginary, they also state that they receive as much companionship, help, and affection from such friends as real ones (Gleason and Hohmann 2006; Gleason 2002) and also experience similar levels of conflict (Taylor 1999). The most common reason given for the appearance of an imaginary friend is that a child felt lonely. However, such friends need not arise out of preexisting deficiency. Consider one Swedish child's tale of how she found her imaginary friend: "It was when I was building a snow sculpture. Then, I made a small house and then it struck me that someone could live there and then I pretended that a mouse fell down from the sky . . . which was supposed to live there" (Hoff 2004). In some societies, adults also engage with supernatural partners. Among Tapirape gardeners of South America, shamans cultivate friendships with animal, forest, or sky spirits who help them in curing patients. The shamans state that they regularly visit these spirits, eat with them, and go on hunts with them (Santos-Granero 2007).

The capacity for language also permits researchers to access children's conceptions of friendship by asking them to reason about difficult friendship dilemmas and to tell stories about their friendships—even with imaginary friends (see box 15). One line of research has focused on the terms and phrases that children use as they talk about their friendships. Based on studying children and adolescents in the first two decades of life, a surprisingly consistent trend emerges. Younger children, about five to nine years old, more often describe concrete aspects of interactions, such as playing, common activities, helping and sharing, physical attributes of their friends, and global qualities of being nice and good. Around nine years of age, children begin to talk more abstractly, in terms of concepts such as intimacy, loyalty, and trust. They also begin to describe their friends' observed behaviors in terms of underlying psychological concepts, such as intentions, needs, and wants.[13]

These recurring observations have led the psychologist Brian Bigelow and colleagues to propose a three-stage model of the development of friendship expectations.[14]

1. In the first stage, children focus on common activities, physical proximity, and the superficial rewards and costs of interaction (e.g., "He
   helps me" or "I like playing with him"). They do not describe the
   interactions as part of a relationship but rather as independent events
   of something they enjoy doing.

2. In the second stage, normative expectations become important, and
   their violation leads to disapproval, guilt, and loss of admiration.
   Children have strong feelings that friends should take turns, that
   they should help one another, and that they should be nice to one
   another. Moreover, not doing so will make a friend unhappy and
   perhaps angry. This is all the more reason to follow the rules.

3. In the third stage, adolescents think increasingly in terms of more
   abstract expectations, such as loyalty, commitment, intimacy, empathy, and unconditional positive regard. Moreover, these expectations
   are seen as important as a way to make friendship work, rather
   than as a norm that is adhered to for fear of punishment or approval.
   Adolescents explain that they are acting in a certain way not only
   because of the partner, but also because of the demands of the
   relationship.[15]

Throughout these three stages, mutual affection and positive regard are consistently mentioned at high levels, suggesting that these are important precursors to even the earliest stage of friendship.

A parallel line of research has focused on how children respond to the following dilemma developed by Robert Selman and his colleagues at Harvard: Sally promised to meet her best friend on their special meeting day. Later, Sally received a more attractive invitation from a third child (e.g., a movie or pop concert depending on age) who had recently moved into the neighborhood. This invitation happens to be at the same time Sally had promised to meet her best friend, and the best friend has prob-lems he or she wants to talk about.[16]

Following the presentation of the dilemma, a researcher interviews the child about six major friendship issues: friendship formation, intimacy, trust, jealousy, conflict resolution, and the termination of the friendship.

Based on children's answers to these questions, Selman and his col-leagues identified five developmental levels that children go through in talking about their friendships (with important similarities to the previous stage model).

1. At level 0, children understand friendship as repeated instances of coming together to play, where friends are valued for what they bring to play or because they have admirable physical attributes. In this stage, physical and psychological closeness are one and the same.

2. Children at level 1 talk about trust as the predictability of a friend's concrete behaviors, not in terms of shared expectations or a friend's trustworthiness. "We play together. I share with her. We take turns."

3. At level 2, children talk about trust in terms of shared expectations. For example, one will keep secrets and promises because that is what one knows a friend expects.

4. At level 3, children see trust as based on a general norm that one follows for the ongoing stability of a friendship. One does things because one wants to be a reliable and trustworthy friend.

5. Finally, at level 4, children understand that one can have competing norms and obligations, and that one must balance the norms of a particular friendship with other social demands. In short, friends are seen as autonomous agents within the relationship with their own needs and wants. This is referred to as "autonomous interdependence."

These two taxonomies of friendship stages, one based on open descriptions of friendship by young people and one confined to reasoning about a friendship dilemma, map closely to each other in many respects (see table 3). The overarching progression captured by these two systems is that children move from talking in terms of concrete behaviors and self-interest, to talking in terms of meeting partner expectations, to talking in terms of general norms of friendship, specifically those based on loyalty, trustworthiness, and mutual aid. The one exception is Selman's level 4, "autonomous interdependence," apparently the most advanced stage of understanding, which does not have a clear correlate in the first taxonomy and which I will discuss shortly.

It is important to realize that these levels are extracted from what children say about friendship, and that what children (or adults, for that matter) say may be different from what they do or even how they think. For example, children who typically describe friends as momentary physical playmates (Selman's level 0) may also show the ability to work out a compromise, which is a characteristic of Selman's level 2.[17] Children who do not verbalize concepts, such as turn taking, may still act as if they understand them, suggesting that children can embody or enact the concepts before expressing them.[18] Nonetheless, despite the limitations of relying solely on what children say, these levels of friendship appear to have some

TABLE 3. Comparison of Bigelow's stages and Selman's levels of friendship

Bigelow's Stage	Main Characteristic	Selman's Level	Main Characteristic
1	Self-interested play	0	Physical play
1	Self-interested play	1	Predictable, fun play
2	Shared expectations	2	Shared expectations
3	Relational norms	3	Relational norms
n/a	No corresponding level	4	Autonomous interdependence

functional consequences, such as predicting how children of different ages or levels of understanding will behave, as well as which children in the same age group will become friends (i.e., those at similar levels).[19]

Development across Cultures

Both Selman's and Bigelow's theories were originally developed with children in the United States and Europe, raising questions about how well they generalize to other cultural settings. The most thorough study to date that assesses the cross-cultural validity of these theories focuses on children and adolescents in four societies that were strategically chosen to represent different kinds of cultural and social histories. The populations were drawn from both "individualist" societies, where value is placed on self-reliance and individual expression (Iceland and East Germany), and "collectivist" societies, where adherence to group norms is emphasized (China and Russia). The countries were also chosen to represent societies with both a communist political history (China, Russia, East Germany) and a capitalist one (Iceland).[20]

In each of these societies, students at ages seven through fifteen were asked about Selman's friendship dilemma. They were also interviewed in more depth about the meaning and importance of friendship. The researchers found that children in all countries followed a similar developmental trend, with older children reasoning about friendship at high levels (figure 23). There were a few subtle differences in these trajectories. Seven-year-olds in the four countries started at different levels, with Russian children already advanced on average to Selman's level 2 (shared expectations) and children from other societies closer to level 1 (predictable, fun play). However, by age fifteen the other cultures had largely caught up, and

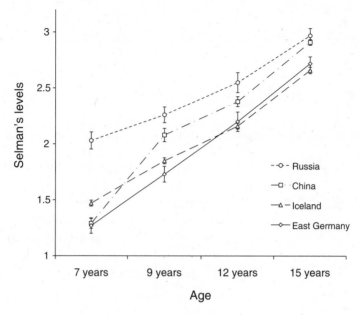

FIGURE 23. Trajectories of friendship reasoning in Russia, China, Iceland, and East Germany. Error bars are standard errors. Reprinted from M. Gummerum and M. Keller, "Affection, Virtue, Pleasure and Profit," *International Journal of Behavioral Development* 32, no. 3, pp. 218–231, copyright © 2008 by The International Society for the Study of Behavioural Development. Reprinted by permission of SAGE.

the average for all children had become much more similar, approaching somewhere between level 2 and 3, a level very close to adults. Therefore, apart from differences in the starting point and the speed of development, children from the different cultures on average approached a similar level of friendship reasoning by age fifteen.

This groundbreaking study has provided an important first look at how reasoning about friendship develops in different cultures. However, questions remain about the cross-cultural uniformity of this progression. For example, are the similar trajectories due to some underlying similarity among the children in these apparently diverse cultures? China, Iceland, East Germany, and Russia differ in terms of their cultural and political histories, but all children in the study had already begun formal schooling. Ample research has shown how formal schooling changes basic patterns of thinking and reasoning.[21] More important for studies of friendship, schools provide frequent and intense contact with same age non-kin peers, providing new opportunities for practicing friendship.[22] Therefore, the similarities observed in these diverse cultures may be a result of common experiences with formal schooling.

Of course, schools are not the only social contexts that foster such intense social interaction with peers. Among the Muria Gond of India, children and adolescents move to common dormitories at the age of six and live in relative isolation from adults. Among Ik foragers of East Africa, children are turned out at age three to form age-grade bands that roam together. These are usually groups of six to twelve kids in which children seek a partner close in age for defense against the other children.[23] Such frequent and dependent interactions may lead to a different pace and perhaps different trajectory of development in friendship reasoning than one would find among children who grow up full-time with their families or in formal schooling or both. Only studies in non-schooled populations will be able to determine to what degree peer interactions in childhood shape the development of friendship reasoning, and therefore how universal such trajectories might actually be.[24]

Socializing Friendship

Existing cross-cultural studies suggest that the psychological motivation and capacity for friendship develops in quite similar ways across cultures. From major U.S. cities to small villages in Papua New Guinea and elsewhere, doting parents and elders encourage this process as well. In the United States, parents annually buy thousands of books, with titles eerily reminiscent of Dale Carnegie's best selling *How to Win Friends and Influence People*—to help *their kids* make friends.[25] Among Ju/'hoansi foragers of Botswana and Namibia, training in the cultivation of relationships through gift giving begins early. At six months to a year of age, parents remove a child's decorative beads and place the beads in the infant's hands to give to a relative outside the nuclear family. Anthropologist Polly Wiessner explains: "From this point on, whether the child agrees or not, parents or grandparents periodically remove the child's beads and give them to a distant relative who takes an interest in the child, explaining carefully the kin term for that relative, how he or she cares about the child, how generous or beloved he or she is, and so on."[26]

This training continues into adolescence as children begin to give to more and more distant kin based on their own initiative. By teaching and encouraging the act of gift giving, parents lay the foundation for a social network of distant kin and eventually non-kin for their children while teaching their children how to create new ties.

Such emphasis on friendship education is common in many societies. Among the Bangwa of Cameroon, for example, elders impress upon children the importance of having friends through folk tales and descriptions of what will happen to them if they don't have friends. Among Kwoma horticulturalists in Papua New Guinea, where children are generally raised to fear non-relatives, especially foreigners, they are taught to be polite and friendly around family friends. Kwoma elders tell children that they can visit and play with these children without fear of sorcery, and that they should be nice to them when they come to visit.[27] The high priority parents place on their children's friendship skills is not universal, however. In one Canadian study of parents' goals for their children, parents of European descent prioritized helping their children to make friends. Aboriginal Canadian parents, on the other hand, placed much more value on teaching children respect for elders and devotion to their families. Similar cultural differences have also been found among Asian American parents, who encourage non-familial ties less, and parents of African American and Latino youth, who encourage non-familial relationships more. However, even when parents don't prioritize making non-kin friends, they often still value it to some extent.[28]

Adolescence

In many societies, adolescence is a time when friendship gains new importance and takes on novel functions. Around this time, friendships last longer, and children start talking about friendship in terms of relational norms and expectations, reaching a level of understanding comparable with most adults, as described earlier in this chapter.[29]

At this time, friends also begin to provide support related to novel needs and desires that emerge in adolescence. For example, as adolescents become interested in sex and marriage, friends gain a new role as go-betweens in courtship—whether they like it or not. In many societies, visiting friends in other communities is an important way to meet boyfriends and girl-friends, and friends are recruited to deliver messages between potential lovers or spouses.[30] Among San foragers in southern Africa, for example, when a young man desires to marry, he first tells his best friend (/hosub), who then approaches the parents of the prospective bride.[31] In some ethnographic descriptions, this role in courtship is the only mention of friends made for a specific culture.[32] Moreover, in cultures where it is not appropriate to discuss sexuality with parents or family members, friends can be the only people with whom adolescents discuss questions of sexuality.[33]

Cultural norms also influence what novel functions friendships begin to fulfill in adulthood. For example, children from the four cultures considered earlier—China, Russia, East Germany, and Iceland—reached the same level of friendship reasoning, involving relational norms of mutual support, sharing, and trust, by age fifteen. However, the fourth level of friendship reasoning proposed by Selman, "autonomous interdependence," may be less important in certain societies where individual autonomy and self-reliance are less valued as cultural goals. Specifically, a comparison between the friendship reasoning of fifteen- versus eighteen- to nineteen-year-old Icelandic and Chinese young adults documented the transition to autonomous interdependence in both societies.[34] As postulated by Selman's theory, Icelandic youth emphasized how friendship provides a way to develop their individual selves. However, Chinese youth emphasized close friendship as a means to integrate into the wider social system, something different from Selman's description of autonomous interdependence, in which friends are seen as autonomous agents within the relationship with their own needs and wants. Therefore, the first stages of friendship may follow a similar developmental pattern across cultures, but how friendship becomes most useful in adulthood may depend on the particular goals that are most valued in a society. At the moment, more research is needed to understand how this happens.

Quantitatively studying the way that humans make use of time is an interesting way to understand cultural as well as age differences in friendship. Many accounts of adolescence describe the large amount of discretionary time that teenagers have to interact with friends, as they are still free from many of the obligations to community and family assumed by adults.[35] This is clearly true in the United States and Europe, where youth have between 30 and 50 percent of waking hours at their discretion and may spend as much as a third of their waking hours with friends.[36] In such contexts, friends (and romantic partners) begin to replace parents as the preferred source of emotional support and affiliation.[37]

Quantitative studies of time use in other parts of the world paint a different picture. Youth in many East Asian countries, for example, have much less discretionary time (only 20 to 35 percent of waking hours), and in some rural semiliterate populations the proportion drops to less than 10 percent.[38]

These differences in leisure time also map onto differences in time spent with friends. For example, in one comparative study of eleventh graders in the United States, Japan, and Taiwan, U.S. teens spent 18.4 non-classroom hours per week with friends, compared with 12.0 for Japanese (d = 0.50) and 8.8 for Taiwanese adolescents (d = 0.89). Such cross-cultural differences have been confirmed by experience sampling method studies, which show

---

BOX 16 Experience Sampling Methods

A researcher hoping to measure how people spend their day and how they feel through the course of it faces a daunting task. A common strategy is to ask individuals such questions in a single survey. However, it is not clear how accurate people are at remembering many daily events, such as how many times they went to the bathroom or how long they spent talking on the phone. Another possibility is to follow participants with a clipboard, regularly recording what they do and asking how they feel. However, this obviously could be awkward for participants and tedious for researchers.

Experience Samping Methods (ESM) were developed in the late 1970s to collect this kind of data more efficiently with cheap and portable digital devices. Like the clipboard method, ESM involves participants responding to repeated momentary assessments as they go about their (otherwise) natural daily life (Scollon, Kim-Prieto, and Diener 2003; Hektner, Schmidt, and Csikszentmihalyi 2006; Larson and Csikszentmihalyi 1983). Participants carry a small device, such as a pager, alarm watch, or PDA, which beeps at random times during the day, asking participants to answer questions about their current mood and activities.

ESM potentially provides a fine-grained, detailed picture of human experience. However, the method has several limitations. It is generally restricted to literate populations and to participants who are willing to tolerate such a strange intrusion into their daily lives, though the device is less strange than a clipboard-wielding tagalong. People are less likely to respond in some contexts, such as in the evening, in the home, when the device must be removed (e.g., at swimming pools), or when the signal might be disruptive (e.g., in church). Moreover, it is still subject to many of the same problems of self-reporting, such as responding in socially desirable ways (Scollon, Kim-Prieto, and Diener 2003). Despite these limitations, however, ESM permits an efficient measurement of how people spend their days and provides more ecologically valid data than can be attained in single surveys.

---

that time spent with friends accounts for 18 to 30 percent of U.S. white and black young adolescents' time, as compared with 19 percent of time for Korean adolescents and 10 percent for middle-class Indian eighth graders (box 16).[39] Although comparable quantitative evidence does not exist for most societies, one cross-cultural study of adolescence in one hundred and seventy-six societies provides a coarse-grained picture of the degree to which peers matter for socialization, a crude proxy for how frequently they spend time together. In the study, peers were judged to be important agents of socialization in less than a quarter of all societies (forty societies for boys and twenty for girls).[40] Therefore, the experience of friendship in adolescence may vary a great deal, depending on the opportunities available for interacting with friends outside the immediate family and how much time is spent taking advantage of these opportunities.

Adulthood

As surveyed in chapter 2, cross-cultural descriptions of friendship in adulthood have a common set of core features. Close friends feel affection and goodwill toward each other, and they help one another when needed. They commonly exchange gifts and are expected not to keep accounts of favors and help. By adulthood, the core behaviors and expectations of friends appear to stabilize.

This is indeed true in the U.S. and European context, where several studies have examined the conceptual and social characteristics of friendship at different stages of the life course. Despite the numerous life transitions that can occur during adulthood, such as marriage, settling down, raising children, developing a career, and retiring, people have a striking consistency in their notions of friendship over the course of adulthood.[41]

These same studies have also revealed that important life transitions in Western contexts (e.g., entry into the workforce, marriage, child rearing, retirement, widowhood) can indeed change time pressures and social mobility, and thus affect interactions with close friends.[42] Therefore, notions about friendship may be relatively static throughout adulthood, despite the dynamic nature of particular friendships.

The most studied transition in the United States and Europe is that associated with dating, cohabitation, and marriage. During this period, couples are hypothesized to go through "dyadic withdrawal," in which partners spend less time with specific friends, and couples develop joint friendships with other couples.[43] Substantial evidence from cross-sectional and longitudinal studies supports the dyadic withdrawal hypothesis. Romantic partners, and especially spouses, generally replace mothers and friends as the primary attachment figure—the person with whom one feels most safe and secure and with whom one most wants to affiliate.[44] And this change in affiliation can become a source of conflict with friends and family.[45] During this time, the sheer number of friends listed in social network questionnaires actually increases as couples combine their social networks. However, the frequency of interactions with these friends drops dramatically. After several years, the number of friends also reduces to pre-marital levels, with substantial turnover in the actual set of friends.[46]

A smaller set of studies also suggests that other life transitions, such as child rearing, divorce, widowhood, and retirement, also affect a person's interaction with friends, often in gender-specific ways.[47] However, the effects of these transitions are relatively small or temporary, with people reporting a great deal of consistency in the prevalence of old friends and in the perceived ability to make new friends.[48]

Although we might surmise that life transitions in other non-U.S. and non-European countries also influence interactions with friends, there are very few systematic, quantitative studies that permit us to test the dyadic withdrawal hypothesis or the influence of any other life transition on friendships. It is also important to remember that notions and expectations of friendship may remain relatively stable despite the changing quantity and quality of friendships over the life course.

DIFFERENCES IN FRIENDSHIP

Despite a common set of core features that adults come to expect of their friendships, individuals also differ in how they make and maintain friends. When asked to name their close friends, for example, people regularly report anywhere from zero to twenty. And some individuals do not feel a strong motivation to make friends at all.[49] People also differ in what they value most in their friends, with some preferring friends they can talk to about personal matters, and others valuing friends they can trust as reliable sources of support.[50] Several reasons have been proposed for such differences in friendship, most notably early childhood experiences with caregivers that influence people's templates for close relationships and gender differences in interpersonal styles. In this section, I review these two proposals.

Friendship and Attachment

A common explanation for differences in friendship derives from a long-standing theory in child development—attachment theory—proposed and extended by psychiatrists and psychologists over the last half-century (box 17). The theory postulates that children learn a template for social relationships, or an *internal working model*, early in life, which is based largely on their interactions with a primary caregiver, who may or may not be their mother. The internal working model is a set of ideas, feelings, and expectations about relationships and about how both partners will and should behave. Since it is a working model, it continues to develop with age and social experience, but as it adapts to novel relationships with romantic

BOX 17   John Bowlby and Attachment

Living in the 1950s, the English psychiatrist John Bowlby was witness to several kinds of broken bonds. The recently concluded world war had left many children in England homeless and orphaned. Hospital visitation practices of the day dictated that children who came for surgery or treatment could be separated from their parents for days at a time. And the American psychologist Harry Harlow was publishing the results of his experiments with rhesus macaque newborns removed from their mothers and given surrogates made of terrycloth and wire.

Drawing from observations about these disrupted social bonds as well as studies of mother-infant dyads and diverse reading in evolutionary theory and ethology, John Bowlby developed a theory of attachment that proposed that human babies have an innate urge to make emotional attachments. Bowlby proposed that this need for social bonds was an evolved predisposition, having increased infants' chances of survival by ensuring that they were protected and fed. Moreover, such early attachments influenced later development, providing a template for cultivating loving relationships later in life.

Although it is impossible directly to *see* an attachment or bond, the theory proposed that attachments were manifest in four kinds of behavior: (1) proximity maintenance, the desire to be near attachment figures, (2) separation distress, anxiety in the absence of attachment figures, (3) secure haven, returning to the attachment figure for comfort in the face of threat, and (4) secure base, exploring the surrounding environment using the attachment figure as a safe home base. Problems with attachments could be observed in these behaviors with primary attachment figures. And events that interfered with nurturing attachments, such as the abrupt separation of infants from their caregivers, could lead to mental health and social problems down the road.

The theory has been the starting point for many studies of the physiology of human social bonds and the role of early experience in adult mental health. It was also more than purely academic in its scope. Bowlby's work helped change common care practices in orphanages and visitation rules in hospitals. It also led to heated debates about the role of mothers in childcare, specifically and more generally the role of women in society (Bretherton 1982).

Despite its influence both in academic and public spheres, Bowlby's theory only goes so far in understanding friendship. Since it was developed from the perspective of an infant, it did not focus on the centerpiece of friendship—that is, providing aid and motivations to help. Moreover, despite occasional nods to the importance of attachment in friendship, most studies have focused rather on parent-child bonds or on relationships with romantic partners. For a review of the few studies that have examined the relationship of attachment to friendship, see Mikulincer and Shaver 2007.

partners, spouses, close friends, and offspring, attachment theorists also argue that this maintains an identifiable form that reflects early childhood experiences.[51]

In the 1960s, American psychologist Mary Ainsworth developed an experimental protocol called the "the strange situation," which she and many other researchers have used to assess how infants interact with caregivers and strangers in a novel environment—in short, to tap into an infant's internal working model of social ties. The strange situation begins as a researcher takes a mother and her child to a novel room with toys. In a series of eight episodes, a stranger enters the room and the mother leaves the infant on two separate occasions. The researcher records how the infants respond to their mothers' absence, how they explore the room, and how they seek proximity to others. Based on infants' behaviors, especially upon the return of their mothers, the researcher classifies infants into one of three categories. *Securely attached* infants, about two-thirds of all infants, are distressed at their mother's absence but are easily soothed upon return. About one in ten infants are *anxious-avoidant*. They do not become distressed at their mothers' departure, and they avoid her upon return. In short, they do not take much notice of their mother. Finally, about one-quarter of infants are *anxious-resistant*. They become highly distressed at their mothers' departure, but when the mother returns, the infant is ambivalent, crying and reaching to be held but then wriggling free.[52]

This classification scheme has since been extended to adults and elaborated, using mostly survey questionnaires to capture two dimensions of behavior and thought in relationships (see figure 24). The first dimension deals with anxiety about the relationship, while the second deals with avoidance in the relationship.[53] Those who are low on anxiety and avoidance (i.e., securely attached) find it relatively easy to become emotionally close to others and to depend on others, and they do not worry about being alone or being rejected by a partner. Those who have high anxiety and low avoidance (i.e., anxious-preoccupied) want to be emotionally intimate with others, but they find that others are reluctant to get as close as they would like. They also worry that others don't value them as much as they value others. Those who have low anxiety and high avoidance (i.e., dismissive-avoidant) feel comfortable without close emotional relationships, valuing independence and self-reliance. Finally, those with high anxiety and high avoidance (i.e., fearful-avoidant) feel uncomfortable getting close to others. They want emotionally close relationships, but they find it difficult to trust others completely or to depend on them, worrying that they will be hurt if they allow themselves to become too close to others.

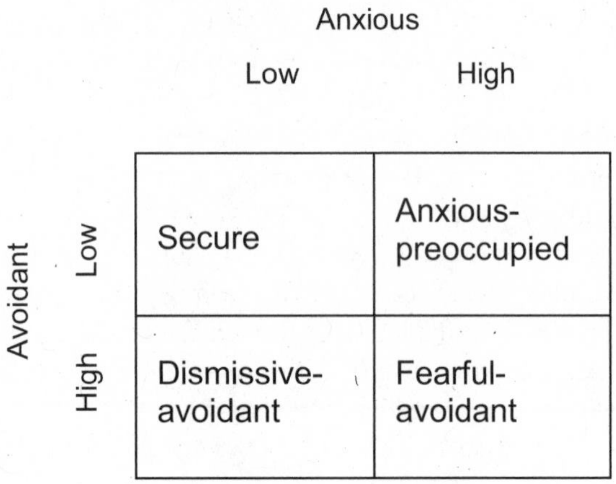

Anxious

	Low	High
Low	Secure	Anxious-preoccupied
High	Dismissive-avoidant	Fearful-avoidant

Avoidant

FIGURE 24. One classification of attachment styles. This two-dimensional classification adds a distinction between two kinds of avoidance—dismissive and fearful—that were not distinguished in Mary Ainsworth's original classification.

Far from a stable personality attribute, people's attachment styles can also change over time and with particular relationship partners. For example, nearly a third of individuals switch attachment styles over the course of months.[54] And adults often have different attachment styles with different partners, such as parents, offspring, romantic partners, and best friends. You can be securely attached with a close friend but fearful-avoidant with a parent and dismissive-avoidant with a romantic partner.[55] Finally, people who have a global attachment style that is not secure may still find such secure attachments in some of their closest relationships. For example, when people were asked to respond to a survey of adult attachment styles, many more were classified as secure (98 percent for a friend and about 84 percent for others) when asked specifically about their closest attachments—to parents, friends, and current romantic partners—compared to when they simply filled out a survey for a global measure of attachment (around 60 percent). Moreover, researchers have been able to experimentally manipulate attachment styles, suggesting that these styles are even sensitive to simplified experimental treatments.[56] These diverse results raise serious questions about the long-term stability of attachment styles.

Based on these findings, some relationship researchers have argued that people do not even have one global attachment style, but rather have different styles for different partners, contexts, or relationships.[57] A recent study tested the three following possibilities: there is a single global style; there are many independent relationship-specific styles; there are relationship-specific styles that draw from a common global template.

The analysis indicated that the last model best fits the data. Therefore, individuals appear to have a global style, but it may be adapted to particular kinds of relationships.[58] There is also emerging independent evidence for the validity of a global attachment style, as it strongly predicts different physiological responses to stresses and social rewards. For example, people who are more anxious and uncomfortable in their relationships have larger jumps in cortisol when confronted with a stressful situation—in this case random, annoying sound recordings of an electric shock (d = 0.84). Moreover, the amygdala region in the brain of people with anxious attachment uses more blood when confronted with angry faces. On the other hand, people classified as avoidant show less activation in reward circuits in the brain—the striatum and ventral tegmental area—when confronted with a smiling face, interpreted by the study authors as finding positive social stimuli less rewarding. These results suggest that people's approach to their relationships may reflect a more general way of responding to the physical and social world.[59] However, the root of these individual differences is still not clear. Are these attachment styles learned during infancy, as standard attachment theory proposes? Or do they reflect genetic differences or developmental differences that are somehow set in utero?

More generally, physiological studies of social bonding indicate that a common cast of characters in the brain, including oxytocin, vasopressin, dopaminergic reward circuits, and opioids, plays an important role in a wide range of attachments, from mother-infant bonds to romantic relationships and perhaps friendships. Such observations are tantalizing given that some of these players—oxytocin and dopaminergic reward circuits—may also influence benevolent and trusting acts (which are different from simple affiliation and attachment). However, how these diverse players interact in the cultivation of attachments and ultimately benevolence is still admittedly quite uncertain and requires further study.[60] Indeed, the feature of attachment theory that most limits its application to friendship at the moment is its original theoretical focus—how infants develop attachments to caregivers. Infants are generally not in a position to provide aid to their partner, and therefore attachment theory has focused on affiliation seeking rather than on motivations to help a partner—goodwill and benevolence. How people seek proximity to others is obviously an important part of making friends, and there is some evidence that children and adults with secure attachment styles are slightly better at making and keeping friends and have slightly smoother interactions with friends than those with anxious attachments.[61] Nonetheless, affiliation is only one part of friendship, and therefore classical attachment theory only takes us so far in understanding the signature behaviors of helping and sharing among friends.

Friendship and Gender

Judging by book sales, people in many parts of the world are obsessed with gender differences. The iconic self-help manual *Men Are from Mars, Women Are from Venus,* which claims a dramatic psychological divide between men and women, has now sold more than thirty million copies in over forty languages. Deborah Tannen's *You Just Don't Understand: Women and Men in Conversation,* which argues that men and women belong to different cultures of talking, enjoyed *New York Times* bestseller status for nearly four years.[62]

These and many other books on sex differences feed on a deep-seated interest about the opposite sex and on what C. S. Lewis described as the enjoyment that women and men gain from laughing at each other.[63] Women muse at men's golf games, fishing trips, and inability to talk about anything other than cars and sports. Men joke about women's phone conversations, coffee dates, and frequent chats.[64] At times the caricatures have not been so benign. Writing in sixteenth-century France, Michel de Montaigne claimed that women possessed neither the constancy of mind nor the communicative abilities to maintain a durable friendship. A quote from Calcutta in the 1960s echoes this sexist viewpoint: "Men have friends, women have only acquaintances."[65] While the vast majority of recorded claims about friendship have been biased toward the superiority of male-male friendships, in the last four decades the pendulum has swung, with frequent claims that men's friendships are less close, less supportive, and less satisfying than those among women.[66]

Although some of these caricatures may possess a kernel of truth, the willingness to accept reports of sex differences often leads to the spread of factoids with little or no scientific basis. For example, in her popular 2006 book *The Female Brain,* Dr. Louann Brizendine reported that women on average utter twenty thousand words a day compared to seven thousand for men, a statistic picked up by numerous mainstream news sources, including CBS, CNN, National Public Radio, *Newsweek,* the *New York Times,* and the *Washington Post.* However, a glance at the book's copious footnotes reveals that the sole basis for this claim was the self-help book *Talk Language: How to Use Conversation for Pleasure and Profit.* When an interested phonetics professor tried to follow the chain of citations to an academic source, the trail vanished.[67] Indeed, only one study has quantified the natural conversations of a large number of people over extended periods of time. In the study, 396 men and women from the United States and Mexico were tracked with special recording devices over the course of their daily routines. When their natural speech was transcribed and counted, women held only a slight numerical, and statistically insignificant, advantage—16,215 words per day, compared to 15,669 for men. The claim of large sex differences regarding language usage was not only unsubstantiated, it was wrong.[68]

This anecdote does not in any way diminish the importance of the differences that do exist between men and women. It is clear that women and men strongly differ in many ways, most apparently in physical measures such as height and voice pitch.[69] The problem is that many claims of difference, especially those that fit our caricatures of men and women, such as in moral orientation and helping behavior, are often accepted without strong or consistent empirical evidence.[70]

Research on friendship follows a similar pattern. There are numerous recent claims about the ways that men and women differ in their friendships: that women's friendships are more intense, intimate, and of higher quality than are those of men; that men's friendships are less personal but more stable; that women value talking with their friends, whereas men prefer hanging out and playing sports; that men have special-purpose friendships, each with a specific function, while women build all-purpose friendships that encompass a broad range of needs and activities; and finally, that women cultivate a few close friendships, whereas men relish acquiring ever more friends.[71] Such apparent differences have led many scholars to agree with one prominent relationship researcher that "there is no social factor more important than that of sex in leading to friendship variations."[72]

Numerous theories have been proposed for such sex differences in friendship. In her book *The Female World*, sociologist Jessie Bernard argued that most men and women live in single-sex worlds that foster different ways of engaging with friends. According to some accounts, girls are socialized to be more relationship-oriented than boys, whereas boys are socialized to be independent, competitive, and self-reliant.[73] What unifies such cultural arguments (i.e., arguments that men and women differ due to different socialization) is that the attitudinal and behavioral differences that men and women exhibit in their relationships are so vast that these differences represent separate and distinct worlds or cultures.

Another set of arguments based on evolutionary theory makes similar predictions but for very different reasons. According to these arguments, the different life circumstances of men and women over human evolution led to the selection of different psychological and social capacities. One theory, coined "tend and befriend," suggests that the cultivation and maintenance of social relationships were particularly important for human mothers as they sought food and care for their offspring. Therefore, women have greater motivation and capacity to form social relationships than do men.[74] Another theory focuses on the different ways that women and men moved between communities in early ancestral environments. Some evidence suggests that human males tended both to stay in their natal group and to form strong, kin-based coalitions. Females, on the other hand, tended to immigrate to their husband's (or sometimes kidnapper's) kinship groups when they reached reproductive maturity. Whereas males had kin on hand for help and support, females were forced into interactions with more distantly related kin or with non-kin. The theory argues that females would have faced unique selection for investing in and cultivating non-kin relationships, and that this would predict, among other things, that women would make a greater investment in a smaller number of relationships.[75]

A third class of theories argues that men and women have the same underlying motivations for cultivating friendships, but that sex differences in social roles, related to work, marriage, and child rearing, provide different opportunities for cultivating friendships. Anthropologists have proposed that such sex or gender differences are prominent in areas of southern Europe and the Middle East, where women are prohibited from forming non-kin ties, especially after marriage. In many of these and other cases, women may be so residentially isolated from other women that forming friendships is neither practical nor feasible, whereas the geographical mobility and material resources enjoyed by their male counterparts render it relatively easy for men to make friends.[76] Despite the relative economic and social equality of men and women in the U.S., some scholars have argued that structural factors can also account for sex differences in friendship in the U.S. For example, the unequal burden of childcare can limit women's interactions with friends, and different job conditions can be more or less fertile situations for cultivating friendships.[77]

What unites these diverse social, cultural, and evolutionary theories is the idea that women and men have different ways of making and having friends. However, there is a deep problem with these theories. When investigators have compiled the hundreds of quantitative studies of behaviors and expectations among friends, such as self-disclosure, empathy, helping, or the sheer number of close friends, they have found that these theories have very little to explain in the first place. Men and women are actually quite similar in how they engage with friends.

Consider the common activity of sharing personal details and concerns through talk. Caricatures in the U.S. suggest that this is much more common in women's than in men's friendships. According to one early study on the topic, women are "blabbermouths" with their friends, whereas men are "clams."[78] However, in the 1990s, two psychologists compiled fifty quantitative studies on how men and women share personal details with their friends. They found that women were indeed slightly more likely to discuss personal matters with friends, but that the sex difference was very small (average $d = 0.28$). Based on this statistic, if you randomly picked one man and one woman on the street, there would be only a moderately better than fifty-fifty chance (59 percent, to be exact) that the woman shared more personal details with her friends than did the man.[79]

When an effect is so small, it is important to identify other subtle differences between men and women that might account for this observation. In self-report studies, is it possible that men are simply less likely to recall what they talked about with their friends?[80] Are the measures biased toward certain kinds of personal issues or self-disclosure? Are reports of men and women biased by the caricatures themselves? For example, in one interview study, when women and men answered broad questions about friendship, they repeated common stereotypes about sex differences—women focused more on talking and men on shared non-verbal activities. However, when asked focused questions about specific friendships, women's reports of non-verbal activities increased, as did men's reports of personal talk.

It is also not clear that these minute sex differences arise in other cultures. In one study of college students in India, where men are expected to be more expressive and interdependent than in the United States, no gender differences were observed in reported self-disclosure among friends.[81] Even in the United States, sex differences might be restricted to particular populations. For example, in one study of U.S. high school students, black students showed no gender difference in talking with school friends about personal issues, though a difference did arise among white students.[82]

A thorough review of quantitative studies reveals similarly small and inconsistent effects for other ostensibly gendered qualities of friendship, namely empathy and helping.[83] In most cases, the similarities between men and women far outweigh the differences. Both men and women report

---

BOX 18 Meta-analyses

Attempting to navigate the large number of studies on sex differences in friendship can be a daunting task. These studies cover a wide range of topics, using various observational and self-report measures, and they often report contradictory findings. One approach taken to make sense of such diverse findings is *meta-analysis*, a method that combines the results of many studies concerning the same hypothesis to permit more accurate conclusions and to understand how different methods and study designs affect results. It also is a way to avoid cherry-picking only those studies that support one's view.

Recently, two psychologists conducted a meta-analysis of eleven published studies that reported sex differences in self-disclosure among friends in childhood and adolescence (Rose and Rudolph 2006). Depending on how one combines the results of these eleven studies, the average d-statistic is 0.65–0.72. This indicates a relatively strong effect, and much bigger than the small effect observed in a meta-analysis of fifty studies conducted among adults (d = 0.28) (Dindia and Allen 1992). This discrepancy raises important questions. Are children and adolescents more likely to show sex differences in self-disclosure than are adults? Are there methodological differences between the studies? Regardless of the possible cause, the fact that these estimates are derived from a number of independent studies makes it worth investigating.

Meta-analyses also have several limitations. If studies have a consistent bias, such as a person's tendency to respond in socially desirable ways, then a meta-analysis will compound that bias. Also, by relying on those studies in the public domain, which tend to be published based on the "merit" of finding a significant result, a meta-analysis may ignore many studies that could not be published because they did not have statistically significant results. In this way, a meta-analysis may overestimate the size of an effect (Sutton et al. 2000).

---

that they rely on close same-gender friends for companionship more than they rely on siblings, mothers, or fathers.[84] Women and men have similar expectations of close friends, and they are quite similar in cultivating both special-purpose and multifaceted friendships.[85]

Despite the lack of evidence for large, global differences between the friendships of women and men, there may be subtle developmental distinctions. For example, one recent meta-analysis (box 18) of hundreds of studies suggests that some behaviors and expectations of friends exhibit more marked gender differences in childhood and adolescence. If girls' thinking and behaviors with friends develop faster than do those of boys, then this is precisely the time when we would expect larger differences. However, we would also expect these differences to disappear once boys caught up in adulthood. Only more research will reveal whether this is in fact the case.[86]

While occasionally showing weak differences, the fact is that most studies show that men and women cultivate friendships in very similar ways. Both men and women define friends as partners with whom one shares mutual affection, goodwill, trust, and loyalty. Both men and women converse (at least in the United States and Europe) with their friends on a regular and frequent basis. Where sex differences do emerge, they tend to be small and culturally specific.[87] Moreover, these differences pale in comparison to the kinds of differences observed between members of different cultural groups (chapter 7).[88]

The social skills necessary for making and keeping friends emerge and build up over the first two decades of life. According to existing studies, the development of basic reasoning about loyalty, goodwill, sharing, and affection follows quite similar trajectories in different cultures. However, this research has focused exclusively on children exposed to formal schooling, a setting that provides ample opportunities for practicing and learning about friendship. Only more targeted cross-cultural studies will help us understand how child-centered institutions, such as formal schooling and age grading, can influence the ways children learn to reason about friendship.

Even in the same society, individuals can differ a great deal in how they make and keep friends. Some prefer many friends, some want a few close friends, and others want no friends at all. Some people care most about having a friend they can talk to about personal problems, others care more about having someone they can count on, whereas others want someone who will make them laugh. In this chapter, I also reviewed how different attachment styles might play a role in some differences in friendship styles and how putative differences between men and women often do not. In the next chapter, I shift focus from how people learn about friendship over a lifespan to examine general processes by which friendships develop between specific partners.

6   The Development of Friendships

Think of your closest friends and how long you have known them. You may remember some from early childhood and others from high school or college. It is unlikely that you met any for the first time yesterday. Now pick one of these friends and try to identify the precise moment in time when your friendship became close. For some friendships it may be possible to identify a turning point, such as an act of great magnanimity or a personally important bonding event. More frequently, when asked to specify the precise point at which a friendship became "close," people stumble.[1]

In such cases, what happens in the period of time between first meeting and the onset of a close, mutual friendship? Why do some acquaintances become friends, whereas others do not? And how do thought processes and behaviors transform as acquaintances become closer?

In this chapter, I describe the basic elements of friendship's development, from first meeting to the dance of cues, signals, and changes in thinking involved in the transformation from acquaintances to close friends. An important part of this process is a change in thinking in which friends help one another regardless of the balance of accounts or the shadow of the future. Such unconditional aid can be a great boon in times of need. But being on the wrong side of a feigned friendship can also leave one open to exploitation by unscrupulous partners, as numerous examples illustrate.[2]

In thirteenth-century Rome, Boncompagno da Signa wrote down a taxonomy of all the ways that friends could abuse their trust. There are "vocal friends," who attend with words alone; "here and there" friends, who hang out but never help; "conditional friends," who help with quid pro quo expectations; "imaginary friends" borne of infatuation; "shady friends," who show devotion until they get what they want; "false friends," who deceive; "haughty friends," who judge and mock; "fair-weather friends," who leave in tough times; "mercenary friends," who secretly hope to gain from their investment; "turncoat friends," who talk behind your back; "pleasure-seeking friends," who will drag you into vice; "dispossessed friends," who will ignore the friendship for a lover; "futile friends," who can't keep a secret; and "hard-hearted friends," who do not share.[3]

With all of these ways that friends can do us wrong, an important part of cultivating friendships is avoiding and deterring such bad behavior. Indeed, defense against such misbehaviors plays a central role in a friendship's growth, maintenance, and, in case of misplaced or excessive defense, demise. Therefore, in addition to the signals and behavior that can increase levels of closeness and trust in a friendship, I also describe how people avoid (or at least try to avoid) partners who might manipulate these signals and exploit their goodwill. Defenses are an important way to avoid false friends, but defenses that are too strong can prematurely destroy a beneficial friendship, and I conclude by describing three "defenses against defenses"—relational blindness, forgiveness, and conflict resolution—that help friends stay together in the face of inadvertent offenses and violations.

CONTEXT AND ATTRACTION:
THE FIRST STEPS TO FRIENDSHIP

Two things must happen before a pair of individuals can become close friends. First, they must meet. Second, they must have sufficient mutual attraction to continue meeting.

Contexts of Friendship

Friendship cannot arise without meeting, and the available contexts for interaction determine which pairs of individuals might build friendships. In small-scale societies where one interacts on a daily basis with kin, one's friendships will be built largely among existing kin ties. In societies with few opportunities for contact between unrelated men and women, there will be few cross-sex friendships.[4] In the twenty-first-century U.S., by contrast, one can cultivate close friendships in numerous places, including high school, camp, college, work, and increasingly on the Internet. However, despite the abundant opportunities for meeting, even friends in the U.S. are often segregated according to such social variables as neighborhood, wealth, occupation, and ethnicity, reflecting implicit limitations on who interacts with whom.[5]

An especially fertile context for cultivating friendships, and one that appears frequently in the ethnographic record, is forced separation from existing social networks. At Bomana Prison in Papua New Guinea, for example, friendships gained new importance for inmates who came from all parts of the country, spoke diverse languages, and were separated from their families. Inmates adopted the English words *friend (pren)* and *partner (partna)* to describe the strong bonds they form with other inmates from different tribes. Johannis, a convict from the western highlands, says his best friend was a cellmate from the Motuan coast. As *partna*, they shared tobacco and cigarettes, helped fetch meals and clothes if the other was ill, and told stories and jokes when the other was depressed and lonely. But, most important, they cared for each other in any way they could in a place far away from their families. In twentieth-century Cameroon, the children of Bangwa farmers were sent miles away to their chief's palace to learn dancing, games, hunting, and fighting and to serve as court pages. Isolated from their families, youth formed strong friendships at the palace court, cemented by both daily interactions and large-scale cooperative endeavors, such as raids against neighboring villages and head-hunting expeditions. Or consider Korean students who have traveled from their rural village to pursue a college degree in Seoul. They describe intense homesickness, an obsessive need for contact with family members. For some, it feels like a physical craving, such as thirst or hunger. The only remedy or source of enjoyment in this novel context is their newfound friends.[6] Indeed, many kinds of separation from existing social networks, whether due to war, imprisonment, migration, schooling, or marriage, often foster the cultivation of new friendships.[7]

New contexts for friendship can arise from life transitions, social change, and the vagaries of daily life. They can also be purposely created. In many societies, children of the same age take part in communal rituals around the time of puberty, which parents and elders hope will bind them together in special age-mate relationships.[8] In some societies, children are assigned sponsors from the elder generation who are charged with teaching them about their community and culture, and with whom they often form close friendships.[9] In other societies, children formally inherit friendship from their parents (box 9). What unites these cases is the expectation that these imposed relationships will become something like friendship, characterized by mutual aid and feelings of goodwill. Although such arrangements do not always lead to close friendship between partners, they often do, in much the same way that partners in arranged marriages in many cultures often come to love each other.[10]

The Necessity of Attraction

Although contexts and ascribed relationships open (and constrain) opportunities for friendship between particular partners, they do not guarantee that two individuals will become friends. Consider the case of Owiyap, a Kwoma gardener in Papua New Guinea, who inherited a friend from his father. Despite his father's endorsement of the match, the friendship never flourished, and Owiyap ultimately grew very close with another villager. In a similar vein, Kwoma age-grade comrades who participate in the same puberty initiation are expected to have intimate friendships. However, the reality is that only some age-mates become close friends.[11]

As Owiyap's story indicates, being thrust together in the same context is necessary to get to know a potential friend, but it is not sufficient for cultivating a friendship. Partners must also be motivated to interact with each other over the long term. Social psychologists use the metaphor of attraction—the drawing together of two objects—to describe this motivation. There are a number of generic qualities that, in many studies, people find attractive in potential friends. These include a reputation for helping, high status, and interpersonal similarity in social class, ethnicity, and personal attitudes. Indeed, simply knowing that a person likes *you* is often enough to increase your liking for him or her.[12]

The initial attraction that leads to a friendship can also be quite strategic. For example, Azande traders in north-central Africa often entered formal blood brother relationships for purely commercial ends: to permit private trade in their partner's land. Although such relationships often began out of purely commercial motives, they also on occasion developed into full-fledged friendship with mutual affection and goodwill.[13]

People can also become attracted, or attached, to an individual not for any generic quality of that person, but rather because they are attracted to that person in particular. The prototypical example of such attachment is a human infant's attraction to his primary caregiver (who may not be his biological mother). This attachment involves several notable motivations and behaviors, including both a desire to approach the caregiver, especially when afraid or threatened, as well as signs of distress when the caregiver leaves. Similar motivations and behaviors play an important role in many human relationships, including romantic ties and friendships. This phenomenon of human bonding formed the basis of Bowlby's and Ainsworth's attachment theory discussed in chapter 5. As I argued in chapter 5, however, this theory, like other theories of attraction, only describes how we find partners rewarding and attractive and not explicitly how people become motivated to help one another.

Beyond Context and Attraction

Context and attraction are necessary foundations for friendship. Context provides opportunities to meet. Mutual attraction serves as a self-reinforcing context, one in which both partners have an interest in continuing to interact over time. However, these two alone can be very weak bases for a relationship. Casual friends may enjoy hanging out but have very few expectations of mutual sacrifice. Indeed, a common description of the false friend in many treatises on friendship, from Confucius to Boncompagno da Signa, is a person who eats and drinks with you and provides enjoyable company when life is good but who deserts you in a time of great need.[14]

What mutual attraction does provide is the *motivation* for future interactions with which partners can build trust, learn their respective needs and wants, and transform their motivations toward one another (box 19). Moreover, mutual attraction gives some small guarantee of repeated interactions in the future, an expectation that makes people, even relative strangers, more likely to help one another, to trust one another, and to cooperate.[15]

In chapters 1 and 2, I identified signature behaviors, feelings, and kinds of communication that occur commonly among close friends. These include feelings of closeness, love, and trust, gift giving, self-sacrifice, and mutual aid without close account keeping. How do people move from giving tentatively and keeping accounts, the common mode of interaction between acquaintances and strangers, to sharing at high levels of trust and support, without keeping track of past favors? How do people's desires change from self-interest to wanting to fulfill a partner's needs? As I described in chapter 1, a part of this process may involve psychological merging such that caring for a partner feels like caring for oneself. These feelings of closeness can be measured and manipulated in controlled laboratory conditions with interesting results. Compared to strangers, or even acquaintances, partners who feel close are more likely to share benefits, make sacrifices, see the world through their partner's eyes, and even—to a degree—view themselves as being like their partners. But how do people get to this level of closeness? How do they induce partners to feel close, and, more important, how do they guard against manipulations of closeness that might make them feel more willing to sacrifice for an unscrupulous partner? I explore these questions next.

BOX 19 Moving from Conditional to Unconditional Trust

An important part of becoming closer in a friendship is moving from calculated reciprocity to unconditional trust and knee-jerk altruism. A recent experiment suggests that this transformation can happen quite quickly and involves changes in how the brain mediates decisions to trust (at least in the context of a simple economic game). As part of the experiment, a team of economists and neuroscientists asked pairs of strangers to play thirty-six consecutive trust games (see box 4) where the role of investor and trustee was switched between games. As the partners played, a Magnetic Resonance Imaging device recorded the blood flow in their brains. After all pairs played, the researchers divided them into two groups: those who most successfully cooperated over the course of the repeated games and those who had high rates of defection.

Not surprisingly, the pairs who had successively cooperated felt much closer to each other than they had prior to the experiment (d ~ 0.83). The uncooperative pairs did not. When viewing the images of blood flow, the researchers made several interesting observations. In a brain region involved in representing the mental states of ourselves and of others (i.e., the paracingulate cortex), the cooperative dyads had higher levels of activation in the first eighteen rounds, but this activity decreased in the final eighteen rounds. The researchers interpreted this as an initial concern about what the partner was thinking that then decreased once trust became unconditional. Interestingly, the opposite trend was observed among noncooperative pairs. Moreover, in a brain region involved in both social bonding and the release of oxytocin (i.e., the septal area), the cooperative dyads increased activation over the course of the thirty-six rounds. Defecting pairs did not. The authors interpreted decreasing activation in the paracingulate cortex and increasing activation in the septal area as evidence for a shift in decision making among cooperative dyads, away from calculation and toward unconditional, oxytocin-mediated trust. A further finding that supports this view is that cooperative pairs, and only cooperative pairs, became much faster in their decisions as they played more games. This supports the notion that they were engaging in more-spontaneous and less-deliberative decisions (Krueger et al. 2007).

In short, cooperative pairs were beginning to run on autopilot. What is astonishing about these findings is the relative speed with which cooperative pairs moved to unconditional trust. Was this a product of the controlled laboratory conditions, with little room for misunderstanding, or the rapid pace of games, which in real life might be drawn out over very long periods of time? Or is it possible that such automaticity can arise quite quickly in real-life relationships?

ENGINEERING FRIENDSHIPS

In 1937 Dale Carnegie published the wildly successful book *How to Win Friends and Influence People*. The book promised to "enable you to make friends quickly and easily," and as a result, "increase your influence, your prestige, your ability to get things done." Within a few months of its first release, it had gone into its seventeenth printing, and today it has sold more than fifteen million copies in more than thirty languages.[16]

Carnegie saw the powerful potential of friendship as a means of encouraging people to do things for one another and carefully dissected ways that people can induce such friendship in an expedient manner. The book made explicit many elements of conventional wisdom and tacit knowledge about how to make friends. Among the six ways to make people like you: smile a lot in their company; appear to be interested by encouraging them to talk about themselves; use their name frequently. As Carnegie stated, "Remember that a man's name is to him the sweetest and most important sound in any language."[17]

Much research since Carnegie's time has shown that such simple manipulations, if done correctly, can make people feel closer to you and more likely to help you. In studies of tipping behavior, waiters who use a customer's name, smile, or lightly touch the customer receive larger tips than those who don't. And these manipulations are far more important than the quality of service in determining the size of tips. Similarly, giving a small gift or providing a small favor can make even complete strangers much more likely to help with far greater costs to themselves. And simply inducing a positive mood in a person, by invoking positive memories, can increase prosocial behavior. Although the friendship that Carnegie hoped to induce was more akin to simple persuasion, many of his techniques have been shown in experimental settings to make strangers feel closer and be more likely to sacrifice for one another.[18]

Based on these and other insights, psychological researchers have developed several methods to create surrogate friendships in laboratory settings. In one frequently used task, strangers take turns sharing personal information (for nine to fifteen minutes) on a variety of progressively more intimate topics, such as "Is it difficult or easy for you to meet people? Why?" and "Tell me one thing about yourself that most people who already know you don't know." Partners who participate in the task report feeling substantially closer to each other than those who simply sit in a room with a person. They also act in ways that are characteristic of friends. In one study, such induced friends were more likely to claim equal responsibility for both successes and failures on a group task. By contrast, strangers committed what is called the self-serving bias, claiming more responsibility for success while accepting less responsibility for failure.[19] Anecdotal evidence suggests that these closeness induction tasks can also have long-term effects. After participating in such a task, one pair of strangers started a relationship and ultimately married.[20]

Many techniques for making people feel closer and act more prosocially involve framing a relationship in terms of "we" or "us."[21] These include assigning people to arbitrary groups, gossiping about outsiders, or manipulating the perceived similarity between two people (e.g., changing one's appearance or name).[22] For example, people are more likely to cooperate in a prisoner's dilemma game if they think they share the same birthday. They are also more likely to help individuals who share either the same first or last name, a fact that may underlie the common practice in many cultures of friends taking the same name or of individuals with the same name automatically gaining status as fictive kin.[23] For example, in the nineteenth century, Nyasa farmers of eastern Africa would exchange names with friends from other tribes. These friends had special duties to help one another in times of need, and should one by chance visit his comrade's town, he expected to receive food and lodging.[24]

Another way to prime friend-like feelings and behaviors is to simply invoke the names of friends or the word *friend* (box 20). People consistently offer more help to strangers when exposed to the name of a very good friend on a computer screen compared to the name of a colleague or an acquaintance (five studies, average d = 0.65).[25] And addressing others as "friend" is a common strategy to gain goodwill or to pave the way for important requests. Politicians regularly use the "F-word" to garner votes and supporters. Among U.S. presidents, for example, Franklin D. Roosevelt, Dwight Eisenhower, Richard Nixon, and George H. W. Bush have all made repeated calls to "my friends" in major convention speeches. Most recently, 2008 U.S. presidential candidate John McCain used "my friends" on average every four minutes during a ninety-minute televised debate.[26] Charities use the term to encourage giving, nation-states use it to motivate international partners to behave cooperatively, and Internet con artists use it to convince people to enter exchanges and hand over money.[27] A similar strategy is used in a wide range of cultures, whether the word for friend is *cideke* (Apache) or *lopae* (East African Turkana pastoralists).[28] Consider the use of *friend* by one Turkana herdsman before asking a favor of British social anthropologist Philip Gulliver: "He said, 'You are my best friend now.' I agreed. 'Then,' he said, 'you must give me shillings for my tax payment.'"[29]

BOX 20 Priming and Framing Studies

When a politician says "my friends" in a speech or a Nigerian get-rich-quick spammer uses "Dear friend" in an email salutation, these relative strangers are trying (probably unsuccessfully) to change our state of mind—to make us more trusting or helpful. In the language of psychologists, they are attempting to *prime* us so that we respond in desirable ways to what they have to say next. Humans are notoriously susceptible to priming, especially when they are unaware that it is happening. Priming people with the word *professor* makes them perform better at Trivial Pursuit. After seeing achievement-related words, people become more competitive, and after being primed with words related to an elderly stereotype (e.g., *Florida, old,* or *lonely*), people walk more slowly down a hallway than those primed with control words (*thirsty, clean,* or *private*). Priming people with the word *helpful* makes them more helpful, whereas priming them with *conformity* makes them more likely to seek consensus, and *intelligent* improves their performance on a general intelligence test (Bargh 2006; Forster, Liberman, and Friedman 2008).

Much research has focused on the extent to which subtle priming can change people's behavior in improbable ways and what mechanisms underlie these effects. Such experimental protocols can also provide insights into how activating thoughts and feelings of friendship can change behavior. For example, we saw in chapter 1 how framing a trust game as a "friendship" game among Maasai herders in Kenya can change their willingness to reciprocate. And in other studies, subliminally priming individuals with the name of a close friend makes them more willing to forgive offenses, even under time pressure. Moreover, subliminally priming people with the names of close partners makes them more willing to help a person in need (Cronk 2007; Cronk and Wasielewski 2008; Karremans and Aarts 2007; Mikulincer et al. 2005).

Such studies provide exciting insights about how our state of mind (and our behavior) changes when thinking about friends. However, as the Nigerian spamming example suggests, we are not complete dupes, and such primes are only successful in specific circumstances. Understanding how such primes actually change behavior and how we guard against exploitation is a useful area for further inquiry.

This whirlwind tour of examples—sharing personal details, manipulating similarity, invoking friendship, and expressing liking for a partner—shows that people have many techniques at their disposal to frame a friendly situation and to compel a partner to start behaving in a way that might begin a friendship.[30] If two partners can use such signals and cues to persuade each other to invest in a mutual, close friendship, then there are numerous advantages for both participants, most notably the possibility of help in times of need without close monitoring or justification. However, someone can smile, use your name, and share some personal information to induce feelings and behaviors of friendship but have no intent of reciprocating. Indeed, these are common strategies used by con artists to prey on unsuspecting victims, or *marks*.[31]

Talking, smiling, and touching are relatively cheap signals that are easy to fake and thus open to manipulation. And not surprisingly, people have caught on that these are signs of potentially false friends. For example, in his thirteenth-century taxonomy of friendship, Boncompagno Da Signa classified such friends as "vocal friends" (*amicus vocalis*)—or those who attend to friends with words alone.[32]

Signals more costly than talk and flattery are also manipulated to create connections for potentially exploitative purposes. Confidence games frequently involve an initial act of generosity on the part of the con artist, which is intended to make the victim more trusting in the future.[33] Corporations use similar techniques to influence buyers. For example, in 2004 the U.S. pharmaceutical industry spent 7.3 billion dollars on visits and gift giving to physicians and hospital administrators in efforts to increase the use of their drugs.[34] There are also numerous examples from the ethnographic record of unscrupulous individuals using "friendship" to exploit another individual. Consider one story of a Zambian trader who ritually contracted a bond friendship with a local farmer to acquire produce at little cost to himself. The exploited farmer finally cut off the relationship, but only after he had given up many batches of home-brewed beer, snared animals, and produce from his plots.[35]

The danger of manipulation is ever-present in fables and proverbs around the world. An entire genre of fables told from Liberia to Tanzania focus on such trickery in friendships. Often begun with two animals who are best friends, the stories tell the tale of one tricking the other, ultimately leading to the end of the friendship. One common and gruesome story in this genre involves the killing of the duped friend's mother during a famine to feed the two friends after they had explicitly agreed to kill *both* their mothers.[36]

Most people would like to think that they can resist such bald misuses of friendship. However, more insidious forms of inequality, dominance, and even unintentional exploitation can also be hidden in the guise of friendship, as long as partners maintain mutual affection and loyalty toward each other. In Ethiopia in the 1970s, Maale farmers could call on friends to help with clearing, cultivating, weeding, and harvesting their fields by sponsoring a *dabo*. To throw a *dabo*, a household would brew beer and ask friends and neighbors to come and work. Wealthier households with large landholdings generally required more labor and were also in a better position to acquire the beer needed for a *dabo*, which led to notable inequalities in who labored for whom. In one group of farmers in 1975, each of the four richest households enjoyed an average surplus of thirty-seven workdays, while each of the seventeen poorest held an average deficit of seven workdays. By framing the *dabo* as a need-based form of communal cooperation, a select few households benefited greatly compared to others. Indeed, in many societies, the ethic of unconditional mutual aid among friends is used to justify persistently unequal relationships between patrons (those with political power in a community) and their clients or followers.[37]

DEFENSES AGAINST ABUSE

Friends use one another all the time, whether they ask for help with childcare, with moving, or for a ride to work. The challenge in friendship is not to avoid such calls for help, but rather to distinguish between being useful and being used. In the previous section, I described many ways that partners can manipulate the cues and signals of friendship to unfairly exploit a partner. There are also many defenses against such exploitation involved in the cultivation of a friendship. These include increasing the value or stakes of the relationship, decreasing the value of outside options that might tempt one to defect from the relationship, and inferring a partner's view of the relationship, which may reveal if a friend does not have one's best interests at heart.

Increasing the Value of the Relationship

One line of defense against exploitation is to make a friendship sufficiently valuable that a partner will hesitate to behave badly if it risks ending the friendship. Friends can increase the value or raise the stakes of their relationship in many ways, thus making it better for both partners to be good to each other.

Over long periods of interaction, friends often develop more efficient ways of communicating, solving problems, resolving conflicts, reading each other's needs, concerns, and desires, and identifying good ways to help each other.[38] For example, in studies where people watched their friends in a conversation, they were 50 percent more accurate at inferring their friends' self-reported thoughts than they were at inferring the thoughts of strangers.[39] In short, friends are experts at how their partners think and feel, while strangers are novices.[40]

One experiment suggests that friends not only predict each other's state of mind better, but also react more automatically to each other's plights, in this case in the form of blushing in an embarrassing situation. To capture a suitably embarrassing situation, researchers asked U.S. undergraduates to sing "The Star Spangled Banner" in front of a video camera. Some days later, the students were asked to bring a same-sex friend to the lab, where they sat with the friend and a stranger to watch their taped rendition. To measure blushing, the researchers attached a photoelectric sensor to each participant's right cheek. To assess sympathetic nervous response, the researchers used a skin conductance monitor (similar to the kinds used in lie detector machines).

Not surprisingly, the performers themselves exhibited the greatest increase in redness and became significantly redder than did strangers who observed the performance. Friends of the performers, on the other hand, blushed at levels between performers and strangers, suggesting that the friends had empathically reenacted their friends' feelings in their own bodies (friends more than strangers, $d = 0.67$). Moreover, the skin conductance responses were significantly correlated among friends ($d = 1.04$) but not among strangers ($d = -0.30$). The greater knowledge and empathy shared by friends is frequently used in everyday interaction to make conversations faster, to coordinate more efficiently, and to monitor one another's needs.[41]

Multiplexity is another common way to build up the value of a relationship. *Multiplexity* is a measure of the diversity of shared activities, aid, and support engaged in by partners in a relationship. If the relationship between a doctor and a patient involved only annual checkups, then it would not be very multiplex. On the other hand, if the patient was also the doctor's attorney, and the two played poker on Friday nights and took turns driving their kids to daycare, then the relationship would be considered more multiplex. Multiplexity makes it possible to fulfill a broad range of needs in the context of one trusting relationship, rather than requiring a new relationship for every new kind of transaction. Close friendships are frequently more multiplex than casual friendships and other kinds of relationships, and such multiplexity is generally built up over time.[42]

BOX 21   Starting Small and Raising the Stakes

Increasing the value of a friendship is often purposely gradual as friends test one another's intentions and goodwill. A case from Africa illustrates how this strategy can screen potential friends before increasing the stakes of helping. In the mid-twentieth century, Gwembe Tonga tobacco growers of Zambia faced a dilemma. The market was generally very far away, and they needed help selling their tobacco. Foreign traders were abundant, offering to take their tobacco, sell it, and return with the money. When such trade first began, foreign traders and Gwembe farmers frequently relied on an existing institution of bond friendship to ratify their agreements. However, the relationships often proved one-sided, with the Gwembe receiving much less than the value of their tobacco. Often a trader took the tobacco and never returned. Well aware that they frequently lost on balance through newly formed friendships, it became common practice to start any new trading relationship with very small amounts of tobacco and gifts, building up the trust and quantity of exchange over time. When talking with anthropologist Elizabeth Colson, Gwembe tobacco farmers were quite explicit about their raise-the-stakes strategy for building a relationship with foreign traders (Colson 1971).

While many friendships do not begin with such openly acknowledged strategy, they frequently do depend on a tacit raising-of-the-stakes, as new friends in the casual stage start with small acts of trust and aid and build up what is expected over time (Rose and Serafica 1986; Hays 1985; Rose 1984). This approach mirrors the practice suggested in Ecclesiastes, "If thou wouldst get a friend, try him before thou takest him, and do not credit him easily" (6:7). This gradual building of friendship plays an important role in minimizing losses from fair-weather friends and in protecting oneself from other exploitative partners.

The greater ability to infer a friend's thoughts, to empathize with a friend, and to engage in multiple kinds of sharing and exchange are all examples of what economists call *transaction-specific investments*. The important part of such investments is that they are non-transferable. We cannot easily leave our current friendship and take our common knowledge, trust, and subtle ways of communicating to a brand-new partner. Such investments are largely lost if we end a friendship, making it more costly to leave the relationship.

A key feature of these transaction-specific investments is that they are built up over time. Perhaps for this reason, people often use longevity as a proxy for the value of a relationship. Whether in Korea, the Southwest

U.S., or Russia, people describe greater trust for and behave differently toward longtime friends, especially those from childhood or school.[43] Anthropologist Keith Basso describes how Western Apache criticized Anglo friendship as being "like air," precisely because it lacked the kind of longevity required for their closest equivalent to friendship: "There is no word in Western Apache that corresponds precisely to the English lexeme *friend*. The nearest equivalent is *shich'inzhoni* ('toward me, he is good'), an expression used only by individuals who have known each other for many years and, on the basis of this experience, have developed strong feelings of mutual confidence and respect. If the stranger is Anglo, it is usually assumed that he wants to make friends in a hurry."[44]

At one level, the duration of a friendship gives clues to a partner's long-term intentions in the relationship, and it gives more opportunities to prove one's commitment to the friendship (box 21). At another level, if the relationship has become sufficiently close over time, with the possibility for very high levels of mutual aid and support, it could be personally disadvantageous to behave in a way that might end the relationship. Therefore, longevity can serve as both a signal of past behavior and a deterrent against future violations.[45] This may be why the handful of longitudinal studies of friendship maintenance has found that the single most consistent predictor of a friendship's future longevity is how long the relationship has lasted already.[46]

Decreasing Outside Options

Transaction-specific investments can make a friendship more valuable and thus deter violations that might end the friendship. Another deterrent is to limit a partner's alternatives or make it more difficult or costly to cultivate new, possibly competing friendships. For example, many of the activities that we expect of a close friend—gift giving, eating together, and just hanging out—are all activities that cannot be scaled up easily to a larger number of partners, thus restricting the ability of a partner to cultivate a large number of friends.[47]

The cross-culturally prevalent practice of ritually confirming friendships is a case in point. In many societies, formally initiated bonds are the outgrowth of already emotionally strong ties that partners want to formalize and further strengthen, in the same way that marriage often arises out of an existing romantic relationship. For example, among the Banyoro of western Uganda, individuals could cultivate close friendships, called *ekimeere*, that involved positive feelings and obligations of mutual aid. Partners could also take such a relationship to another level of exclusivity by engaging in a ritual to confirm them as blood brothers. A ritual clearly and publicly ratified the existence of a relationship, so that betrayal also became public, and a would-be betrayer might find it more difficult to cultivate new relationships.[48]

Jealous reactions against a partner are another way to decrease outside opportunities. Although we often associate jealousy with romantic or sexual partners, it is also common among friends.[49] Consider Melville's description of jealousy among best friends in Tahiti: "Though little inclined to jealousy in love-matters, the Tahitian will hear of no rivals in his friendship."[50] And in the U.S., time spent with other friends or romantic partners can lead to jealous reactions and even to the end of a friendship.[51]

In truth, decreasing the value of outside options or making it difficult to cultivate new relationships is another way of increasing the value of one's relationship. However, such an approach does not increase the intrinsic value of the relationship. Rather, it increases the value of a relationship relative to other possible opportunities.

Inferring a Partner's View of the Relationship

Well before we are adults, many of our decisions about how to behave toward people are governed not by their behaviors but by the intentions we think are guiding those behaviors. For example, people are more willing to reciprocate a favor if they feel it was originally extended without ulterior motives. They are more likely to forgive a transgression if they feel it was unintentional, and they will more readily punish an intended versus an unintended offense.[52] In short, people evaluate exchanges and behaviors at two levels: the concrete outcomes of a behavior, and the symbolic or intentional level—what the outcomes tell them about the intention of the other party.[53]

The first behavioral level is much easier to evaluate and forms the basis for many theories in economics and behavioral ecology. However, the second symbolic level takes center stage among friends. In short, it's the thought that counts. Although close friends regularly disregard the precise balance of give-and-take, it is entirely legitimate to be concerned about how a friend is feeling and thinking about the relationship. Especially at early uncertain stages of a friendship, people are intensely interested in a partner's mindset and motivation and also in the potential friend's consideration of their needs. As a friendship progresses, clear evidence that a partner is not considering one's own well-being is grounds for ending the relationship. Contrast this with ending a relationship because a friend was unable to reciprocate a favor immediately.[54]

Focusing on the symbolic level requires reading a friend's mind, which is not a straightforward task. We can only see what people do, including what they say, how they behave, or how they inadvertently express their feelings. From these pieces of information, our brains must somehow infer their feelings, thoughts, and intentions. Psychologists refer to this ability as a theory of others' minds, a capacity that develops in humans over the first few years of life. At an early age, most human infants learn that some things in the world behave as if they have intentions (such as people and animals), and at around one year, they begin to interpret the behaviors of others as intentional and goal directed. Finally, around four years of age, children begin to understand that other people have not just intentions that are manifested in behavior, but also thoughts and beliefs, which may or may not be expressed in actions. They also come to realize that people sometimes mask their true intentions and attempt to deceive. More important, at some point along the way, they are able to piece together observations of another's behavior to reconstruct a model (perhaps only partially correct) of what that person is thinking and feeling. This model of what we think they are thinking in turn influences how we interpret and respond to their behaviors, for example, whether something given is a gift, a bribe, a repayment, a loan, or an insult.[55]

Although humans have the capacity to develop theories of others' intentions and feelings, they are not always very good at it. For example, in one study of nearly ninety thousand U.S. high school students in the 1990s, only about 57 percent of boys and 73 percent of girls named a best friend who returned their nomination as a close friend. And these findings mirror those for U.S. adults.[56]

Initially low agreement about the qualities of a friendship can increase in longer lasting relationships. For example, in the 1960s a psychologist at the University of Michigan followed two groups of college freshmen who lived together in two houses. At the beginning of the year, how much one student liked another bore almost no relation to how much the second student liked the first. However, over a fifteen-week period, as people's friendships became more defined, students became more reciprocal in their affection. After fifteen weeks, these correlations between pairs of students were moderate ($d = 0.74$), but for those friends who have been together for many years, the correlations get much stronger. In studies of friendship pairs that have lasted just a few years, friends agree at moderate to high levels about aspects of their friendship, including positive qualities, competition, and conflict ($d = 0.63$ to $d = 1.31$).[57] In one longitudinal study of friendship that lasted over nineteen years, however, partners who were still friends at the end of the study strongly agreed on the strength of, commitment to, and closeness of the relationship ($d = 1.86$ to 32.8).[58]

What causes this increase in agreement about the state of a relationship? What are the signs and cues that people use to learn about their friends' intentions and feelings toward the relationship and whether their partners value the relationship? The most obvious indication that a friend does not value the relationship is failing to help when one is in need. Willing and obvious violations of this norm can produce a sense of betrayal and are a common cause of relational breakups. As anthropologist Steven Piker observed of friendships among Thai peasants, "So long as neither party suspects the other of withholding help when it is needed, the relationship can and does continue indefinitely without goods or services being rendered in either direction."[59]

However, focusing solely on failures to help poses a problem. The sporadic rhythm of needs can leave long periods where the willingness of friends to sacrifice goes untested. Therefore, people often rely on other signals that a partner may not value the relationship. For example, a too-close attention to accounting in a relationship can be a warning sign that a partner cares too much about the balance of inputs and outputs in the friendship or feels that the relationship may not last. Therefore, attempting to pay back a debt too quickly can be perceived as a mild violation of friendship, and charging the market rate for a service performed for a friend can cause great offense or even end the relationship.[60]

Paradoxically, not asking a friend for help can also be a signal that one does not sufficiently value the friend or the relationship. Indeed, in one recent study in Japan and the United States, college students were asked how they would feel if a close friend needed help—with taking care of a dog, fixing a computer, or having a place to stay for the night—but didn't ask them for help, going rather to another friend or a market-based service. In both cultures, students said they would feel sadder, more disappointed, and less close with the friend than they would have if the friend had come to them for help.[61]

Another sign that a friend does not value a relationship is a failure to devote enough effort to the relationship, in the sense of spending time, giving gifts, or simply devoting attention to a partner. Spending time with a partner is a very honest way of communicating how much one values that partner. Time cannot be hoarded and spent like money or material resources, and so by spending time with one partner as opposed to another, one demonstrates an exclusive interest in that partner.[62] In studies that have examined the causes of conflict in friendships, disputes over time commitments are by far the most common.[63]

In lieu of focusing on equity or the strict balance of favors in a relationship, friends often focus on how much their companions value a friendship, not only in terms of failures to help, but also with such cues as how partners keep accounts, how they ask for help, and how they devote time and effort to the relationship.[64]

RESPONSES TO VIOLATIONS AND
DEFENSES AGAINST DEFENSES

Benjamin Franklin advised, "Be slow in choosing a friend, slower in changing." As we all know, betrayal hurts. The discovery of a friend's abuse of trust or failure to help in a time of need can cause profound psychological wounds, including persistent and debilitating sadness or anger, confusion about the event and its future consequences, and obsessive rumination about the events surrounding the betrayal. These feelings are captured vividly in a song by a Nigerian Hausa woman who found her best friend sleeping with her husband:

> The best friend broke the faith,
> The useless friend, the worthless friend,
> I shall never be an "older sister" again,
> I shall never take another "younger sister."[65]

The anguish evident in this song may also have real links with perceptions of physical discomfort. Recent studies of how betrayal is registered in the brain indicate that it activates networks involved in visceral feelings of pain.[66]

Considering the high standards held for close friends, we might also predict that friends' violations hurt worse than do those by acquaintances or strangers. For many kinds of violations this is true. U.S. college students, for example, say they would feel more betrayed if a close friend failed to defend them against verbal backstabbing than they would if it were simply an acquaintance. They would also feel more betrayed if a close friend tried to "steal" their romantic partner, attempted to immediately balance accounts, didn't defend their reputation, harmfully deceived them, or failed to provide help, such as a much-needed car ride.[67]

In many cases of betrayal, victims may seek vengeance or demand retribution, silently adjust their feelings of goodwill toward the transgressor, or simply walk away from the relationship.[68] The strong emotional reactions triggered by friends' violations are important defenses against exploitation of one's unconditional aid. However, they also pose a danger. People make mistakes; expectations and communication are always imprecise; and there are times when friends are legitimately unable to help. In such situations, a draconian response could end a long-term and mutually beneficial friendship at great cost to both partners. Therefore, people often have defenses against their own defenses, which prevent premature reactions to perceived slights.

Blinded by Friendship

One of the best-documented defenses against defenses is the tendency for people to perceive their friends in a more positive light than they perceive strangers and acquaintances. In general, friends see friends' statements as more truthful (even if they are not) and attribute more positive intentions to their actions, and for this reason they are more likely to overlook or excuse their partners' transgressions, risking the chance of similar misdeeds in the future.[69]

Consider the following statement by a seven-year-old about an incident at school where he accidentally hit another boy: "And I tried to go up to Jim to play with him again, but he won't come near me. And he's not . . . When a kid isn't really your friend yet, they don't know you didn't mean to do it to them."[70] The boy's statement illustrates that even young children can understand how friendship should change the way people make inferences about one another's behavior.

Forgiveness

Despite a tendency to perceive friends' behaviors in a positive light, there are many times when people do feel betrayed by friends. Among U.S. college students, friends rival romantic partners for letting down, hurting, and betraying their partners (34 percent compared to 46 percent for romantic partners and only 16 percent for family members).[71] When people infer that friends are behaving inappropriately or have violated a relationship norm, there is a second line of defense against defenses—forgiveness. Forgiveness involves several changes in thoughts and feelings—a decreasing urge to retaliate against an offending partner, an increasing drive to reconcile, and greater goodwill toward the offender—once an offense has been inferred. It requires that we recognize a violation, but it also requires that we do not retaliate.

Despite the frequently stronger emotional responses to friends' betrayals, people are also more willing to forgive friends to whom they are closer and more committed.[72] One set of experiments suggests that forgiveness is also a more automatic reaction among close friends. In the experiments, college students in the Netherlands were asked to think of someone with

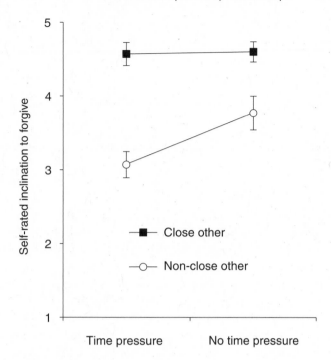

FIGURE 25. Inclination to forgive by closeness and time pressure; data from Karremans and Aarts 2007. Error bars are standard errors.

whom they had a very close relationship and another person with whom they had a non-close relationship, and to type these names into a computer. After some time, the students were then instructed to rate on a computer how willing they would be to forgive specific offenses such as lying, cheating, insulting, and deceiving. Before each rating, the name of either the close or non-close partner flashed on the computer screen so quickly that the students would not consciously realize it had happened. When primed with a close partner's name, as predicted, individuals expressed a greater likelihood of forgiving the transgression, compared to when primed with a non-close partner's name (four studies, average d = 0.73).

To assess how automatic such decisions are, the researchers then assigned students to one of two conditions that changed the time pressure to respond. In the high time-pressure condition, students were asked to respond to questions about forgiving offenses as quickly as possible and within at most four seconds. In the low time-pressure condition, students were asked to think as much as they wanted about each dilemma. When given more time to think in the stranger condition, individuals were more likely to express an inclination to forgive than they were if they had been asked to respond quickly (d = 0.43). However, regardless of the time pressure imposed in the close partner condition, individuals expressed a uniformly high inclination to forgive the offense (see figure 25). In these studies, people automatically felt an inclination to forgive close others regardless of the time given to make a decision, whereas forgiving non-close others required some thought. In other words, for strangers but not close friends, we tend to mistrust first and ask questions later.

Talking versus Walking

Despite a common stereotype that friends shouldn't fight, both behavioral and self-report studies indicate that friends are more likely than acquaintances to get into fights and arguments. The key difference is not that friends engage in fewer or more conflicts, but rather it lies in how they resolve them. In conflicts with acquaintances, people very commonly disengage by walking away, by not thinking about it, or by talking to a third party about the problem. Friends, on the other hand, are overwhelmingly more likely to work at resolving the conflict. They spend more time thinking about the situation and make a greater effort to go talk to the friend. In behavioral studies of disagreement, friends generally offer more explanations, provide more constructive feedback, and show greater tolerance for criticism and argument. Moreover, friends are more likely to perceive comments as productive and positive.[73]

The End of Friendships

In many places, friendship is ideally for life. Zuni farmers in the southwest U.S. expected their *Kihe,* or ceremonial friendships, would last a lifetime and might even be continued by their families after death. Zande blood brothers of north-central Africa ingest their partners' blood to ensure the relationship's longevity.[74] However, despite all of the defenses against defenses that preserve friendships from premature ends, many friendships do not last forever. In one study of U.S. college students, for example, more than half reported losing at least one close friend in the previous five years, and a third had lost one in the past year.[75]

Friendships can end in many ways, including betrayal, interpersonal frictions, and conflicting expectations.[76] They can also end quite abruptly, as in the friendship of Juan and Pedro, two youths in a Guatemalan village. The day after their epic fight, Pedro and his father went to the mayor's office and asked that the mayor officially separate the boys (as in the case of a husband and wife) on the grounds that Juan could not be trusted as a friend and that Juan encouraged Pedro to drink too much. Ultimately, no formal charges were filed, but Juan and Pedro agreed to end the friendship. After that, they would neither greet each other in the street nor talk about each other to third parties. At one point Pedro sent a few messages to Juan to restart the friendship, but Juan had already begun to work out a friendship with another boy of his age, in order to "forget his ex-camarada."[77]

Friendships need not witness such dramatic ends. They can also fade away after physical separation or be supplanted by relationships with other friends or romantic partners. Indeed, in a study of U.S. college students in the 1980s, 78 percent of friendships ended primarily because of such external factors. Betrayals accounted for no more than 22 percent of breakups.[78] Therefore, more friendships among U.S. college students end because friends lose interest in maintaining them or find them difficult to maintain than end due to some obvious betrayal on the part of one partner.

This chapter examined what is known about the development of friendships, from first meeting to attraction to the building of relationships based on mutual aid. In particular, it examined two related challenges in the cultivation of a mutually beneficial friendship. The first challenge is created by the unconditional aid expected of close friends and the many ways to induce feelings and behaviors of friendship. These open up opportunities for exploitation, and an important part of cultivating a friendship is avoiding such a possibility, by making the relationship too valuable to lose, by decreasing a friend's outside options, and by paying attention to subtle signals about a friend's view of the relationship. With so many defenses, a second challenge is to avoid prematurely ending a friendship, and there are three important defenses against defenses—relational blindness, forgiveness, and conflict resolution.

The last two chapters have examined questions of development, both how children learn the rules of friendship and how friendships develop over time. Chapter 7 moves the focus to social, cultural, and ecological influences that shape how people think about and behave toward friends.

7   Friendship, Culture, and Ecology

> You are a passenger in a car driven by a close friend, and
> he hits a pedestrian. You know that your friend was going
> at least thirty-five miles per hour in a zone marked twenty.
> There are no witnesses. Your friend's lawyer says that if
> you testify under oath that your friend's speed was only
> twenty miles per hour then you would save your friend
> from any serious consequences.
>   What would you do?

Although most people around the world agree that it is important to help
friends, the passenger's dilemma tests one's commitment to the proposition. Does the unknown pedestrian have a right to truthful testimony? Is
it immoral to lie under oath to protect a friend? Similar ethical dilemmas
arise frequently in human societies, each with unique norms for dealing
with them. Among Pashtun pastoralists in northern Pakistan, lying violates basic Koranic law, but there are times when one should lie to help a
friend. Orokaiva gardeners in Papua New Guinea frown upon fighting,
but it can be acceptable when a friend needs protection. And among Wolof
herders in Senegal, a true friend should be prepared to help, even if it
means breaking the law. In the United States and Europe, philosophers
have long struggled with such dilemmas and, despite many attempts, have
found no universally acceptable way of resolving them.[1]

To understand how people in different cultures negotiate such dilemmas between loyalty to friends and to society, two Dutch social scientists
presented the passenger's dilemma to thirty thousand white-collar workers in over thirty countries. In the U.S., fewer than one in ten managers
said they would lie to protect their friend, most preferring to tell the truth
under oath. Although this same pattern held in northern European countries, in Japan and France the number who would lie under oath rose to
about three out of ten. The number reached a high in Venezuela, where
seven out of ten managers said they would lie to help their friend.

Regardless of the particular social dilemma, these cross-national differences persisted. If you were a doctor and your friend came in for a checkup
required by his insurance, would you lie about your friend's health to

improve his premiums? If you were a food critic, would you lie about the poor quality of your friend's restaurant? If managers in a country are more likely to lie for a friend in one of these dilemmas, chances are that they would lie in the others as well.[2]

People's choices in these dilemmas reflect value differences that social scientists call *particularism* and *universalism*. Someone who is more universalist would rather uphold the law than protect a friend, seeing the value of general principles of conduct that should be applied to everybody. On the other hand, someone at the particularist end of the spectrum would violate standard codes of good behavior if a friend's well-being depended on it, putting greater value on protecting loved ones and allies. Anthropologists and sociologists have long argued that societies differ along this continuous dimension of particularism and universalism.[3] The large cultural differences observed in the Dutch study confirm the qualitative insights of these generations of social scientists.[4]

In addition to cultural differences in particularism, friendship differs in a number of other ways across cultures, including the degree to which partners value and expect different kinds of support (e.g., material aid and verbal support) and in how people resolve other loyalty conflicts between, for example, kin and friends.

Numerous authors have proposed why friendship might differ systematically across societies. Some claim that friendship is a relatively modern invention that arose with the breakdown of traditional kin systems. Others claim the opposite, that modern institutions have weakened friendship, with insurance companies taking the place of mutual aid, enforceable contracts taking the place of loyalty, psychiatrists replacing confidants, and safe highways, diners, and hotels obviating the need for long-distance hosts. With the rise of the Internet and the many novel forms of social interaction that it supports, social commentators have also proposed numerous and conflicting theories about how friendship will change, for the better or for the worse, as more people spend time on the Internet.

This chapter is divided into two sections. The first explores cross-cultural differences in expectations of friends, in the relative value placed on friends, and in the kinds of aid they provide. The second section critically examines hypotheses about how social, cultural, and ecological contexts influence people's friendships and what they expect of their friends (box 22). In particular, I examine four hypothesized reasons for differences in friendship, including material and social uncertainty, geographic mobility, competition with kin institutions, and changes in the media that people use to communicate and interact.

BOX 22   The Ecology of Friendship

The derivation of the term *ecology* (literally, study of the household) gives little indication of its current definition—the study of interactions between organisms and their environments. By referring to the ecology of friendship, I also take liberty with definitions. I conceive of friendship as an organism—the long-term interaction between two friends—and the ecology of friendship as the study of interactions between friendships and their environment.

The ecology of friendship might ask the following questions: To what degree does economic uncertainty influence the forms and functions that friendships take? How do other institutions in a society influence the importance of friendship? How do cultural norms affect how people think about friendship? And to what degree do friendships influence the broader social system, for example, by circumventing universal rules of behavior or standard codes of conduct?

In this chapter, I will address these specific questions, and more generally I will use the ecology of friendship as a way of thinking and asking questions about friendships in different social and cultural environments.

CULTURAL DIFFERENCES IN FRIENDSHIP

Over the past two decades, sociologists and psychologists have developed numerous measures for comparing friendships across cultures and ethnic groups. These include what people expect of good friends (e.g., loyalty, affection, sharing of personal secrets), how people behave toward friends in hypothetical and real situations, and how people report feelings in existing relationships.

In many cases, the differences across cultures are quite small relative to the vast amount of variation between people from the same culture. Indeed, when researchers have compared friendships across various countries, including Zimbabwe, China, Brazil, Korea, the Netherlands, Turkey, Morocco, Costa Rica, Poland, India, Russia, and the United States, they are quite similar in a number of features, including perceived quality and emotional support, expectations of a good friend, kinds of communication and influence, feelings of closeness, and what counts as betrayal or bad behavior.[5]

Despite these striking similarities, researchers have observed three relatively big differences in how people make and have friends across cultures. First, people in different cultural settings vary in how they prioritize the importance of material aid and the discussion of personal matters in their friendship. Second, as the Dutch study of the passenger's dilemma suggests, members of different societies differ in how they resolve moral dilemmas between helping friends and following other kinds of social norms, such as obeying the law or following a universal code of helping. Finally, people differ in what kinds of material aid are most important among friends. Here, I describe these differences in more detail.

Talking about and Sharing Personal Matters

In 2004, one of the longest-running surveys of social, cultural, and political issues in the U.S. revealed a troubling fact. One in four Americans lacked a confidant. More striking was the fact that the number of Americans without a confidant had increased dramatically, from 10 percent to 25 percent, since the last time the question was asked two decades earlier. The study results were picked up by *USA Today*, the *New York Times*, the *Washington Post*, and MSNBC, providing fodder for pundits' concerns about the increasing social isolation of Americans, their loss of close friends, and the possible causes of this trend.[6]

What these stories failed to report was that two years earlier in the 2002 version of the same national survey, another set of questions was asked about close friends and revealed that more than 95 percent of Americans reported having at least one close friend.[7] Why the big difference? One reason is that the 2004 edition had focused on only one kind of behavior among friends—the discussion of personal matters. Many Americans and U.S. researchers see talking and self-disclosure as an important if not essential part of close relationships. Therefore, it is not surprising that some commentators assumed that the 25 percent of Americans who lacked confidants also lacked close friends.[8]

In the end, the comparison of these two surveys reveals that for many people, in this case at least 20 percent of Americans, talking about personal matters may not be a necessary part of friendship, or even of close friendship, and this may be an increasingly common take on friendship. While Americans have different opinions about the relative importance of talking and personal self-disclosure in their friendship, people in different cultures also value talking and self-disclosure to differing degrees.[9] For example, in one experiment conducted in the 1980s, Korean and U.S. college students were asked to bring in their best cross-sex friends for a study of communication. During a structured conversation, each partner was asked to talk about him- or herself for two minutes on four separate topics. In these meetings,

Korean students reciprocally disclosed far fewer highly intimate details (a 30 percent reduction) with their best friends than did U.S. students.[10]

A reduced emphasis on talking and self-disclosure with friends has also been observed in behavioral and self-report studies in other East Asian countries, such as Japan, China, and Indonesia. Indeed, studies of daily activities among U.S. and East Asian youth indicate that East Asian youth spend less than an hour on average talking with friends and family, compared to two to three hours among U.S. youth.[11]

Some scholars have explained the lower levels of personal discussion among Chinese friends in terms of the concept of *hanxu*. *Hanxu* refers to an ideal of communication that is contained, reserved, implicit, and indirect, both in terms of verbal expression and non-verbal displays of emotion, such as anger and joy. The importance of *hanxu* is reflected in many Chinese proverbs, such as "Talking a lot will lead to personal loss" (*yan duo bi shi*) and "mutual understanding lies in the heart not in words" (*xin zhao bu xuan*). According to this ideal, close friendships are developed and nurtured through actions—mutual aid and care—rather than the open expression of feelings and thoughts.[12]

Researchers recently bet that these observed cultural differences would persist even for people with different cultural ancestry living in the same country. Specifically, they conducted an experiment with U.S. college students of East Asian ancestry and majority white background, to examine differences in how verbal social support reduces or contributes to stress. In the study, students were asked to give a five-minute speech (something known to induce stress in most people) arguing why they would be a good administrative assistant in the university psychology department. After having three minutes to prepare the speech, they were randomly entered into one of three conditions:

1. In the implicit support condition, students were asked to think about a group to which they were close, and then to write about aspects of that group.

2. In the explicit verbal support condition, students were asked to think about people to whom they were close, and then to write a letter directly seeking advice and support for the upcoming speech from one of these people.

3. In the no support condition, students were asked to think about campus landmarks, and then to write about places they would recommend for campus tours.

The students then delivered their speeches. Before and after the task, students filled out a question about how agitated, upset, stressed out, and nervous they felt. They also spit into a cotton swab so that researchers could track the level of a particular stress hormone, cortisol.

For majority white Americans, explicit verbal support led to a reduction in reported feelings of stress (d = -0.56 to -0.52). However, for East Asian Americans, explicit verbal support had exactly the opposite effect, making those who had been assigned to condition two feel more agitated, upset, and stressed than either the intrinsic support (condition one) or the no support (condition three) group (d = 0.71 to 0.83). Moreover, the East Asian American students in the explicit verbal support condition experienced a much greater jump in cortisol than did those in either the no support or the implicit support condition (d = 0.86 to 0.96). These results suggest that verbally mediated support can be either helpful or detrimental, depending on one's culture of origin.[13]

Although the differing emphasis on talking about personal matters among friends in English-speaking and East Asian cultures has received considerable support, there is no dominant explanation for it. One reason may be methodological. If there are a few topics that are taboo in East Asian countries, such as family life and personal feelings, and these change the overall frequency and degree of disclosure, then the difference may not be about *all* verbal expression, but only certain kinds. The relatively public nature of the experiments may also make some people less likely to share personal details, even though they would do so with friends in more private settings.[14]

Researchers have also tried to explain this difference in terms of a cultural dimension called *collectivism,* which is claimed to influence behavior in several ways. According to these arguments, collectivist individuals prioritize the goals of their group and their relationships over their own individual goals. They place greater weight on preserving the harmonious functioning of relationships; sharing personal information and feelings can potentially damage this harmony. Finally, members of collectivist cultures theoretically prefer implicit communication, relying on common knowledge between partners rather than explicit verbal communication.[15] For these reasons, some scholars have predicted that members of collectivist cultures (e.g., China, Japan, Korea, Indonesia) should rely less on talk and self-disclosure compared to members of less collectivist cultures (e.g., the United States, the United Kingdom). This argument works for many East Asian countries and for Russia and Poland, which are also highly collectivist and where friends report less-personal discussion.[16] However, in studies comparing friendships between Americans (putatively individualist) and other cultures classified as collectivist—such as India and

Ghana—friends report sharing private information at comparable levels.[17] Therefore, the commonly invoked explanation in terms of collectivism and individualism fails to hold water.

Then what can account for such cultural differences in the importance of talking among friends? As I mentioned earlier, the reason may be methodological, due to the kinds of topics that researchers use to explore talking among friends or the relatively public nature of the experiments. It may be related to other institutions in society that encourage or discourage talking in general and that have only marginal relationships with coarse-grained variables such as collectivism and individualism. Perhaps the roots are historical, based on recent experiences with an oppressive state where it is generally a bad idea to reveal one's secrets to anyone. While these are all possible hypotheses, only further work can identify the reasons for cultural differences in the importance of sharing secrets and personal talk among friends.

Material Support

Studies of support among friends in the U.S. often focus on the intangibles of friendship—emotional support, talking, companionship, affection, and feelings of closeness—with much less attention given to concrete examples of material support. This bias may reflect the value placed on self-reliance when it comes to material needs in the U.S. or a more general Western notion that the highest forms of friendship are somehow free from material concerns.[18] Regardless of the reasons for this bias, it doesn't represent the views of friends in many parts of the world (or even perhaps in the U.S.), where acts of material help, sharing, and support are crucial elements of good friendship.[19]

Cross-cultural studies have shown clear differences in the degree to which friends value and supply material aid. In the 1990s, for example, researchers asked secondary schoolteachers in Helsinki, Finland, and St. Petersburg, Russia (which are located less than 250 miles apart), to fill out structured diaries about their social interactions over the course of two weeks. Each evening, teachers recorded all kinds of information about their encounters of the day, whether with kin, colleagues, neighbors, friends, or others. They described the time, duration, place, and content of these encounters and attributes of their partners and their relationship. Most important, they recorded what kinds of favors they gave and received.[20]

Although the teachers recorded similar numbers of favors and exchanges with kin (seventy-two for Russians and fifty-two for Finns), the

BOX 23 Lending Money to Friends

"The holy passion of friendship is so sweet and steady and loyal and enduring in nature that it will last through a whole lifetime, if not asked to lend
money" (Twain 1894, p. 93). As Mark Twain, perhaps one of the most notable
commentators on white U.S. culture, notes, people expect friends to help
them, but some forms of aid are more or less appropriate than others. Of
particular relevance in the U.S. is the lending of money to friends, a frequent
target of newspaper advice columns. The warnings are numerous. Lending
money to deadbeat friends only enables their bad behavior. Money matters
can strain a friendship, and one is unlikely to see the money returned.
Exchanging money among friends is often thought to weaken a friendship
(Cohen 1961), and such concerns have led many to believe that lending
money to friends is quite uncommon in the U.S. On the contrary, Americans
seem to lend money to friends quite regularly (Board of Governors of the
Federal Reserve System 2008; Glaeser et al. 2000).

Mixing money and friendship is also a common practice in societies
around the world (23 percent of PSF societies), although its acceptability
and importance differ across cultural groups. For example, a study of white
and Asian shopkeepers in three U.K. cities showed that shopkeepers from
the two cultures raised start-up capital for their shops in very different
ways. Asian shopkeepers raised substantial amounts of capital through
friends; white shopkeepers raised almost no capital through friends, perhaps reflecting a culture-specific concern with mixing money and friendship
or differential access to money markets (Zimmer and Aldrich 1987).

Russian schoolteachers exchanged more than *three times* as many favors
with non-kin friends. The Russian schoolteachers exchanged more favors
in nearly all categories, including in shopping for needed items, arranging
medical aid, lending money, helping with home repairs, arranging useful contacts, sending parcels, arranging a place for one's child at school,
and tutoring another's child. Only in two categories—giving a ride and
lending non-monetary goods, such as books and clothes—did the Finnish
schoolteachers exchange more favors.

These behavioral patterns were matched by the schoolteachers' own definitions of friendship. For Finns, definitions of mutual help between friends
centered on psychological and spiritual support provided by listening to
one another's worries and consoling one another in times of need. Like

the Finns' descriptions, Russian definitions included emotional support, but they also made many more mentions of practical, material aid (box 23).

Far from being an isolated example, such differences in the importance of material aid have also arisen in comparisons of Icelandic and Chinese adolescents, U.S. and Ghanaian college students, and British and South Asian merchants, with members of Chinese, Ghanaian, and South Asian cultures having a far greater tendency to expect and to exchange material aid among friends. Later in the chapter, I will describe one possible explanation for these differences.[21]

Loyalty Conflicts

Friendship does not exist in a social vacuum. As with the Dutch study's passenger's dilemma described earlier, people frequently face tough decisions about competing social obligations—to one's spouse, family, religious tenets, job, codes of conduct, and society in general.[22] People differ in how they resolve these daily dilemmas, and how people resolve loyalty conflicts can tell us a great deal about their values and preferences.

The Dutch study revealed how people in different cultures value helping close friends over obeying standard codes of conduct. Another tension is the relative value placed on friends over kin. In the 1960s, a team of anthropologists examined this dilemma in four different cultures in Tanzania and Kenya, whose members included both farmers and herders. They asked simply, "Is it better to have many friends or many kinsmen?" and found a dramatic difference among and within these four societies in the tendency to prefer friends (from a low of 20 percent to a high of 50 percent). They also found that how people made a living, whether primarily by herding or farming, influenced their relative preferences for friends and kin. Specifically, farmers were always more likely than pastoralists to prefer having many friends over having many kin (see figure 26, which also includes comparable data for a fifth society, Sangu, collected later). The reasons for this pronounced difference are not clear. One argument based on the shadow of the future is that with high mobility among herders, friends will have shorter windows for interaction and therefore will have less incentive to provide aid. In such situations, kin would be more reliable partners and therefore preferable to friends.[23]

In different cultures, friends are prioritized in different ways. Limited measures of these priorities exist for the competition between friendship and standard codes of conduct and friendship compared to kinship. However, conflicts can also arise between friends and religious tenets, friends and mates or spouses, and friends and duty to country, raising

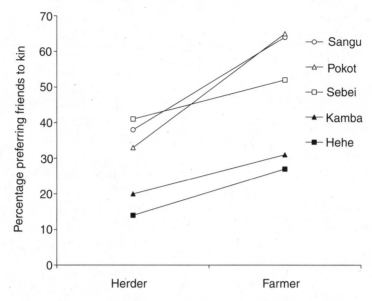

FIGURE 26. Preferring friends to kin in five East African societies by mode of subsistence; data from Edgerton 1971; Sangu data from McElreath 2004. Kamba and Hehe are predominantly herders, but all groups included herders and farmers.

many interesting opportunities for understanding how friendship fits into the diversity of values in any specific society.[24]

The Varieties of Material Aid

When material aid is important, there are often profound differences in the kinds of aid provided. For example, first-grade friends in the U.S. help each other with schoolwork and support one another during conflicts with other students. The *puan daj* (loosely, friend to the death) of a Thai peasant may provide loans without interest, lend water buffaloes, or aid in staging life course ceremonies, while the *bagay magtaymanghud* (blood friend) of a Tausug horticulturalist in the Philippines can be counted on to assist with debts, to loan guns if needed, to provide food and shelter, and to come to his aid in a fight.[25]

Indeed, knowledge about what friends do for one another reveals a great deal about the recurring social needs within a specific society, whether they involve hosting life-course rituals, stocking herds, handling disputes over grazing land, arranging love affairs, providing safe haven when traveling and trading, or helping to find a job or get a promotion.[26] As the way of life in an area changes, the kinds of favors and exchanges can also change, depending on available resources and recurring needs. Consider one friendship between an Ik herding family of Uganda and a neighboring Napore family that had recently transitioned from foraging to farming. The friendship went as far back as their grandfathers. In the old days, the Napore family would give gifts of game, honey, and tobacco, while the Ik provided milk, blood, and goats. However, as the Napore family moved from foraging to farming, their gifts changed to maize, millet, and sorghum.[27]

Although friendship may endure, the changing resources and needs of the partners can influence what goods are exchanged and what aid is provided. This is not surprising. More interestingly, in the cross-cultural record there are several particularly common kinds of help provided by friends, and I will discuss three of them here: hospitality from friends in novel or far-off places, mutual insurance against uncertainty and risk, and help with labor and in disputes.[28]

Having Friends in Strange Places.   Although we take for granted the ability to travel to new regions with relative ease and security, in many places and times, traveling to the nearest settlement could be a dangerous proposition. Even after a successful journey, one might not have a safe place to stay or be able to find food. Hotels with built-in restaurants aim to corner this market in modern settings. But elsewhere and in the past, having a trustworthy friend who would provide food and shelter was a prerequisite for traveling, trading, finding a spouse outside of one's community, and having a place to move in the case of drought or famine. Therefore it is not surprising that friends in many small-scale societies are obliged to feed and protect their visitors. In many of the sixty societies described in chapter 2, this promise of hospitality and protection was an important obligation among friends (43 percent of societies in the PSF sample).

In some cases, trouble at home is the impetus to visit a friend. For example, among Fore horticulturists in Papua New Guinea: "Men take up residence with their *wagoli* [friend] when they are in trouble at home. It is said that if a man kills his brother or injures his father, he lives out his temporary banishment with distant *wagoli* until the anger of his brothers and age mates subsides, a period of about three years or more."[29] In more extreme examples, entire villages find temporary safe havens in the settlements of allies and friends. Among the Yanomamo of South America, when a village is attacked and destroyed, the residents can go to another ally's settlement and stay for a year or longer as they establish new gardens and rebuild their own settlement.[30] More often, however, the guest

---

BOX 24 Travel between Hostile Groups

In the nineteenth century, American novelist Herman Melville spent a month stranded on one of the Pacific Ocean's Marquesas Islands, in a valley occupied by a tribe called the Typees. In the book that chronicles his adventure, *Typee: A Peep at Polynesian Life*, Melville describes how people from neighboring but hostile groups were able to travel throughout the island. One day a stranger belonging to a hostile tribe entered the valley. When Melville expressed surprise at the young man venturing among the Typees, the stranger looked at Melville and exclaimed, "Ah! me taboo—me go Nukuheva—me go Tior—me go Typee—me go everywhere—nobody harm me, me taboo." The stranger was relying on a custom among the Marquesans later explained by Melville as follows: "Though the country is possessed by various tribes, whose mutual hostilities almost wholly preclude any intercourse between them; yet there are instances where a person having ratified friendly relations with some individual belonging to the valley, whose inmates are at war with his own, may under particular restrictions venture with impunity into the country of his friend, where under other circumstances he would have been treated as an enemy. In this light are personal friendships regarded among them, and the individual so protected is said to be 'taboo,' and his person to a certain extent is held as sacred. Thus the stranger informed me he had access to all the valleys in the island" (Melville 1846, pp. 178-179).

---

relationship serves an economic purpose, permitting traveling and trading in potentially hostile communities. In the Marquesas of the Pacific, despite mutual hostilities between adjacent communities, a well-connected individual could travel throughout the hostile valleys on the islands, provided that he had personal friendships in each of the regions (box 24).[31]

In some cultures, marriage is the first time when men or women must move far away from their own families. This can be a tense, lonely, and potentially hostile situation for the new spouse, and friendships made in the new community often serve to buffer this stress and to provide support as needs arise. In some cases, these friendships are made informally. Among Agta hunters in the Philippines, for example, close friendships develop between men or women who have joined a community through marriage. These "co-sibling-in-laws," or *idas*, though not directly related, pull together as mutual outsiders in a potentially hostile community. They work together, support one another in times of need, and may speak one another's names, a privilege generally reserved only for one's spouse and children. In other situations, families facilitate a newlywed's entry into a community by assigning him or her to a friend, often sealed by a formal ritual. Whether such ties are created ritually or informally, they ensure a base of support for outsiders as they adapt to their new living conditions.[32]

Sharing Risk. In many human societies, food supplies are variable: herds can die from epidemics or drought, crops can suffer from blight, pests, and the vicissitudes of rain, and a hunter can have a bad month or become incapacitated by illness or injury. In small-scale societies, one of the most common ways to insure against these risks is through building and relying on social ties. In southern Africa, Ju/'hoansi San hunter-gatherers cultivate and maintain a network of gift-giving partners who can be called on for help and visited if food becomes scarce. Among many East African groups, herders distribute their cattle among stock-friends in far-off places, so that if one's herd is hit by an epidemic or other calamity, the cattle entrusted with stock-friends have hopefully survived.[33] Common forms of pooling risk involve sharing food and other necessary items (63 percent in the sixty PSF societies), giving monetary loans (23 percent of societies), and helping with labor, hunting, childcare, or food preparation when one falls ill (10 percent of societies).

One of the most frequent economic shocks a family can experience is not drought or attack, but rather the expected outlays in food, gifts, and entertainment required to host life-course rituals for a member of the family—for example, funerals, weddings, and circumcisions. The expenses and labor requirements for such ceremonies can exhaust the capacity of even extended families, and friends frequently play a role in preparing food, making loans, and participating in the ceremonies (20 percent of societies).

Help with Labor and Disputes. In 38 percent of the PSF societies, friends were an important source of help in tasks that could benefit from many hands. These included house building, babysitting, clearing and plowing fields, harvesting crops, hauling nets, preparing food, and caring for livestock. Another important contribution of friends is helping to resolve disputes. In societies where there is no centralized legal system for resolving conflicts, friends provide both physical and reputational backup. Friends, rather than kin, become important when the dispute involves in-fighting among kin over inheritance and property.[34] Friends (specifically non-kin friends) also become important when coalition sizes are so great that one's kin are not sufficient. For example, in Chagnon and Bugos's detailed description of a within-village ax fight among the Yanomamo, neither of the two coalitions involved could have achieved a competitive size without relying on allies through marriage (box 13).[35] Such aid in disputes is described as important in 37 percent of societies.

WHAT CAN ACCOUNT FOR CULTURAL DIFFERENCES?

There is ample evidence that people in different societies think about and behave toward friends in very different ways. Talking and sharing secrets with friends is valued in some societies more than others, as is the provision of material aid. And people in different societies prioritize friendship higher or lower when compared to other social goods.

Scholars have proposed four primary reasons for observed differences in friendship across societies. The first argument is that uncertainty—in the ability to acquire necessities, to protect oneself and one's possessions, or to achieve other important life goals—leads to greater investment in and reliance on friends and a greater focus on material aid among friends. The second set of arguments claim that geographic mobility affects the ability to maintain friendships. Depending on the argument, mobility can either hurt friendships or make them more appealing. A third argument is that strong and binding kin institutions make friendship impossible, and friendship only flourishes in societies where kinship has weakened. The final argument is more recent, based on changing modes of interaction on the Internet and how they are transforming the quality and meaning of friendship. I turn to these four arguments now.

Uncertainty and Friendship

In the literature on personal relationships, Russians are legendary for the value they place on their closest friendships.[36] Consider one New York–based former Muscovite opining about his *drugii,* or friends, in the USSR.

> With your friends was the only time you could be absolutely yourself. . . . Friends help friends find something to eat and tell where you can get something black market so to speak. Friends help you find something to wear, places to spend the summer. These are just the practical things of everyday life which are necessary to everyone and theoretically available to everyone but are not, except through connections. So friends provide this.
>
> Then there is the moral part, the emotional part. Friends are the people who you can share political jokes with and that is the primary way of expressing yourself, especially in our circles against the repressive system. But you have to know exactly who your friends are, otherwise you run the risk of getting arrested. . . . Friends then served many

functions: They helped you provide physically for yourself and gave emotional and moral support.[37]

Narratives of Russian friendship from the cold war era generally go hand-in-hand with descriptions of uncertainty as a fact of life: in acquiring daily necessities, such as food and clothing; in gaining access to public services, such as health care, higher education, and housing; and in confiding in partners who may turn them over to the secret police.[38] Moreover, the transition to a market economy in the 1990s created a new set of economic problems that the same institutions of friendship helped to resolve. In the early 1990s, for example, only one in eight Russians reported earning enough from their official job to meet basic needs, with the majority surviving through the help and favors of friends and relatives.[39] Given this daily reliance on friends, it is not surprising that in the passenger's dilemma described earlier, more than half of Russians expressed a willingness to help a friend rather than follow universal codes of conduct.[40]

While the Soviet system might have been the crucible for deep, mutual friendships in their home country, immigrants to the U.S. in the 1980s described a persistence of this approach to friendship, with favors including babysitting, co-signing loan applications, assisting with home repairs, lending substantial interest-free sums, helping with big purchases, and finding apartments and employment.[41]

Such friendships provided an effective insurance system, but they also imposed obligations and high expectations. Consider the ambivalent words of a thirty-eight-year-old former Muscovite living in New York for six years, which also provide an interesting outsider's perspective on friendship among Americans born in the U.S.:

> I told you about the good and bad sides of friendship according to Russian ways. This is one of the bad sides: very demanding, obligations, expectation. This I do not like at all. The kind of [American] friendship, without obligation, where we see each other without being pushed into it, this to me is real friendship, different from that in Russia where there was no other way or no other choice. . . . I am beginning to understand the American version of friendship—a tie. A feeling of warmth without expectations and obligations, a sharing of spirit. I told you that sometimes I miss the old way of absolute friendship, living not really for yourself but for your friends. Other times, and now more often than not, I thank God that those days are over and that friendship has reached a new level.[42]

In Russia, during both the Soviet and transition periods, strong friendships were a necessity. However, in a relatively functional market system where most basic material needs could be met through anonymous interactions, the trade-offs changed. In a more certain world, the burdens of such absolute friendships could outweigh the costs.

This trade-off of friendship—between the benefits it provides and the obligations it entails—is not unique to Russian émigrés. In many societies, friends provide an important buffer against risky events: food shortages, traveling in foreign lands, disputes, and possible cheating in exchanges. For this reason, social scientists have argued that friends will be most prized in societies where uncertainty is pervasive, in the form of food scarcity or inaccessibility due to high prices, material unavailability, violence, or lack of legal sanctions.[43]

While strong, long-term relationships can solve problems created by uncertainty, they also have a downside. As the Russian émigré expressed, strong friendships can be very demanding, and so people often find a middle ground that fits their particular situation. Consider the statistic from a classic study of midwestern America in the 1920s, where nearly a third of wives said they had no intimate friends.[44] While potentially horrifying to those who value friendships, consider the view of one woman: "I like this way of living in a neighborhood where you can be friendly with people but not intimate and dependent."[45]

Given the burden of diffuse obligations of strong friendship, when circumstances permit, we would expect people to lighten their friendships by expecting fewer kinds of aid, focusing less on material support, and generally limiting help to small favors. This is exactly how the anthropologist Paul Baxter explained the demise of one form of blood brotherhood among Azande farmers in Sudan. The Azande institution of blood brotherhood included strong obligations to help when a friend was in trouble and to provide protection when a friend visited. However, with the introduction of colonial rule in the early twentieth century and the ensuing *Pax Britannica*, people needed fewer protections while traveling and trading. Moreover, the introduction of money and commerce meant that most people could dispense with circuitous means of acquiring and storing goods and wealth. In short, many of the social situations that formerly made the blood brother bond so useful had ceased to exist. Furthermore, the possibility of large debts permitted by money and commerce made the unlimited responsibility for a friend inherent in these blood pacts especially dangerous. In such a setting, a profligate blood brother could become a serious drain on resources.[46]

Such changes in bond friendships and blood brother ties have been described in other cases when long-distance trade ceases to be dangerous,

BOX 25   Cross-Group Analysis

When a researcher wants to understand how ecological factors, such as social institutions or resource availability, influence individual behavior, one approach is to compare populations that are subject to differing ecological conditions. This can be done at the level of any relatively bounded group, such as neighborhoods, cities, cultural-linguistic groups, or nation-states.

For example, it is possible to assess the claim that greater social, economic, and political uncertainty at a national level is associated with an increased probability of supporting a friend over following the rules of the legal system. For each of the countries included in the Dutch passenger's dilemma study discussed earlier, the World Bank also collects data on the degree of economic, social, and political uncertainty, measured according to several concepts. *Rule of law* is the extent to which citizens have confidence in the rules of society, in particular the quality of contract enforcement, the police, and the courts as well as the likelihood of crime and violence. *Government effectiveness* is the quality of public service provision, bureaucracy, and civil servants. *Control of corruption* is the extent to which exercise of public power for private gain is curtailed. *Political stability* is perception that the government in power is safe from overthrow or destabilization. These are all highly correlated (correlation > 0.80), representing a general yet multi-dimensional uncertainty in daily life, and one can average these to make a composite measure of social, economic, and political uncertainty.

Figure 27 shows the strong correlation between this measure of uncertainty and the probability of breaking the law to help a friend (correlation = 0.70, d-statistic = 1.96). This strong correlation is thought provoking, yet it calls for further study to confirm the actual relationships and isolate the precise mechanisms. For example, it is not clear to what degree uncertainty makes people value friends more or whether the tendency to break the law to help friends is the cause of broader social uncertainty. This can only be

(continued)

necessary, or profitable, when agriculture becomes less risky, and when new or upgraded welfare systems reduce individual uncertainty.[47] For these reasons, some sociologists have proposed that people have lighter friendships in the modern West, because when legal and market systems flourish and when money permits one to acquire most material necessities from strangers, it can substitute for many of friendship's functions. As I

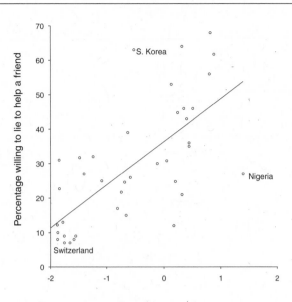

FIGURE 27. The correlation between uncertainty and the willingness to lie to help a friend

determined by a study that takes place over time. Moreover, there may be some underlying mechanism, such as the tendency to break the law, that drives the correlation. Further analysis suggests that a tendency to break the law—such as illegally claiming government benefits, avoiding a fare on public transport, or cheating on taxes—cannot account for this relationship (uncertainty with law breaking, correlation = 0.05, helping friend with law breaking, correlation = 0.02 [World Values Survey 2005 Official Data File 2009; Worldbank 2009]). However, there may be other confounding variables, such as poverty or cultural history, that deserve further scrutiny.

write this in the U.S. in the midst of the 2008–2009 recession, it will be interesting to see if friends have gained new importance in terms of job seeking, making ends meet, and finding new ways to cut costs, such as ride sharing and bulk purchasing.[48]

The analysis in box 25 indicates that these theoretical arguments are consistent with data from a cross-national sample. Specifically, in those nation-states with greater political, legal, and economic uncertainty, individuals are much more likely to value their friendships over following universal rules. Meanwhile, in those states where life is relatively certain (e.g., Switzerland), most people prefer to obey universal rules rather than help their friends.

At least one scholar has extended this argument to say that in well-run economic and legal systems, the "economic benefits derived from friends are in most cases quite minor and do not radically affect an individual's economic position."[49] However, even in the best-run economies, one finds that friends still help in diverse ways, such as lending money, babysitting, providing sick care, giving rides, helping with household chores, proofreading, dog-sitting, house-sitting, and hosting, some of which often arise during emergency situations (chapter 3). Only studies of the dollar equivalent of these services will be able to tell the extent to which these services become "minor" in well-run economies.

Geographic Mobility

Certainty in a modern society can reduce many of the benefits of friends. Other factors can also make friendships more difficult to cultivate and maintain. One commonly proposed deterrent to friendship is competition for time from other activities, such as working for money or watching TV, which whittle away the time available to spend with friends.[50]

Perhaps the most heated debate in the twentieth century in the social sciences revolved around geographic mobility and whether it deters the cultivation and maintenance of close social ties. According to the argument, when people move, they leave behind relationships that physical distance renders difficult to maintain. Moreover, movers must cultivate new ties that lack the same history and depth as their previous relationships. In short, every move tears the fabric of our social networks, leaving it irreparably damaged.[51]

The argument has intuitive appeal and feeds from nostalgic notions of a happier, stable past. However, when researchers have left their theoretical armchairs and asked people about their experiences, the results regarding friendship are quite contradictory. Mobility does reduce the number of friendships within one's immediate neighborhood. In representative surveys of U.K. residents in the 1960s and 1980s, for example, long-time residents indeed listed more friends. However, these studies focused on friends within a ten- to fifteen-minute walking distance, and when asked about friends outside the local community, recent migrants reported many more.[52] Similarly, a recent large-scale study of high school students in the

U.S. shows that mobile students are less likely to have a best friend in school (61 percent versus 71 percent), but since many students maintain best friends outside of school, it is not clear if there is any overall difference between mobile and non-mobile students.[53]

If we focus on what friends actually do for each other, the results are strikingly similar. In a recent study of social support among American families, mobile adults continued to receive comparable levels of emotional and financial support from friends and other non-kin, regardless of moving distance or frequency. Although they experienced a statistically significant reduction of help with small tasks and companionship, the differences were relatively slight.[54] In another U.S. study, families that lived in a different state than the household head's home state actually had *greater access* to help from friends, partly as compensation for the lack of kin living nearby.[55]

There are a number of reasons why mobility might not be as bad for friendships as some scholars have suggested. Migrants often specifically choose to move near existing friends and family, and they actively cultivate new friendships.[56] Moreover, geographic distance is not so lethal to old relationships as might be expected, as migrants frequently maintain their former relationships, thanks in part to modern communication and transportation systems that permit long-distance social interaction.[57] Finally, migrants may be *more* likely to rely on new friendships as a way to compensate for kin ties that were left behind.[58]

Rather than suffering from a loss of social ties, movers on average adapt quite quickly to new locations, maintaining some social ties from their place of origin while cultivating new ones where necessary.[59] Indeed, the institution of friendship appears to be a social solution to the problem of geographic mobility, by providing what the sociologist Vance Packard has referred to as a quick way to "plug in" to a new location—a way to re-root oneself.[60]

The utility of friendship as a way to "plug in" to new communities is not a use of friendship unique to modern society. For example, from the fourteenth to nineteenth centuries, when Serbia was at the edge of the Turkish Empire, family groups frequently migrated away from advancing armies and into liberated zones. In this context of population flux, ritual ties based on godparenthood helped small and weak groups move into new regions. A newly arrived family that had traveled a long distance generally offered a child for baptism to a resident family, and the newly arrived family might later become incorporated as a genealogical segment of the sponsoring group.[61]

Geographic mobility does not appear to be as damaging to friendship as has often been proposed. People actively maintain ties from their places of origin and make new friends in their destinations. Indeed, friendship may be one solution to the problem of geographic mobility, by providing a way to quickly cultivate new social ties in a novel place.

Kinship and Friendship

Social scientists have often painted kinship and friendship as opposites, with kinship a powerful organizing principle in small-scale societies and friendship a residual category that can only flourish when kinship weakens (chapter 3). A common prediction from this line of thinking is that the importance of friendship in a society would wax as kin institutions unravel and would wane as kinship gains strength. Other anthropologists have argued precisely the opposite: when kin institutions are *too* strong, people will develop friendships as a way out or as an "emotional release and catharsis from the strains and pressures of [kin-based] role-playing."[62]

A final argument proposes that friendship becomes more important than kinship in situations of social mobility—whereby individuals are able to enter new social roles and achieve new social statuses. In such situations, social climbers have an interest in avoiding the traditional, diffuse obligations to kin who remain at lower rungs on the social ladder, so as not to quickly give away the hard-earned capital that comes with new and higher social roles and statuses. Anthropologist Elliott Leyton described this situation among emerging urban elites in Ireland, who were constantly "bombarded" with requests for aid from kin—to hire incompetent kinsmen, to play favorites in times of layoffs or in the distribution of easier or better-paying jobs, and to provide loans. In such cases, elites avoided economic obligations to less fortunate kin, who were considered a drain on resources, by cultivating friendships with equally fortunate elite, thereby alienating their kin and making themselves less accessible.[63]

Unfortunately, there have been no systematic cross-cultural or longitudinal studies showing that when kin ties weaken, friendships flourish, or vice versa. Indeed, it is quite possible that both can rise in tandem as a society becomes more particularist. In such a case, people would value *all* kinds of social relationships (whether kin or friends) relative to following standard codes of conduct. It is also possible that in some universalist societies, neither friends nor kin are particularly important compared to obeying the legal system or following a standard code of conduct. Therefore kinship and friendship may not compete for importance in an either-or fashion, but rather vary along quite independent dimensions.[64]

Medium of Interaction

In the words of John Donne, "Sir, more than kisses, letters mingle souls; for, thus absent friends speak."[65] For most of human history, friends could interact in two ways—in the face-to-face presence of each other or via the long-distance exchange of things, such as gifts and letters, the latter coming with the advent of writing. With the invention of the telegraph and then the telephone in the nineteenth century, the number of opportunities for interaction increased, permitting friends to have instantaneous conversations around the world. These novel technologies supplemented other kinds of interactions among friends and permitted the maintenance of friendships in an increasingly mobile society. These new technologies also raised concerns about their potentially negative effects on society. In 1926, for example, the Knights of Columbus proposed that its group meetings around the U.S. discuss, among other topics, "Does the telephone break up home life and the old practice of visiting friends?"[66]

In the last three decades, the introduction of a new suite of technologies, referred to as computer-mediated communication, is transforming the ways that people not only interact with friends, but also how they cultivate new relationships. Aficionados of topics as varied as ABBA, bodybuilding, pets, snowboarding, and *Star Wars* can meet like-minded fans and share their interests in specially designated online chat rooms. Online gamers can enter three-dimensional virtual worlds, such as World of Warcraft, and interact via lifelike avatars with thousands of others in fantasy adventures. Children, adolescents, and adults can keep track of old acquaintances, meet new friends, and showcase their personal interests on social networking sites, such as Ravelry, a website devoted to sharing and discussing knitting and crocheting projects. And both offline and online friends can email, send instant messages, keep up-to-date through social networking sites, and engage in video chat via webcams.[67]

These new technologies are also spurring debates about their effects on society, and on friendship in particular. Social critics have raised concerns that people will become socially isolated as face-to-face social relationships are replaced with time spent interacting with anonymous strangers over media that limit emotional expression to happy-face emoticons.[68] As one prominent critic of the Internet warned, "Human kindness, warmth, interaction, friendship, and family are far more important than anything that can come across my cathode-ray tube."[69] Opponents of this view argue that the Internet opens up new opportunities for social relationships, freeing people from the constraints of geography, removing superficial biases based on gender, race, and class, and permitting people to meet others based on common interests.[70]

Since these debates began over a decade ago, numerous surveys and longitudinal studies have shown that Internet use does not eat away at time spent with family and friends. Indeed, most of the time spent on the Internet is shifted from time spent watching TV or reading newspapers.[71] Rather, the Internet seems to be a way to maintain existing relationships (via email and instant messaging) and to stay in touch with long-distance ties.[72]

Instead of displacing face-to-face friendships, the Internet has permitted individuals to extend their social networks beyond existing offline ties.[73] In a survey of over five hundred members of randomly selected popular Internet newsgroups, more than one in seven people had developed a close friendship with a person they had met through the Internet (about the same number who had developed a romantic relationship). Moreover, many friendships progressed from initial Internet meeting to communication by letter or phone to face-to-face meeting, actually becoming stronger and closer over a two-year period.[74] The authors of the study argued that interacting over the Internet focuses partners on mutual self-disclosure, common interests, and revealing their "true selves" rather than superficial features such as physical attractiveness, race, or gender, and thus provides a more stable and durable basis for relationships.[75] The anonymity of most Internet interactions makes people engage in "hyperpersonal" talk, where they can reveal deeply personal aspects of their lives that they would normally reserve for much later stages of non-anonymous interactions.[76] It may also make cross-sex and cross-ethnic relationships easier to develop by removing the structural and normative constraints existing in offline settings.[77]

The Internet provides a context for meeting people with shared interests and trying out acquaintances within the safe confines of an anonymous space. These interactions often migrate to face-to-face encounters as the relationship develops. Indeed, face-to-face interactions with online friends may be a necessary part of becoming close. In a comparison of online and offline friendships of 162 Hong Kong newsgroup users of similar duration, online friends expressed less commitment to the friend ($d = -1.12$) and less integration ($d = -0.75$) with a friend's significant others. It is difficult to judge from this finding whether exclusively online relationships are less close because of the nature of communication or rather that people only permit online relationships to become offline once they have reached a certain level.[78] Moreover, the difference between offline and online friendship is larger for casual friends and less pronounced for close friends, suggesting that once one becomes close, the specific means of communication don't matter.[79]

By providing new avenues for meeting people, forming relationships, and keeping track of old acquaintances, the Internet also has the capacity to change the very meaning of the word *friend*. Two decades ago, when people in the U.S. or U.K. were asked to name their close friends, they would generally name no more than ten, and when asked to name their other friends, they would name somewhere between ten and a hundred. They might send Christmas cards to 120 friends, family, and associates.[80]

With the advent of social networking sites, where one can electronically "friend" other individuals, share personal information, keep in touch, and read newsfeeds about what these friends are doing, this number has changed. Now users regularly amass friends in the hundreds, with mean friend counts for U.S. college students between two hundred to three hundred.[81] Notably, in a burgeoning group of highly friended individuals on Facebook, Facebook Whales, each boasts over a thousand friends. Many of these people use Facebook as an electronic rolodex to keep track of anyone they have ever met. Others view these friends as potential contacts for business networking. For minor celebrity bloggers, friends are an admiring audience.[82] In any case, these novel technologies, by permitting such easy tracking and communication, are expanding the definition of *friend*.

This is not a new phenomenon. Consider the following by Hugh Black, a Scottish preacher at the turn of the twentieth century: "The sacred name of friend has been bandied about till it runs the risk of losing its true meaning."[83] Indeed, the stretching of *friend* to encompass links on Internet social networking sites is part of a much longer historical process in the U.S., whereby the stand-alone term *friend* has come close to meaning *acquaintance*, requiring new verbal modifiers to differentiate close friends from the rest (*true, real, best*).[84] On blogs that grapple with the most recent semantic expansion, it is common to make distinctions between friends from social networking sites and offline friends by using scare quotes or modifiers such as *virtual, Internet, online,* or *MySpace*. The same process is also happening in other languages, with terms like *wang you* (Net friend) in Chinese and *Cy-Ilchons* (online buddies) in Korean.[85] Thus, as new opportunities have arisen for cultivating friend-like social ties, people have already developed new ways of differentiating

---

BOX 26   "She's not my friend—she's just my Friendster"

Social networking sites, such as Friendster, MySpace, and Facebook, have provided new technologies for interacting with friends and acquaintances— for keeping abreast of the lives of friends from college and high school, for making plans, and for posting memories and preferences for others to see (Boyd 2006, p. 1). Due to the use of *friend* to signify mutual links on their systems, such sites have also provoked debates about what it means to be a friend. It turns out that, as the quotation above suggests, people are still quite capable of distinguishing between close friends and "friends." For example, when asked about their reasons for accepting people as "friends" on Friendster or MySpace, people in one study gave many different reasons, including (1) they are actual friends; (2) they are acquaintances, family members, colleagues; (3) it would be socially inappropriate to say "no" because you know them; (4) having lots of friends makes you look popular; (5) it's a way of indicating that you are a fan (of that person, band, product, etc.); (6) your list of friends reveals who you are; (7) their profile is cool, so being friends makes you look cool; (8) collecting friends lets you see more people; (9) it's the only way to see a private profile; (10) being friends lets you see someone's bulletins and their friends-only blog posts; (11) you want them to see your bulletins, private profile, private blog; (12) you can use your friends list to find someone later; (13) it's easier to say "yes" than "no." Thus, despite the apparent expansion of the term *friend*, people are quite aware of the fact that it means something very different from a real, close, or true friend (Boyd 2006).

---

these ties and clarifying what kind of friendship they are talking about (box 26).

Friendships share many features across societies, including positive affection, feelings of goodwill, and expectations of mutual aid. At the same time, how we view and value our friends can change, depending on the society and culture we live in. In some places, verbal expression and explicit emotional support are important elements of friendship, whereas material aid and concrete favors matter more in others. Moreover, how much we value friendship compared to other social institutions, whether kinship or the law, differs substantially across cultures.

A number of theories have been proposed to explain how our social and cultural environment might influence how we think about and act toward friends. Many do not stand up to further inspection. Despite concerns about geographical mobility and new ways of communicating, people who move or who have access to the Internet are not at a loss for friends. And the new approach to "friending" on Internet sites is not necessarily changing people's close friendships. Moreover, the explanation for differences in how friends talk about personal matters as a difference between collectivist and individualist cultures does not neatly fit the facts. For some theories, such as those that pit friendship against kinship, there is too little support one way or the other. The strongest candidate among these theories for cross-cultural variation in friendship, and the one backed up by good qualitative and quantitative data, posits that material help among friends becomes more important in societies where daily life is more uncertain. However, a great deal of work still remains in order to understand the various reasons for cross-cultural differences in how people cultivate and maintain friendships.

8 Playing with Friends

A friend may well be reckoned the masterpiece of nature.
RALPH WALDO EMERSON

When Emerson wrote these lines, he was musing about what he called the paradox of friendship—that a good friend lives in a separate body but can also feel like a reflection of one's own self. This is certainly one of the most intriguing qualities of close friends, but Emerson could have justified his claim of friendship's masterpiece in several ways. Friendships are striking in their potential longevity. They can last decades, surpassing the lifespan of many living organisms. And in those societies where children formally inherit the friendships of their parents, such ties can even outlive their original hosts. Friendships are remarkably ubiquitous, and despite some cross-cultural differences, are surprisingly consistent in their basic form across human communities. Compare this to the many other social institutions we know today, such as banks, schools, and specialized law enforcement, which are limited to a small set of complex societies.

Perhaps the most notable quality of friendship is the diverse and profound ways that friends make sacrifices for one another. Simple acts of kindness, such as taking a Saturday afternoon to help with a move, preparing a dinner for a sick friend, or giving a lift to the airport, are fundamental to good friendships. Friendships can also be a context for extreme acts of selflessness. A seventeen-year-old in the Philippines drowns after jumping into a strong current to save his best friend. A sixty-five-year-old commercial painter in Burbank, California, is stabbed to death while trying to protect his friend of forty-five years from an unknown assailant. A ten-year-old boy in South Africa rescues his best friend from a crocodile's grip and dies himself.[1]

So far in this book, I have documented self-sacrifice as a key expectation and behavior among close friends, not only in modern, industrialized nation-states but also in a wide range of small-scale societies. Moreover, close friends avoid the kind of accounting procedures—following a norm of reciprocity, maintaining a balance in favors, or considering the shadow of the future—that evolutionary theorists have proposed as ways to avoid exploitation in the social world. Indeed, close friends appear to help, give, and forgive automatically, with very little thought at all. How could such unconditional willingness to sacrifice have survived the gauntlet of natural selection? More specifically, how was it advantageous, in an evolutionary sense, to spend time and resources and perhaps risk one's life to help friends?

Two evolutionary psychologists, John Tooby and Leda Cosmides, propose that friends have an incentive to help (even at great cost to themselves) because they have become mutually irreplaceable to each other. Originally a friend might be irreplaceable because of a special skill, resource, or attribute. The friend may be a good cook or very good at making you laugh. Over time, however, close friends can also become irreplaceable in a much more subtle and mutual way. Specifically, the fact that a friend thinks *you* specifically are irreplaceable (and thus worthy of support) in turn makes that friend irreplaceable.[2] In chapter 6, I described several ways that people can become mutually irreplaceable by building the value of a relationship and decreasing partners' outside options. In this chapter, I describe economic models that suggest that many of the features of friendship-like relationships—a "courtship" period, gift giving, gradually raising the stakes, and positive affect—play an important part in making the unconditional help observed among friends particularly robust (though never failsafe) against cheating and exploitative behaviors. To do this, I draw from mathematical approaches in economics and game theory to understand when we would expect a behavioral system like friendship to be stable against false friends. Specifically, when does behaving like a friend make it more likely that one's partner will also want to behave like a friend? When friendship is a self-reinforcing system and thus robust against exploitation, participating in friendships can make evolutionary sense, in that it provides advantages over not cultivating friendships.

In this chapter, I will first outline the basic elements of game theory, a body of analytical techniques used to make predictions about behaviors in strategic situations. Then, I describe a particular game (the prisoner's dilemma) and strategy (tit-for-tat) often used to understand the evolutionary stability of cooperation in a game theoretic framework. I describe the limitations of the standard game for understanding helping among friends and outline an extension, the favor game, that captures the variable nature of favors among friends. I then outline how a simple friend-like strategy can permit the cultivation and long-term maintenance of a mutually beneficial relationship at high levels of support. The strategy involves three stages: starting with small levels of calculated help, moving to higher levels of calculated help if a friend's actions warrant it, and finally a shift in decision making to knee-jerk altruism based not on past behaviors or the shadow of the future but on simple decisions about whether a partner is a friend. I conclude by discussing how the gradual cultivation of a relationship and the subsequent shift in thinking is a special case of a more general process—niche construction—by which people change their environment in ways that benefit them in the future.

GAME THEORY

When most people decide to help a friend, choose a birthday present for a spouse, or send a gift to charity, they don't think of these decisions as occurring within a game. Calling these actions part of a game can sound callous and, depending on the audience, downright sociopathic. In most people's view, good friends shouldn't play games with one another. Spouses shouldn't have strategies. Close relationships are not like chess.

Although many people do not consciously analyze their close relationships as games, an entire branch of economics, appropriately called *game theory*, does precisely this. Game theory focuses on situations where an individual's success in achieving his or her goals is affected by the actions of others. Robinson Crusoe did not have to deal with such contingencies in his island exile, but in most real-life situations, the success of our actions depends on how others behave. Whether one can get a job, earn a raise, buy gas at $2.00 a gallon, or get married and have kids depends in many ways on what other people do. In each of these cases, game theorists see the world through game-colored glasses.

A game theorist often begins an analysis by asking a number of questions about a given situation: Who are the major players? What are the kinds of actions that players can take? What do players know? What kinds of outcomes do players care about, and how much do they value these outcomes?

Consider the case of two traders who live in different villages. Each trader can visit the other village and sleep under a tree on the outskirts of town, thus risking robbery at a cost of all his goods. Or he can stay safely at the other trader's house for free, but the host must agree to part with some of his food to feed the traveler. At another point in time, the host will return the visit, and the first trader must decide to part with *his* food or turn the traveler out. With this description, we can answer all of the above questions

TABLE 4. The potential costs and benefits to trader 1 given different possible actions

	Trader 1 hosts next	*Trader 1 doesn't host next*
Trader 2 hosts first	Trader 1 saves goods from robbers but parts with food when trader 2 visits	Trader 1 saves goods from robbers and keeps food rather than sharing with trader 2 later
Trader 2 doesn't host	Trader 1 loses goods to robbers and parts with food when trader 2 visits	Trader 1 loses goods to robbers but keeps food rather than sharing with trader 2 later

in a precise way. With such information, game theorists often distill such two-person dilemmas to a two-by-two matrix, showing the costs and benefits to players when they and others act in different ways (table 4).

With these tools, we are also in a position to make some predictions about what the traders would do. In this game, if neither trader hosts the other, then it is likely they will both lose their goods to robbers in the night, leaving them their food at home but no goods to trade with. This is such a great cost that it might make people avoid trading between villages altogether. On the other hand, if both traders agree to host the other, then they only incur the cost of feeding the other, making this a mutually better option than refusing to host (since the cost of food is much less than the value of one's goods). Although this is a mutually beneficial option, once trader 1 has enjoyed the hospitality of trader 2, he can do better by refusing to host trader 2. In mathematical notation: goods < goods + food. Trader 2 knows that trader 1 has this incentive, and this makes him reluctant to provide hospitality in the first place. Therefore, we are back to the asocial starting point where neither trader hosts the other and each receives a payoff of nothing.

When traders only have the opportunity to interact once (an admittedly unrealistic assumption), then the asocial outcome is self-reinforcing in a very simple way. Regardless of what one trader does, the other trader will always do better by not hosting. Neither has an incentive to decide independently to host. This self-reinforcing state of affairs is referred to as a *Nash equilibrium*, a central concept in game theory named after John Nash, a mathematician known by many as the protagonist of the Oscar-winning film *A Beautiful Mind*. Trying to find the self-reinforcing states, or *equilibria*, is a primary task in game theoretic analyses. Specifically, they permit one to make predictions about the kinds of behaviors that one would expect to find in the real world, as these are the kinds of behaviors that should maintain themselves over time.

Such game theoretic models raise questions about the currency that people use when making decisions, whether it is food, money, status, reproductive success, or some other culturally determined goal. In the example above, we can only add food to goods if we assume some under-lying common currency that can be measured across these two kinds of things. This is a necessary simplification when making predictions from game theoretic models. If we frame the game as an economic model, then we might compare food and goods either in terms of their subjective value to the decision maker or in terms of some exchange equivalent, such as their rough cash value. If we frame the game in evolutionary terms, then we might compare food and goods in terms of how they contribute to an individual's fitness or reproductive success. Of course, such quantities are exceedingly difficult to measure. Here, I assume that people make choices with the highest subjective value to themselves, and that this bears some (perhaps weak) relation to the reproductive value of their choices. In the case of the trading game, this means that people prefer having both goods and food to just having goods. People prefer keeping their food to giving it away. Although such assumptions are simplifications, they also let us use another kind of reasoning based on formal mathematics to hone intuitions and test insights about behavior in strategic interactions.

The Prisoner's Dilemma and Tit-for-Tat

The trader's dilemma is a special case of a more general dilemma called the prisoner's dilemma, a game that has dominated the study of cooperation in economics and evolutionary biology for the last thirty years.[3] In the prisoner's dilemma, there are two possible moves—defect or cooperate—with the following restrictions on how actions translate into payoffs (see table 5, which sums up the game's basic outcomes).

1. Whatever a partner does, one's best move is to defect. In table 5, the temptation to defect (T) is greater than the reward from mutual cooperation (R), and the punishment for mutual defection (P) is at least better than being a sucker (S).

2. But players can achieve more if they both cooperate than if they both defect. In table 5, the reward for mutual cooperation (R) is greater than the punishment for mutual defection (P).

TABLE 5. The potential costs and benefits to player 1 given different possible actions

	Player 2 cooperates	*Player 2 defects*
Player 1 cooperates	Reward for mutual cooperation (R)	Sucker's payoff (S)
Player 1 defects	Temptation to defect (T)	Punishment for mutual defection (P)

The prisoner's dilemma is appealing because it captures major elements of many cooperative dilemmas and it frames them in a simple yet general mathematical framework. For this reason, it has figured prominently in landmark studies of the evolution of cooperation. Robert Trivers made direct reference to the prisoner's dilemma game when he introduced the concept of reciprocal altruism in his groundbreaking 1971 paper "The Evolution of Reciprocal Altruism."[4] Nearly a decade later, Robert Axelrod chose a repeated version of the prisoner's dilemma as the basis for his famous tournament comparing the success of different strategies in a cooperative environment.[5] The dilemma's simplicity has permitted clear mathematical proofs about the conditions under which cooperation can get started and be sustained.[6] Over twenty thousand articles and books have been published about the prisoner's dilemma, with nearly one in every thousand published works in 2007 including the phrase.[7]

As in the trader's game, the prisoner's dilemma has only one Nash equilibrium—neither trader hosts the other. However, if two partners play the prisoner's dilemma many times, the results can change. Consider a player who starts by cooperating and then does whatever her partner did in the previous round. This strategy is often called tit-for-tat and is based on a strictly balanced form of reciprocal altruism. If your partner plays this hair trigger strategy, then you can do no better than to also play this strategy as long as the expected number of interactions is sufficiently large.[8]

Since the early 1980s, when it was declared the victor of a repeated prisoner's dilemma tournament run by Robert Axelrod at the University of Michigan, tit-for-tat has enjoyed celebrity status and is included in nearly as many scholarly references as the prisoner's dilemma itself.[9] Since that time, a wide range of mathematical and simulation studies have confirmed that three properties make tit-for-tat and its many variants so successful. Foremost, tit-for-tat starts off being nice and thus never defects first. However, tit-for-tat is no chump, and it can be provoked to defect if another player does so. Finally, it is willing to repair a relationship, if the partner begins to cooperate again. Indeed, most successful strategies in the standard repeated prisoner's dilemma game possess these same properties: nice, provokable, and forgiving.[10]

The great success and academic celebrity of the prisoner's dilemma and tit-for-tat has made one-for-one, or more generally balanced, accounting over outcomes the primary mechanism proposed for cooperation with non-kin partners, including friends. According to this notion, individuals keep track of help given and received. This can be the number of times help was provided, the costs of help, or the benefits of such help. Individuals calculate the difference between help given and help received, and when this is too great, they try to restore balance. When a partner has benefited too much, then one stops helping until the balance is restored. When a partner has benefited too little, one tries to restore balance by providing benefits to the partner. Tit-for-tat does this very simply by doing what a partner has done previously. If the partner has helped, then one helps. If the partner has not helped, then one doesn't help.

One can complicate the repeated prisoner's dilemma in many ways, and these can influence which cooperative strategies perform best. For example, when people make occasional mistakes, a pair using pure tit-for-tat strategies would break into long bouts of retributive defection. In such cases, strategies that forgive some defections and also forgive retribution against their accidental defections perform better than pure forms of tit-for-tat. In another complication of the model, individuals can stay with partners for as long as they want or leave to search for other partners. In such environments, people who stay with cooperating partners but leave non-cooperative partners (sometimes called an out-for-tat strategy) perform well.[11]

Complications to the repeated prisoner's dilemma influence the kinds of strategies that ultimately perform best. However, people using these variations of tit-for-tat, whether contrite tit-for-tat, forgiving tit-for-tat, or the partner-hopping out-for-tat, all base their decisions on the behaviors of their partners in a relatively short time frame. Thus, they violate basic empirical findings about friendship: that close friends' decisions to help cannot be explained in terms of a norm of reciprocity or a longer-term balancing of accounts (chapter 1). Therefore, the strategies that work well in the standard repeated prisoner's dilemma game cannot account for the kind of reciprocal altruism observed among close friends.[12]

Another problem with the standard specification of the repeated prisoner's dilemma is that it lacks an important feature of the help shared by friends—the favor needed at any moment may be either small, very large, or somewhere in between. The variable size of favors creates new kinds of decisions—do I maintain a friendship of small favors, or do I try to raise the stakes when the opportunity arises? And do I raise the stakes to some level of helping but then plateau? Favors of variable size and currency also make calculations exceedingly complicated, putting pressure on individuals to find shortcuts in how they make decisions to help.[13] In the next section, I will describe a game that captures the variable size of favors and argue that a friend-like strategy that begins with small calculated favors but then raises the stakes, eventually moving to unconditional aid, can avoid exploitation and solve the problem of making decisions when the benefits and costs of a relationship are exceedingly difficult to calculate.

THE FAVOR GAME

Consider two people, Sally and Ann, living in a risky world. At any time, one may benefit from the other's help. Sally may need some food to tide her over for a few days, care while she is sick, or simply a ride to the market. The degree of need changes from time to time, as does the potential cost to the would-be helper. Assume for now that there are two kinds of favors—small and large—with the cost and benefits of large favors (let's say, loaning a large sum of money) substantially bigger than those for small favors (let's say, loaning money for lunch). In all of these cases, if Sally decides not to help Ann, then she incurs no cost, and Ann receives no benefit.

If Sally and Ann met only once in their lives, and Sally was in need of a small favor, Ann would lose by helping Sally. As long as there is no reputation, repeated interactions, or relationships (an admittedly unrealistic assumption), if Ann regularly helped strangers like Sally, she would be worse off than if she never helped. The same is true for Sally, and in such a dystopian world, we would expect both to live alone and never help each other.

The tragedy of this situation is that if Ann and Sally could somehow trust each other and agree to help each other over time, then they would both do better than if they lived alone. Luckily, in this game, individuals do have the opportunity to interact over time. They also have the capacity to move from their partner, and either to live alone or try to find another partner. Moreover, they are able to expand the amount of help they are willing to provide over the course of their relationship.

The analysis follows several logical steps based on recent economic models of behavior in repeated games that progressively open novel possibilities for action (see appendix B for the formal models). By gradually increasing the kinds of decisions open to players, the analysis shows how a number of practices, such as costly "courtship" and the willingness to bear increasing costs as a relationship progresses, are crucial for making it personally beneficial for individuals to cultivate and maintain a high-stakes helping relationship.[14]

First, if both Sally and Ann could agree (and trust each other) to help when help was needed, then they would on average do better than if they lived alone trying to make do by themselves. The reasoning behind this first step relies on the fact that a long-term cooperative relationship can be much more valuable than the cost of providing a single favor. One way that Sally can give Ann an incentive to continue helping is to threaten to end the relationship if Ann doesn't help—essentially a tit-for-tat strategy.

Such a solution works well as long as the only outside option for Sally or Ann is to go it alone. However, people often have the possibility of cultivating friendships with other people, and the solution above could easily be exploited in the following way. Sally accepts favors from Ann until Ann needs help, and then Sally simply moves to another partner. One safeguard against such flighty friends is for both Sally and Ann to require an early courtship consisting of costly but intrinsically worthless gifts—such as hanging out, small presents with costly wrapping, and perishable goods. Such a courtship period makes it difficult for someone to simply enjoy the benefits of a friend and then move on when asked for a favor. Another important part of this solution is that the required cost of a gift is never explicitly stated and furthermore that it is impolite to ask how much a gift should cost. Otherwise, we would find partners agreeing to give less and less until no courtship gifts were required, leading us to the problem we started with in the first place. Therefore, in a world where friends require costly gifts of time and intrinsically worthless gifts of unstated cost, both individuals have an incentive to stay in the relationship rather than to simply cultivate new relationships when a partner needs a favor.

Courtship costs raise a final concern. To be effective in the formal model, they must be higher than the cost of providing a very large favor. However, if the large favors are really large and rare, then it could be a waste to spend such large amounts on an initial courtship gift. A solution to this problem can be found in a simple strategy that starts with a small courtship cost and attempts to expand the cost of favors over time. If a partner reciprocates by helping with big favors in the future, then the relationship ratchets to a new level. If the partner doesn't reciprocate, the relationship isn't lost. Rather, the friendship stalls out at the previously low level of favors.

Until now, I have assumed that individuals will follow something like a tit-for-tat strategy to enforce favors. If Sally fails to help with a small favor at the early stage of a relationship, then Ann ends it. If Sally fails to provide a large favor, then Ann returns to helping only with small favors. However, once Sally and Ann have built up their friendship to providing large favors, they can monitor each other at quite low levels, be very forgiving, and still maintain an incentive to help each other. Indeed, with reasonable assumptions about the length of relationships and the rates of small and big favors, Sally would only need to retaliate against Ann's small betrayals 5 percent of the time to give Ann an incentive to help with small favors (appendix B). For large favors, Sally would still only need to retaliate by moving back to small favors 15 percent of the time to give Ann an incentive to provide big favors. These are far lower levels of retaliation than derived for optimal generous tit-for-tat strategies in the standard repeated prisoner's dilemma games.[15] Therefore, when individuals can cultivate a relationship at different levels of helping, even very low levels of monitoring and retaliation can provide an incentive to help a partner. If monitoring has any cost, then it would make sense to monitor at such low levels.

What these economic models show is that under a number of conditions, a raise-the-stakes strategy that begins with calculated helping permits the cultivation of relationships where partners do better than if they were asocial, followed a sneaky court-and-leave strategy, or limited their help to a maximum determined by the initial courting cost.

The raise-the-stakes strategy exhibits several properties that are commonly observed (or at least expected) among friends. First, it involves an initial series of gifts that are costly and surprising but also worthless in material terms. Second, it permits the maintenance of a relationship with long-term imbalances dictated by the occurrence of needs and involves raising the stakes of helping far above the original courtship costs.[16] When the game is extended to involve a continuous distribution of helping costs, there is theoretically no limit to the kinds of help that partners have an incentive to provide as long as the relationship lasts long enough. However, at any point in time, there may be favors that are "too big" for the specific friendship.

An important part of the raise-the-stakes strategy is that the value of a relationship can increase over time. This gives partners an incentive to maintain the relationship and can also account for a puzzle raised by Trivers in his original framing of reciprocal altruism. He suggested that people should be less likely to maintain reciprocal ties in old age because the shadow of the future is shorter. And yet elderly people continue to maintain strong friendships. According to the models described above, older individuals can have much more valuable friendships in terms of the potential help provided, thereby outweighing the higher probability of the relationship ending.

FROM CALCULATION TO KNEE-JERK ALTRUISM

A raise-the-stakes strategy, by increasing the future value of a relationship, can account for many of the stylized facts about friendship—an initial courtship, a gradual raising of the stakes, and relatively high levels of forgiveness. However, a glance at the mathematical model in appendix B reveals that this still involves very detailed calculations about the costs and benefits of future interactions (i.e., the shadow of the future) and at least some reactions to a partner's past behaviors (i.e., tit-for-tat accounting). What further changes in decision making are required to reflect the way that close friends make relatively unconditional decisions to provide aid? I contend that when the relationship becomes suitably valuable and the calculations become too computationally complex, the best way of making decisions is simply to rely on a judgment about whether someone is a friend.

This transformation involves a shift between two kinds of decision making—what some psychologists call deliberation and habit. When people deliberate, they usually weigh future consequences more carefully, take a longer-term perspective, and think more rationally according to traditional economic models. When people act habitually, however, they follow automatic heuristics and preferences, regardless of the future consequences.[17]

Many of the decisions described above involve calculations about the future value of relationships, which would require recruiting the deliberative system. However, relying on the deliberative system for such decisions poses a problem. The deliberative system is very costly, and it breaks down frequently when we do not have sufficient time and energy—when we are tired, hungry, or stressed.[18] As described in chapter 6, one aspect of relationship preservation, the tendency to forgive, is low in strangers. Moreover, if we rely on time-limited habitual decisions, the tendency to forgive is even lower. If our initial impulse with others is to be defensive, then relying on the deliberative system to make decisions about something as valuable as a long-term relationship means that during times of weakness, we could react over-defensively and thus prematurely end a friendship. Similarly, in moments of weakness, we may also fail to help a friend. Interestingly, in the study of forgiveness described in chapter 6, the tendency to forgive when primed with a close friend's name was very high and also insensitive to time pressure, suggesting that it had become habitual and automatic.

It is possible that positive feelings reflect this shift in decision making toward friends, whereby we want to help friends and attempt to safeguard the relationship habitually rather than relying on the deliberative system to calculate the future benefits of the relationship. By shifting toward the habitual system, the calculations described in the model are no longer made. Rather, decisions are made automatically because a person is a friend.[19]

Of course, this system of decision making will also lead to errors—a tendency to forgive and to help when one shouldn't. However, one theory, error management theory, can help understand when it would be advantageous to make more errors in favor of friends than against them. The premise of error management theory is that when making decisions in uncertain situations, errors are quite common, but some errors are more costly than others. Accordingly, the best way to make decisions is not to minimize the number of errors, but to minimize their *costs*.[20]

The standard example of this is a fire alarm. The costs of a false alarm are much smaller than the costs of a failure to detect a fire, so fire alarms are generally designed to give false alarms much more frequently than they fail to detect a fire. This same approach can be used to understand how people might learn to make decisions toward close friends. Consider that one can follow one of two rules when someone, let's say Frank, asks for help: always help Frank, unless there is some reason to believe he is not a friend; don't help Frank, unless there is some reason to believe he is a friend. The first rule describes a habitual tendency to help Frank, unless upon further deliberation there is a reason not to help him. The second rule describes a habitual tendency to *not* help, unless upon further deliberation there is a reason to.

According to the first rule, there will be very few cases where one fails to help Frank if he is a friend, but there will be more cases where one helps him even if he is not really a friend. Therefore, there is a bias toward helping Frank. Conversely, according to the second rule, there is a bias against helping Frank.

We can mathematically analyze these rules to determine when one should be biased in one way or the other (appendix B), and what we find is a simple inequality describing when we should help Frank:

$$p(friend) > \frac{C_h}{C_n}$$

Here, C_h is whatever it costs to help Frank, C_n is the cost of ending a mutually beneficial relationship if Frank is a friend, and $p(friend)$ is the probability that Frank is a friend. In words, if the probability that Frank is a friend is sufficiently high and the cost of not helping is sufficiently large relative to the cost of helping, then it makes sense to follow rule number one—be good, and ask questions later. In short, in a sufficiently long and valuable relationship, it makes sense to treat Frank with the adage "innocent until proven guilty" rather than "guilty until proven innocent." Positive feelings toward Frank may reflect changes in decision making necessary to make this shift.[21]

An important implication of this kind of decision making is that if people are uncertain about the cost of *not* helping—which requires estimating the probability of the relationship ending as well as the cost of losing the relationship—they may focus more on judging whether their partner is a friend rather than on the costs and benefits of the relationship. Therefore, the most efficient kind of decision making can be transformed from a concern with the costs, benefits, and consequences of helping to judgments about a person's intentions toward oneself, which changes the focus to such signals of friendship as the longevity of the relationship and costly signals about a partner's intentions.[22]

The importance of signaling in such a scenario can also account for another common feature of close friendships—that partners *continue* to give costly but worthless gifts, not just at the beginning but throughout the course of a relationship. Of note is that such exchanges, unlike the provision of help, should generally be balanced. The continual, reciprocal giving of costly but intrinsically worthless gifts could be one way to honestly communicate to a partner that one is still a friend (i.e., *p(friend)* is high). Conversely, gift giving that becomes too unbalanced can cause the disadvantaged partner to question the other's intentions.[23]

In the last two sections, I outlined a simple friend-like strategy that can permit the cultivation and long-term maintenance of a mutually beneficial relationship at high levels of support. The strategy involves three stages: (1) starting with small levels of calculated help, (2) moving to higher levels of calculated help if a friend's actions warrant it, and (3) finally a shift in decision making to knee-jerk altruism based not on past behaviors or the shadow of the future, but on simple decisions about whether a partner is a friend. This trajectory reflects many of the observations made in earlier chapters about how close friendships develop from first meetings, and how close friends help one another for different reasons than do casual acquaintances or strangers.

This trajectory also has implications for how we study friendship and its development. Most experimental work on how people help and share—two of the defining behaviors of friends—has focused on strangers. One primary reason for this is the quest for control in laboratory settings. Strangers come to the lab with a blank slate. Friends, on the other hand, could share any number of different histories that define their current state of interactions. Therefore, experimenters are rightly concerned that bringing friends into the lab shuttles in historical baggage that defies experimental control. This problem has led some researchers to try "inducing" friendships in the lab, as a cleaner surrogate for messy, real-life friendships. However, if the transitions described previously generally occur over extended periods of time, then findings about "friendships" that are induced in time-limited laboratory settings may reflect only the initial stages of a relationship and may not accurately capture the kinds of thinking and behavior that arise in later stages. This problem is most notable in the differences between close friends and acquaintances in the use of a norm of reciprocity, the balancing of accounts, and a shadow of the future in deciding to help. For this reason, work that builds on the innovative studies described in chapters 1, 3, and 6 that examines how real friends behave in laboratory settings will hopefully improve our understanding of how friends behave at all stages of their development.

FRIENDSHIP, HABIT FORMATION, AND NICHE CONSTRUCTION

The models described previously suggest that when needs are variable, a friend-like strategy that builds a mutually valuable relationship by gradually raising the stakes can give both partners an incentive to maintain a friendship, both in terms of helping when needed and by being highly forgiving. Once such a mutually beneficial relationship is created, it can also be advantageous to switch decision making from deliberation to a more habitual urge to help (and to forgive).

In this way, building friendships shares many similarities with the general process of habit formation observed in non-human animals. The

---

BOX 27   Can Non-human Animals Be Friends?

Biologists occasionally use the term *friendship* to describe interactions of affiliation and support between non-human animals (Silk 2002; Smuts 1985). But do such animals really participate in friendships? A challenge to answering this question is the fact that friendship in the human case is deeply tied with internal states, such as emotions, expectations, and theories of the other's mind. In experiments with humans, researchers often ask people about these internal states. Consider the Facebook experiment from chapter 1, in which people helped friends with little regard for the shadow of future consequences. Not only did the students have to name their best friends for the experiment, the researchers also needed to let the students know that their friends wouldn't find out who provided help. It is difficult to imagine how such an experiment could be conducted with non-human animals, or how researchers could determine that non-human animals give preferentially to friends without concern for the shadow of the future.

Despite these challenges, recent work is investigating how feelings involved in human friendship, such as wanting to affiliate, feeling secure in a partner's company, or wanting to maintain a relationship, are expressed through gestures and behaviors. Among wild chimpanzees, for example, behaviors such as visual monitoring or rough self-scratching behavior can indicate anxiety, and the absence of such behaviors suggests an individual feels secure around a partner (Kutsukake 2006). Reconciliation after aggressive conflict, through either grooming or other affiliative behaviors, may signal the motivation to maintain a relationship (Aureli and Schaffner 2002). However, until researchers can show that non-human "friends" help one another and share with the same lack of accounting as observed among human friends, it will be difficult to claim definitively that non-humans have friendship.

---

first phases of decision making toward a partner involve careful deliberation about which actions lead to which outcomes and are very sensitive to these contingencies. However, once one has had a series of successful interactions with a partner, one engages in habitual action that is driven by more proximal stimulus-response behaviors. In the case of friendship, the stimulus is a friend in need and the response is to help.[24]

There is one problem with thinking of friendship purely in terms of habit formation. Forming a habit of brushing one's teeth is an efficient way to maintain dental hygiene, with few deleterious consequences. However, in a strategic environment, it is possible that a partner could act as a friend just long enough to get you in the habit of helping, and then start to exploit your goodwill. Therefore, friendships also involve numerous other defenses, such as regular signaling and gradual raising of the stakes, which can ensure that both partners have an incentive to keep up the habit.

The practice of gradually cultivating a friendship also shares many similarities with the general process of niche construction, whereby individuals invest in modifying or maintaining their environment, thus changing the resources that are available in the environment. The prototypical example of such niche construction is an American beaver investing time and energy building and maintaining its dam and lodge—felling trees, weaving branches, patting mortar, and digging canals for transporting trunks. Although initially costly, the beaver's effort pays off, providing protection from predators, such as wolves and bears, a warm lodge in the winter, and slow-moving water for storing food. The beaver's ecological niche is composed of these physical resources. People, and other animals, can also build social niches that are composed of social resources—such as friendships and alliances.[25]

In his aphorism "The bird a nest, the spider a web, man friendship," William Blake implies that friendship is a quintessential form of niche construction. Birds build nests and spiders weave webs. Humans make friends.[26] As with other kinds of niche construction, the time, effort, and risk spent in cultivating a friendship often pays off in the long run as partners change the "game" to expand the possibilities for stable prosocial interaction.[27] In friendship, the game moves from one where the largest risk is exploitation by a partner to one where the largest risk is losing a mutually beneficial relationship. More than simply reacting to their environment in a tit-for-tat manner, friends make decisions to invest in a niche, and once a niche is created, make a change in the mode of decision making toward maintenance of the niche. If the raise-the-stakes strategy is considered as a rudimentary model of friendship, it provides some insight into key game changes that can occur. By increasing the stakes, there are larger rewards from interaction, and by increasing the future value of the relationship, violating the rules becomes more costly.

The important point here is that neither niche construction nor habit formation are uniquely human, suggesting that the basic psychological requirements for building friendships may not represent novel selected features. Rather, they could represent a suite of general adaptations for forming habits and building niches that are also available to other non-human animals. A crucial question here is whether other non-human

BOX 28   Identifying Different Kinds of Reciprocity

On most winter days at the hot springs in Jigokudani Park, Japan, a visitor will observe a curious scene—human-like faces poking out of the misty pools, lounging, eating, playing, and sleeping. These are Japanese macaques, and among the bathing monkeys, some will be in pairs, with one partner picking at the other's puffy gray fur. These monkeys are grooming—a behavior that acts like an external immune system and is common to many social animals. In this case, one partner meticulously identifies louse eggs, loosens the adhesive ring that glues the egg to a hair, and then carefully pulls the egg off the end of the hair (Tanaka 1998).

Without such grooming, one researcher has estimated that the number of louse eggs on an adult Japanese macaque would increase from about 500 to nearly 15,000 in just one month (Zamma 2002). While beneficial to the receiver, grooming is a time-consuming process and impedes the groomer from pursuing other activities, such as foraging and being groomed (Tanaka 1998).

If one visited the park over many weeks and carefully recorded these grooming pairs, one would notice an interesting regularity. When one monkey, let's say Ann, frequently grooms another, Sarah, then it is also very likely that Sarah will groom Ann. Moreover, this correlation between giving and receiving occurs among kin and non-kin alike (Schino and Aureli 2008).

Such correlations between giving and receiving are frequently interpreted as evidence for reciprocity, but what kind of reciprocity? Are Sarah and Ann following a tit-for-tat-like strategy, where each day they give and receive grooming in equal amounts and stop the relationship if one fails to reciprocate? Do they keep accounts, so that too great an imbalance ends a relationship? Perhaps they simply have an affinity for being near each other, and when it comes time to groom, they choose a close associate. Or do they freely groom a specific "friend" without a concern about past behaviors or future consequences, as is observed in human friendships?

Each of these mechanisms could create the kinds of correlated helping observed among the Japanese macaques, and so we need some other way to determine which mechanism is at work. Examining the time sequence of grooming would help to some extent. If we observed anything other than a lockstep alternation in helping, then we could rule out a strict tit-for-tat-like mechanism. On the other hand, a lockstep alternation could not rule out any of these other mechanisms. Another option is to conduct experiments that manipulate the balance between benefits provided and received or the ordering of help (de Waal and Brosnan 2006). However, even such experiments would have a difficult time ruling out a concern about the shadow of the future, as has been done in human experiments (chapter 1). In many cases, it may be very difficult to extract which mechanism is actually underwriting an observed pattern of reciprocity. And care should be taken in interpreting any such pattern as evidence for a specific mechanism.

animals also engage in something like friendship. I explore this in further detail in box 27, arguing that much more research is required before we can claim that other animals have something like human friendships.

In this chapter, I have verbally described several game theoretic models that capture many of the basic elements of friendship. I argued that a strategy based on initial courtship and a gradual raising of the stakes can create mutual incentives for helping a partner and maintaining a relationship. Moreover, once someone has cultivated a suitably valuable friendship and has sufficient confidence in a partner, one can move from deliberative decision making about the costs and benefits of interactions to habitual decisions based on whether one judges a person to be a friend. In short, it is possible, and can be adaptive, to move away from decisions based on tit-for-tat–like algorithms or on maintaining balance to those based on a knee-jerk, stimulus-response action, in which the stimulus is a friend in need and the response is to help.

According to this algorithm, over a lifetime, many friendships will *look* balanced, suggesting that partners might be following an algorithm that tries to seek balance over outcomes. However, the appearance of balance does not imply that the balance is at all important to the actual decision-making algorithms (box 28). Rather, it is possible that the most important inputs to such algorithms are not solely instances of aid, but also signals about the partner's intention and whether that person is a close friend.

Conclusion

If you are right, there is no point in any friendship; it all comes
down to give or take, or give or return, which is disgusting, and
we had better all leap over this parapet and kill ourselves.
      Dr. Aziz to Cyril Fielding in E. M. Forster's *A Passage to India*

In this passage, Dr. Aziz expresses indignation at his English friend's
materialistic approach to relationships. According to Aziz, Fielding pays
too much attention to the *observables*—what is said and what is done—
and not enough to what motivates people's behaviors. Aziz's reaction to
Fielding's materialism is shared by many who first come across economic
and evolutionary theories of social exchange. Whether couched in terms of
inputs, outputs, costs, benefits, or behaviors, these theories often posit that
people make decisions based solely on the concrete, measurable outcomes
in a relationship. Someone following a tit-for-tat strategy need know nothing of her partner's intent. She simply reacts to past behaviors. Someone
maintaining balance keeps tallies of past costs and benefits and behaves
accordingly. But there is no need to know a partner's motivation.

Such accounting practices have their place in social exchange, but they
also miss many of the ways that people make decisions in their relationships. Frequently, people care about more than just behaviors and outcomes.
They act according to what they think the *other* person is thinking. People
reciprocate gestures they perceive as favors more than those perceived as
bribes. And they are more likely to forgive an unintentional slight than
an intentional one.[1] In these cases, people do not simply make decisions
based on past behaviors and the costs and benefits of current options. They
also rely on judgments about a partner's intentions and make knee-jerk,
stimulus-response decisions based on how close they feel to a partner.

Close friendship is a prime example of how people think about more
than behaviors when making decisions. Close friends help each other at
much greater levels than do acquaintances or strangers. But models of
exchange based solely on behavioral outcomes—such as a norm of reciprocity, a drive to balance accounts, or a concern about the shadow of the

future—do a poor job of explaining such behaviors. Rather, what seems most important is a psychological state—how close we feel to a person or whether we think that person is a friend. Of course, such judgments are based on myriad concrete behaviors, such as past help, gift giving, conversations, careless sharing, and hanging out as well as the longevity of the relationship. However, in making the decision to help a friend, people do not always rely directly on so many past behaviors. Rather, they focus on the singular judgment of how close they are or whether a person is a friend.

Notably, these judgments are embedded in a much longer process by which partners cultivate friendships—increasing the value of the relationship, decreasing outside options, and signaling and inferring feelings and intentions. During this process, many psychological changes also occur. Feelings of closeness toward a friend become an important basis for decisions. Paralleling this change in feeling, judgments move from deliberation about costs and benefits to impulsive, habitual urges to help. Friends become knee-jerk forgivers, partially blinded to their friends' faults and more willing to constructively resolve conflicts when they arise. In short, friendship reorganizes how we think about and act toward others.

Substantial cross-cultural evidence indicates that this transformation to knee-jerk goodwill is not confined to a subset of humanity. In the lush highlands of Papua New Guinea, gardeners rely on the goodwill of friends for help with planting and for a safe haven when visiting other villages. Foragers in the southern African desert depend on good friends to share scarce water and food in times of need. And East African pastoralists avoid losing their herds with the help of friends.

In the face of this remarkable cross-cultural regularity, social and ecological circumstances also influence the specific functions of friendship and the relative importance of friendships compared to other institutions in a society. In situations of chronic resource uncertainty, whether among Ju/'hoansi foragers in southern Africa or Russian workers during the Soviet era, friendships and the material aid they provide gain particular importance. In such cases, friends can be worth defending, even in the face of countervailing laws or obligations. Among corporate managers in the United States, on the other hand, friends may be important, but not so important that they would be willing to break the law for one (chapter 7). When confronted with the specific ecological and cultural conditions of a given society, friendships change in form and function. Nonetheless, the underlying system based on goodwill and mutual aid remains.

This system is simple and elegant, leading Emerson to make his claim that "friendship is a masterpiece of nature." However, many mysteries remain. We still have much to learn about how friendships reorganize thinking toward a partner, how this arises in the brain, and how this turns us into knee-jerk (but discriminating) altruists. We know next to nothing about how people in different cultures make difficult choices about helping friends when faced with all the other appeals to their altruism that arise in daily life. And we are only beginning to understand how the psychological machinery underlying friendship can be used for both good and bad. I finish the book by exploring these issues.

FRIENDSHIP'S PHYLOGENETIC ROOTS?

Why does something like friendship recur with such regularity across human cultures? And how does it recruit preexisting adaptations—such as those underlying kin-biased helping, pair-bonding, habit formation, and niche construction? In chapter 3, I argued that friendship is not simply an application of kin-biased feelings and behaviors, since friends and kin respond very differently to both feelings of subjective closeness and the costs of helping when sacrificing for one another. In chapter 4, I compared and contrasted friendship with sexual relationships. Superficially, friendship is different from these relationships because friends need not engage in sexual behavior or reproduction. But I also argue that the kinds of attachment observed in long-term human mate bonds can exist independent of sexual desire or behavior, opening the possibility that friendship recruits the same systems involved in such long-term relationships. Some scholars have emphasized the importance of pair-bonding as the template for human friendship, whereas others have focused on the mother-infant bond or on ties with immediate kin. At the moment, however, there is insufficient evidence to make a strong case one way or another. One of the biggest challenges in distinguishing these alternative claims is that they rely on many of the same physiological systems (e.g., oxytocin and dopaminergic reward centers) and involve many of the same psychological constructs (e.g., subjective closeness and love).[2] Ultimately, more research that compares how people think about, respond to, and help biological kin, non-kin friends, and romantic partners will be necessary to disentangle the psychological and physiological processes involved in these different kinds of relationships and to understand *how* the capacity for friendship emerged from earlier capacities for social bonding with kin and mates. Returning to the debate between Aziz and Fielding, such research will require careful measurement of *both* psychological states, such as feelings of closeness, and real behaviors between individuals.[3] Moreover, given the Western bias in current research, a fuller understanding will also require cross-cultural investigations that use the precise techniques of social psychology and experimental economics to understand the micro-behaviors and motivations among friends in diverse cultural settings.

In chapter 8, I briefly discussed the possibility that friendship may recruit nothing more than primitive adaptations for niche construction and habit formation. Humans around the world build shelters for protection from the elements, but it is unlikely that we have a specific adaptation selected for the purpose of building a roof over our heads. Rather, our propensity and capacity to build homes draws from a number of other adaptations, including the ability to manipulate materials, the propensity to plan ahead, and the capacity to teach and learn. Humans are consummate niche constructors, and the cultivation of friendships may be a generalization of the same suite of abilities that permits humans to craft tools, fashion clothes, make fire pits, and build homes across so many diverse societies. Only more research will determine whether the psychological systems involved in cultivating a friendship and in guiding behaviors toward friends are the same as those used in more general kinds of niche construction and habit formation. A comparative perspective that examines the social behavior and brain organization of our mammalian and avian relatives and compares it with our own will be essential in this endeavor.[4]

FRIENDSHIP IN THE BRAIN?

An important part of identifying the roots of friendship will be careful study of how psychological states and behaviors observed among friends are mediated in the brain. At the psychological level, subjective closeness plays an important role. How does subjective closeness overlap and interact with other psychological constructs, such as love and commitment? And how does this extend to cultures where a metaphor of closeness is not used to differentiate relationships?[5] At the physiological level, we have only fleeting clues about the role that chemical messengers, such as oxytocin, might play in the willingness to accept social risks. By desensitizing people to social risks and to partners' previous behaviors, oxytocin may mediate the kinds of unconditional support provided by friends. Further research that builds on the exciting experiments described in chapters 1 and 6 will hopefully shed light on these potential links between oxytocin and behaviors among friends. At the level of brain organization, we know even less— that some parts of the brain activate differently when seeing a close friend as compared to an acquaintance.[6] The neuroimaging techniques employed in such studies have already shed new light on how human brains work as their owners perceive the faces of loved ones, solve social problems, and engage in elementary social interactions.[7] Extending these to interactions with close friends and comparing this to other kinds of partners—that is, immediate kin, romantic partners, acquaintances, and strangers—will hopefully provide insight into the psychological and physiological systems that govern behaviors among friends.

FROM DELIBERATION TO STIMULUS-RESPONSE (AND BACK AGAIN)

A crucial part of friendship is moving from deliberative, contingent decision making to a habitual kind of knee-jerk altruism. How does this transformation occur? As many researchers have suggested, psychological (and neural) merging of self with another may play a part in this process. If this is true, we should see a number of psychological and behavioral changes as friends become closer. For example, if close friends care about a combination of their partner's and their own outcomes, then they should be more willing to sacrifice other norms (e.g., equality, equity) for a higher group payoff. In negotiations, close friends should see one another's concessions as being bigger than if they were strangers. And such psychological merging may also lead to uniquely maladaptive behaviors. Consider the case of Della and Jim, the two poor lovers in O. Henry's short story *The Gift of the Magi*. Della sells her prized waist-length hair to a wig shop to buy Jim a chain for the watch he was given by his grandfather. On the same day, Jim sells his watch to buy Della a set of jeweled combs for her hair. In this case, too much sacrifice led to a lesser material outcome (although the affection expressed in their acts led O. Henry to call them the "wisest").[8]

Habits are meant to be broken, and friendships frequently end. So how and under what circumstances do people move out of habitual behaviors toward friends and return to their contingency-based deliberation? In short, how are the habits of helping and forgiving among friends broken? A critical difference between helping friends and other kinds of habits, such as brushing one's teeth, is that once a habit is formed, a partner may take advantage of that habit. With this added potential for exploitation, habit formation in strategic situations may involve special kinds of vigilance. Friends appear to be less sensitive to friends' past behaviors. They are automatically willing to forgive and less sensitive to a norm of reciprocity.[9] Given that friends appear to enter a state of autopilot with little feedback, what are the cues that might lead one to exit autopilot? Answers to these questions will require carefully controlled experiments like those described in chapters 1 and 6 that consider the behaviors of individuals in real friendships and their decisions when subtly primed to think of a friend.

CONFLICTS OF ALLEGIANCE

The vast majority of writing on altruism has contemplated how individuals sacrifice their own self-interest for the good of others. But humans, with so many potential outlets for their altruism, face an even greater challenge: how to choose between competing appeals to their goodwill. People not only tithe to charity; they must choose specific charities among a universe of good causes. Greek drama frequently hinged on the conflict between such competing appeals, between duty to god and to the state, between protecting friends and adhering to the law, between helping family and serving society. Most questions in life are not about being naughty or nice, but about *how* to be nice.

The passenger's dilemma described in chapter 7 captures one such dilemma, between protecting a friend and obeying the law, and shows that members of different societies resolve such conflicts in markedly different ways. This is a pressing problem in international development, since favoring friends is a common form of corruption. When resources are unpredictable or difficult to acquire, people often resort to friends and friends of friends to ensure access to them. While friends can be a local antidote for social uncertainty, a heavy reliance on friends can also contribute to the very uncertainty that makes friendship necessary. During the Soviet period in Russia, for example, friends not only provided a stopgap for a failing centralized economy (chapter 7), they also helped each other beat the system, thus further weakening any system that might have existed. The obligations of friendship, as well as those of the family, tended to undermine objectivity, so that personal ties were more important than professional performance for success.[10] Thus, a reliance on friends can lead to a vicious feedback cycle, in which friends become more and more valuable as broader social institutions break down. And this requires no explicit self-interest. Even if people would never break the rules for themselves, they may be more than willing to break the rules to help a friend.[11]

There are many other kinds of conflicts—with family, religion, and other friends—and we know very little about how people learn to resolve such conflicts in different cultural conditions. How does this differ in societies where children don't attend school and therefore don't have chronic exposure to non-kin peers? When and how do kinship and friendship come into conflict?[12] Only further studies can illuminate how and why members of different cultures negotiate such trade-offs between different appeals to their altruism.

HARNESSING THE POWER OF FRIENDSHIP

The power of friendship goes far beyond its ability to regulate mutual aid between two people. Friends connect society, serving as informal match-makers and job finders, as conduits for gossip, and as ambassadors between otherwise indifferent or hostile groups.[13] Close friendships are often the seed crystal for social change; many cultural movements, political revolutions, and business start-ups have begun as groups of friends.[14] Friends motivate us; people with friends at work are more willing to exert effort for their company.[15] Friends also limit us, as commitments to them may make us blind to advantageous outside options.[16] Friends play a hidden but significant role in the economy through informal exchange, help, and caregiving. And friends are a lucrative business; a substantial portion of consumer spending goes to gifts among friends.[17]

People have long sought ways to harness the power of friendship, especially the trust, goodwill, and loyalty that influence behaviors among friends. Dale Carnegie's *How to Win Friends and Influence People* focused on efficiently creating friendships for personal gain, a concern that goes at least as far back as politicians in ancient Rome. Social movements, from Al-Qaeda to millenarian cults, explicitly recruit new members through existing friendships.[18] And for decades, corporations have made use of friendships to sell their wares. A prime example is the Tupperware party, in which a host invites friends and neighbors to a gathering centered on selling the product (at a commission to the host).[19] A more recent strategy, word-of-mouth marketing, recruits large networks of "influentials" who are given free product samples and encouraged to promote those products to friends, family, and acquaintances. One of the most prominent recent programs, founded by Procter and Gamble in 2001, includes about two hundred thousand teenagers chosen for their extensive social networks and their ability to spread the word about upcoming products. As Steve Knox, the CEO of one such word-of-mouth program, has observed, "We know that the most powerful form of marketing is an advocacy message from a trusted friend."[20] We also know next to nothing about how such uses of friendship change people's perceptions of and feelings toward friends (or the organizations that exploit them).

Social movements have also attempted to harness the power of friendship for the good of others. When people are asked about the people who make them the happiest, friends are often at the top of the list, whether one asks middle-class Americans or middle schoolers in Saudi Arabia, Israel, or the United Kingdom.[21] And in experimental settings, people react less to stressful situations when accompanied by a good friend. One U.K. economist recently calculated that meeting with friends more frequently was equivalent to a wage increase of 85,000 British pounds in terms of its effect on life satisfaction.[22] Building these insights into interventions, several organizations have developed programs to match volunteer "friends" with individuals faced with mental health challenges in attempts to improve their well-being.[23]

Perhaps the most ambitious attempts to harness the power of friendship have been calls to extend the same feelings of compassion and goodwill felt toward friends to all of humanity. Jesus said to love your enemies, and the Buddha aspired to treat all sentient beings with equanimity, without distinguishing among friends, enemies, and strangers. In the twentieth century, the influential psychologist Lawrence Kohlberg proposed that favoring friends is a lower grade of moral development than thinking that all people deserve equal treatment.[24] Numerous methods, including meditation, visualization, and exposure to suffering, have been used to cultivate such equanimity in loving others.[25]

How far can people extend feelings of goodwill and compassion beyond their close attachments?[26] Experiments described in chapter 1 show that simply priming people with the names of friends can make them more willing to help strangers. But the effect of such priming is small in comparison to the influence of a real friend. People also regularly volunteer for noble causes. But such altruism often relies on recruitment through close relationships. And people generally favor causes when a close friend or family member is directly affected by the issue at hand.[27] Consider Nancy Reagan's support for Alzheimer's disease research or Rob Lowe's support of breast cancer research. These findings suggest that close relationships may be an important stepping-stone for the broader equanimity extolled by Jesus and the Buddha. Further study of how feelings and thoughts toward friends play a role in helping behavior will hopefully shed light on how we might harness the power of friendship for the greater good.

Ethnographic Data and Coding

There are 396 societies coded in the Human Relations Area Files (HRAF) database, with 60 societies selected for the Probability Sample File (PSF). Another sample of societies is the Standard Cross-cultural Sample (SCCS), selected for quality of documentation and relative independence of observations. Of the 186 SCCS societies, 150 have equivalents in the HRAF database, and I use this sub-sample in addition to the HRAF and PSF to examine the probability of finding friendship by pages of text.

NUMBER OF PAGES

In most cases, the HRAF lists how many text pages are devoted to the description of a particular society. However, for some societies, the dataset only lists the number of microfiche for each culture. In these cases, I estimate the number of text pages by using a formula based on number of fiche. The formula is estimated from those societies where we know both the number of pages and fiche. In those cases where there is an electronic version and microfiche version for ethnographies of the culture, I estimate the number of pages from the electronic version, if it is greater, and the sum, if it is less than the microfiche version.

In the electronic HRAF, a search revealed 2203 paragraphs coded for *friendship* (Outline of World Cultures [OWC] code = 572) and an additional 2035 paragraphs coded for *artificial kinship* (OWC code = 608). There were 14927 paragraphs coded for kin relations (OWC code = 601) and an additional 9572 coded for kin terminology (OWC code = 602). Figure 28 shows the probability of finding a referent to friendship (OWC code = 572) by the number of pages in the text.

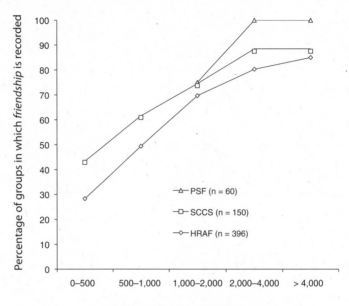

FIGURE 28.   Probability of finding references to friendship by number of pages in HRAF database

CODEBOOK

Inclusion criteria for references to friendship

Any text coded as *friendship* (OWC code = 572), any text coded as *artificial kinship* (OWC code = 608), any text including reference to *friend, friendship, friends,* or *friendships.*

Exclusion criteria for references to friendship

The following references to friendship were excluded: relationships with an anthropologist or with European colonists, discussions of *friendship* as a peace between two groups (e.g., "they entered a pact of peace and friendship between the two groups"), a brief mention of how "friendly" people were, and references to friends that contained no information about the function, obligations, or roles of the relationship (e.g., simply *friends and relatives*).

Codes for aspects of friendship-like relationships

For each of the features of friendship, I developed two codes referencing whether the description included statements related to the feature (e.g., voluntariness, privateness), and whether such statements confirmed or disconfirmed the feature as a part of the relationship.

BEHAVIOR—MUTUAL AID

Description: References to aid, support, or help provided between partners (or lack thereof).

Confirmation: References to partners helping, supporting, protecting, caring for, assisting, aiding, hosting, or sacrificing in times of need.

Disconfirmation: References to the absence of help or support among partners.

BEHAVIOR—GIFT GIVING

Description: References to giving gifts or exchanging presents (or lack thereof).

Confirmation: References to giving gifts or exchanging presents.

Disconfirmation: References that partners do not exchange gifts or presents.

BEHAVIOR—RITUAL INITIATION

Description: References to a ceremony, ritual, rite, or initiation that marks a new stage in the relationship (or lack thereof).

Confirmation: References to the existence of a ceremony, ritual, rite, or initiation that marks a new stage in the relationship.

Disconfirmation: Reference that no ceremony is involved.

BEHAVIOR—SELF-DISCLOSURE

Description: References to the sharing of sensitive personal information between partners (or lack thereof).

Confirmation: References to partners sharing secrets, discussing personal problems, and sharing personal dreams, plans, and designs.

Disconfirmation: References to partners intentionally not sharing secrets or personal information.

BEHAVIOR—FREQUENT SOCIALIZING

Description: References to the frequency of socializing and barriers to frequent socializing.

Confirmation: References to the fact that partners frequently engage in common labor or activities, that commonly doing things together is part of the relationship, or that partners frequently visit each other.

Disconfirmation: References to seeing partners infrequently (seasonally, only on certain occasions, or infrequent trips) and barriers to frequent socializing, including living in distant areas, groups, or islands.

BEHAVIOR—INFORMALITY

Description: References to looser or stricter adherence to normal rules of conduct between partners (other than rules about mutual aid).

Confirmation: References to looser adherence to normal rules of conduct between partners (other than rules about mutual aid).

Disconfirmation: References to stricter adherence to normal rules of conduct between partners (other than rules about mutual aid).

BEHAVIOR—TOUCHING

Description: References to sustained bodily contact between partners.

Confirmation: References to sustained bodily contact between partners.

Disconfirmation: References to the inappropriateness of sustained bodily contact between partners.

FEELINGS—POSITIVE AFFECT

Description: References to what partners feel toward each other, the relationship, or both.

Confirmation: References to positive feelings, such as love, closeness, affection, amity, warmth, passion, liking, deep emotions, and positive sentiments.

Disconfirmation: References to apathy between partners or consistent enmity.

FEELINGS—JEALOUSY

Description: References to feelings of jealousy between partners.

Confirmation: References to feelings of jealousy between partners.

Disconfirmation: References to the fact that partners are not jealous of each other.

ACCOUNTING—TIT-FOR-TAT

Description: References to helping a partner based on tit-for-tat accounting.

Confirmation: Helping a partner based on tit-for-tat accounting.

Disconfirmation: References that helping a partner is not based on tit-for-tat accounting.

ACCOUNTING—NEED

Description: References to helping a partner based on a partner's need.

Confirmation: Helping a partner because a partner is in need.

Disconfirmation: References that helping a partner is not based on a partner's need.

FORMATION—EQUALITY

Description: References to the relative social standing of the partners in a relationship. Criteria for judging social standing can be culturally salient dimensions of status or authority.

Confirmation: References to the fact that partners are similar in status, that they are "equals," or that there are no authority relations between them.

Disconfirmation: References to one partner being senior or junior relative to the other, to partners coming from different-ranked castes or classes, to one partner having authority over the other, and to asymmetric behavior between the two.

FORMATION—VOLUNTARINESS

Description: Any reference to individual or group decisions or circumstances leading to the formation or dissolution of the relationship. This includes references to individual choices, the influence of others in such choices, heavy circumstantial limits on who can be partners (i.e., limited to fewer than five others), and formal or informal sanctions against ending a relationship.

Confirmation: References to the relationship as voluntary, limited only by mutual consent, and voluntarily dissoluble.

Disconfirmation: Any references to pre-arrangement by others, permission from others, ascription or inheritance of the relationship, or ritual or social control over beginning and ending the relationship.

FORMATION—PRIVATENESS

Description: References to control and sanctioning of behaviors in the relationship by partners and outside sources. This is different from voluntariness in that voluntariness involves control over decisions to enter and leave a relationship, whereas privateness involves control over behavior within the context of a recognized relationship.

Confirmation: References to the fact that only partners know about the relationship; violations of the relationship are dealt with by the partners alone.

Disconfirmation: References to external sources of control over the relationship, including engagement in public rituals, public or legal enforcement of violations, the social ostracism of betrayers, and religious or supernatural sanctions.

Characteristics of Friendship by Society

Culture	Mutual aid	Gift-giving	Ritual initiation	Self-disclosure	Frequent socializing	Informality	Touching	Positive affect	Jealousy	Tit-for-tat	Need
Dogon	Y			Y	N	Y		Y			Y
Akan	Y	Y		Y				Y			
Tiv	Y		Y					Y			Y
Ganda	Y	Y	Y	Y	Y			Y			Y
Maasai	Y	Y	Y		N			Y		N	Y
Mbuti	Y	Y	Y		Y			Y			Y
Azande	Y	Y	Y		N			Y		N	Y
Bemba	Y	Y						Y			
Lozi	Y		Y		N						Y
Somali	Y	Y		Y		Y					Y
Amhara	Y			Y				Y			Y
Hausa	Y	Y		Y				Y			Y
Kanuri	Y			Y		Y	Y	Y			
Wolof	Y	Y	Y	Y						Y	Y
Libyan Bedouin	Y	Y		Y		Y		Y	Y		
Shluh	Y					Y		Y		N	Y
Korea	Y	Y	Y		Y	Y		Y			
Taiwan Hokkien	Y	Y	Y	Y	Y	Y	Y	Y			Y
Central Thai	Y	Y			N	Y	Y	Y		N	Y
Garo	Y							Y			Y
Khasi	Y							Y			
Santal	Y	Y	Y	Y			Y	Y			Y
Sinhalese	Y					Y		Y			
Andaman	Y	Y		N	Y	Y	Y	Y			
Ifugao	Y										

Equality	Voluntariness	Privateness	Hospitality	Food/material sharing	Money loans	Help when ill	Ceremony help	Labor help	Aid in disputes	Courting aid	Inherited from parents	Gender described*	Between sexes (non-kin)
			Y	Y				Y		Y		M	
N												B	N
N			Y					Y	Y			M	
				Y	Y				Y			M	
	N		Y	Y			Y		Y		Y	B	
	N			Y				Y			Y	B	
	N	N	Y	Y		Y			Y			B	
N	N											B	
			Y	Y		Y					Y	M	
				Y					Y	Y		M	
				Y					Y	Y		M	
N	N			Y	Y		Y	Y				B	Y
Y			Y	Y	Y		Y	Y				M	
N				Y	Y			Y	Y		Y	B	N
N			Y	Y								B	
N	N		Y	Y				Y	Y		Y	M	
	Y							Y	Y			B	N
Y				Y	Y		Y	Y	Y			B	
N			Y	Y	Y		Y	Y				B	
						Y		Y				B	
			Y	Y	Y		Y				Y	B	
N				Y	Y	Y	Y	Y				B	N
								Y				M	
			Y	Y		Y					Y	M	
			Y					Y				M	

(continued)

Characteristics of Friendship by Society *(continued)*

Culture	Mutual aid	Gift-giving	Ritual initiation	Self-disclosure	Frequent socializing	Informality	Touching	Positive affect	Jealousy	Tit-for-tat	Need
Iban	Y	Y	Y			Y		Y			
Yakut	Y	Y	Y					Y			Y
Chukchee	Y		Y					Y			Y
Serbs	Y	Y	Y					Y			Y
Saami	Y	Y	Y	Y				Y			
Highland Scots	Y			N							Y
Tarahumara	Y										
Tzeltal	Y	Y	Y					Y			Y
Kuna	Y	Y	Y								
Kurds	Y						Y				Y
Tlingit		Y						Y			
Copper Inuit	Y			Y	N	Y		Y			
Blackfoot	Y	Y				Y		Y			Y
Ojibwa	Y	Y	Y					Y			Y
Iroquois	Y	Y	Y	Y				Y			Y
Pawnee	Y	Y						Y			
Klamath	Y										
Hopi	Y	Y	Y					Y		Y	
Eastern Toraja	Y	Y						Y			
Aranda	Y	Y				Y	Y				
Kapauku	Y	Y					Y	Y			Y
Trobriand	Y	Y					Y	Y			
Lau Fijians	Y										
Chuuk	Y		N	Y		Y	Y	Y		N	Y
Tikopia	Y	Y	Y	Y				Y			Y

Equality	Voluntariness	Privateness	Hospitality	Food/material sharing	Money loans	Help when ill	Ceremony help	Labor help	Aid in disputes	Courting aid	Inherited from parents	Gender described*	Between sexes (non-kin)
												B	
												B	
			Y	Y				Y				M	
N			Y	Y					Y		Y	B	Y
			Y				Y	Y				B	
				Y	Y							B	N
			Y										
				Y	Y			Y				B	
			Y	Y								B	Y
Y				Y	Y		Y					B	N
												B	
			Y	Y				Y	Y			B	
				Y					Y	Y		B	
				Y					Y			B	
N	N	N				Y			Y			B	N
N												B	
				Y					Y				
N			Y	Y				Y					
									Y			B	
Y			Y										
				Y	Y		Y		Y	Y		B	Y
				Y	Y						Y	B	
			Y							Y			
	Y		Y	Y				Y	Y			B	Y
Y	Y	Y	Y	Y			Y			Y			

(continued)

Characteristics of Friendship by Society *(continued)*

Culture	Mutual aid	Gift-giving	Ritual initiation	Self-disclosure	Frequent socializing	Informality	Touching	Positive affect	Jealousy	Tit-for-tat	Need
Kogi								Y			
Aymara	Y		Y					Y			
Ona	Y							Y		Y	
Guarani	Y							Y		Y	
Bahia Brazilians								Y			
Bororo			Y								
Yanoamo	Y	Y						Y			
Tukano	Y	Y	Y	Y		Y	Y	Y			
Ndyuka	Y	Y				Y		Y			Y
Mataco	Y	Y									

*M = only descriptions refer to males, B = described for both females and males.

SOURCES:

Dogon: Parin, Morgenthaler, and Parin-Matthey 1963, pp. 136–139; Paulme 1940

Akan: Rattray 1927; Warren 1975; Field 1960; Sarpong 1977; Meyerowitz 1974; Fortes 1950

Tiv: Abraham 1933; Malherbe 1931; Bohannan and Bohannan 1968; Keil 1979; Bohannan 1957, 1954

Ganda: Robbins 1979; Southwold 1965; Roscoe 1911; Robbins and Kilbride 1987

Maasai: Cronk 2007; Merker 1910; Spencer 1988; Huntingford 1953; Hollis 1905; Fosbrooke 1948; Bernardi 1955

Mbuti: Turnbull 1962, 1965; Putnam 1948

Azande: Anderson 1911; Baxter 1953; Calonne-Beaufaict 1921; Casati 1891; Czekanowski 1924; Lagae 1926; Larken 1927; Evans-Pritchard 1933; Seligman and Seligman 1932; Hutereau 1909

Bemba: Hinfelaar 1994; Richards 1939

Lozi: Gluckman 1967; Turner 1952

Somali: Helander 1988, 1991

Amhara: Young 1975; Levine 1965

Hausa: Cohen 1969; Hassan, Shuaibu, and Health 1952; Hill 1972; Smith 1957; Smith 1954; Salamone 1974

Kanuri: Cohen 1960, 1967; Peshkin 1972

Wolof: Irvine 1974; Gamble 1957; Lasnet 1900; Magel 1984; Venema 1978

Equality	Voluntariness	Privateness	Hospitality	Food/material sharing	Money loans	Help when ill	Ceremony help	Labor help	Aid in disputes	Courting aid	Inherited from parents	Gender described*	Between sexes (non-kin)
					Y			Y				B	
												B	N
					Y			Y				B	
									Y				
N				Y				Y					
			Y	Y								M	
	Y			Y				Y	Y			B	Y
			Y		Y					Y		B	Y
			Y	Y									

Libyan Bedouin: Abu-Lughod 1986; Davis 1988; Peters 1990; Cole 2003; Behnke 1980; Obermeyer 1968

Shluh: Hatt 1974

Korea: Beirnatzki 1968; Bishop 1898; Kang 1931; Brandt 1971; Chun 1984; Han [1949] 1970; Hurlbert 1910; Janelli 1982; Kennedy 1977; Kim 1992; Knez 1960; Osgood 1951; Yi 1975

Taiwan Hokkien: Barnett 1971; DeGlopper 1974; Diamond 1969; Gallin and Gallin 1974; Gallin 1966; Gates 1973; Gould-Martin 1976; Harrell 1974; Jacobs 1977; McCreery 1974; Wolf 1978

Central Thai: Phillips 1966; Kaufman 1960; Mulder 1996; Preecha 1980; Sharp 1978; Terwiel 1975; Hanks 1962; Foster 1976; Piker 1968; Bunnag 1973.

Garo: Burling 1963; Marak 1997

Khasi: Nakane 1967; Assam 1963; Gurdon 1907; McCormack 1964; Stegmiller and Knight 1925

Santal: Skrefsrud 1942; Archer 1984; Culshaw 1949; Mukherjea 1962; Orans 1965

Sinhalese: Yalman 1971; Leach 1961; Tambiah 1965

Andaman: Radcliffe-Brown 1922; Cipriani 1966; Man 1932

Ifugao: Beyer 1911; Villaverde 1909; Barton 1930, 1919

Iban: Kedit 1991; Sandin 1980; Gomes 1911; Howell 1908–1910; Low 1892; Sutlive 1973

Yakut: Sieroszewski 1896

Chukchee: Bogoraz-Tan 1904–1909; Sverdrup 1938

(continued)

SOURCES *(continued)*:

Serbs: Halpern 1986; Denich 1977; Filipovic 1982; Hammel 1968; Pavlovic 1973; Lodge 1941; Kemp 1935; Simic 1973

Saami: Anderson 1978; Ingold 1976; Pehrson 1957; Whitaker 1955; Pelto 1962; Itkonen 1948

Highland Scots: Parman 1990, 1972; Walker 1974; Stephenson 1984; Coleman 1976; Ducey 1956; Ennew 1980; Vallee 1955

Tarahumara: Bennett 1935; Champion 1963; Fried 1952

Tzeltal: Hunt 1962; Nash 1970; Metzger 1964; Villa Rojas 1969

Kuna: Marshall 1950; Sherzer 1983; Stout 1947; Nordenskiold 1938

Kurds: Masters 1953

Tlingit: Oberg 1973; Olson 1967; Kan 1989; De Laguna 1972

Copper Inuit: Condon 1987; Jenness 1922, 1959; Damas 1972; Pryde 1972

Blackfoot: Goldfrank 1966; Wissler 1911; Hanks and Hanks 1950; Ewers 1958; Hungry Wolf 1980; Kane 1925; Lancaster 1966; Schultz 1930

Ojibwa: Landes 1937; Vennum 1988; Bishop 1974; Hilger 1951; Theriault 1992; Tanner 1830; Kohl [1860] 1985

Iroquois: Wallace 1969; Shimony 1961; Fenton 1953; St. John 1981; Tooker 1970; Weaver 1972

Pawnee: Dorsey 1940; Lesser 1933; Weltfish 1965

Klamath: Spencer 1956; Stern 1963

Hopi: Titiev 1972, 1967; Beaglehole and Beaglehole 1937; Talayesva 1942; Beaglehole 1935; Aberle 1951; Bradfield 1973; Clemmer 1995; Cox 1969; Dennis 1940; Eggan 1950; Lowie 1929; Schlegal 1989; Sekaquaptewa 1969; Stephen 1969

Eastern Toraja: Adriani and Kruijt 1950

Aranda: Morton 1992; Basedow 1925; Chewings 1936; Strehlow 1947; Spencer 1927; Roheim 1945

Kapauka: Pospisil 1978

Trobriand: Malinowski 1922, 1929, 1935, 1926; Weiner 1988

Lau Fijians: Hocart 1929; Thompson 1940

Chuuk: Gladwin and Sarason 1953; Bollig 1927; Fischer 1950; Marshall 1977; Mahony 1971

Tikopia: Firth 1936; Spillius 1957

Kogi: Reichel-Dolmatoff 1951

Aymara: Tschopik 1951; Hickman 1963; Cole 1969; Buechler 1971

Ona: Gusinde 1931; Cooper 1917

Guarani: Reed 1995; Watson 1952; Watson 1944

Bahia Brazilians: Borges 1992; Hutchinson 1957; Azevedo 1969

Bororo: Fabian 1992

Yanoamo: Barker 1953; Chagnon 1967

Tukano: Goldman 1963; Hugh-Jones 1979

Ndyuka: Bilby 1990; Lenoir 1973

Mataco: Alvarsson 1988; Karsten 1932

Mathematical Models for Chapter 8

Consider two people, Sally and Ann, living in a risky world. At any time, one may benefit from the other's help. Also, assume that there are two kinds of favor—small favors and large favors. Small favors require Sally to incur some cost, c_S, to provide a benefit to Ann, b_S, where the benefit is greater than the cost ($b_S > c_S$). Large favors also require Sally to incur some cost, c_L, to provide a benefit to Ann, b_L, where the benefit is greater than the cost ($b_L > c_L$). As one might expect, the costs and benefits of large favors are bigger than those for small favors. In each of these cases, if Sally decides not to help Ann, then she incurs no cost, and Ann receives no benefit. For each participant, at any point in time, the opportunity for a large favor arises with probability p_L, and the opportunity for a small favor arises with probability p_S. And there is some probability ($1 - p_L - p_S$) that Ann does not need a favor at all.

If Sally and Ann met only once in their lives and Sally were in need of help costing c_S, Ann would lose by helping Sally. Specifically, Ann would bear a cost and Sally would achieve a gain, b_S. As long as there is no reputation, repeated interaction, or relationship, if Ann regularly helped strangers like Sally, she would be worse off in terms of payoffs than if she never helped. The same is true for Sally, and in such a world, we would expect both to live alone and never help each other.

The Value of Repeated Interactions

In this game, if Sally and Ann met only once then they would lose by helping the other. However, if they were able to credibly agree to always help each other in the future, then they would be much better off. Here I show how much better off.

Consider a world where time occurs in steps (let's say days). On any given day, there is a probability, p_S, that Ann will need a small favor and a probability, p_L, that she will need a large favor. On that same day, Sally will need favors with the same probabilities. If Sally and Ann need favors on the same day, then assume that they are still able to help each other.

If Sally needs help and Ann decides not to help, she incurs no cost and Sally gains no benefit, meaning the benefits and costs at a given time are 0. However, if both partners agree to help regardless of the cost, the expected net benefit to a partner at any time is the benefit they might accrue from being helped minus the costs they would bear for helping:

Expected Benefit $= p_S b_S + p_L b_L - p_S c_S - p_L c_L$ 　　　　Equation 1

Since the benefits of help are always greater than the costs (b > c), the expected net benefit to Sally and Ann is always greater than 0.

This benefit at each time step can add up over a lifetime. Suppose Sally cares more about help in the next few days than she cares about help far into the future. This is consistent with a discount factor that people's expected payoffs from future events are less important because they are less likely to happen (due to either death or some other change in circumstances). We can state this precisely in terms of a decay factor so that she cares less about future help on each consecutive day (0 < D < 1). The lifetime benefit of the future relationship can be written as:

$$Lifetime\ Value = \sum_{t=1}^{\infty} D^t \left(p_S b_S + p_L b_L - p_S c_S - p_L c_L \right)$$ 　　Equation 2

$$= \left(\frac{D}{1-D} \right) \left(p_S b_S + p_L b_L - p_S c_S - p_L c_L \right)$$

This means that if both Sally and Ann could agree (and trust each other) to help when help was needed, then they would on average do better than if they lived alone trying to make do by themselves. One way that Sally can give Ann an incentive to continue helping is to end the relationship if Ann doesn't help—a kind of tit-for-tat strategy. In this case, Ann would be better off helping, if the cost of helping at a given time point is less than the expected lifetime benefit from the relationship (Neilson 1999):

$0 <$ *Lifetime Value* $- c_L$ 　　　　　　　　　　　　Equation 3

The Incentive to Help with Outside Options

The above result holds as long as both partners have only two options—to cooperate or live alone. However, in everyday life, people often have the option to leave one partner for another. We see this most strikingly among con artists, who quickly build up relationships, exploit them, and then drop their partner to find another victim. In the favor game, this is exemplified by an individual who develops relationships, enjoys the benefits of a partner's aid, and then leaves when the time comes to bear the cost of helping. An individual who simply follows the tit-for-tat strategy of helping until one's partner fails to help will be exploited by such a con artist.

Consider the case where Sally is in need of a small favor and Ann has the option of incurring cost, c_S, or leaving and starting a new relationship. Starting a new relationship is costless, and the lifetime value of one relationship is equivalent to another. In this case, the lifetime value of a new relationship (LV_n) is greater than the lifetime value of the old relationship minus the cost of helping in the current round ($LV_o - c_S$). Therefore, Ann can do better by betraying Sally and simply starting a new relationship with another person. This would be a prime opportunity for con artists.

One ingenious way to deter strategies that "exploit and run" is to institute a round of courtship or gift giving at the beginning of a relationship, where each partner must bear some start-up cost, c_c (Carmichael and MacLeod 1997). With such an institution of courtship, the gain to a con artist of ending a relationship and starting a new one is $LV_n - c_c$. As long as this is worth less than staying with the old relationship (as long as the following inequality holds), a partner's best option is to bear the cost of helping and stay in the relationship:

$$LV_n - c_c < LV_o - c_S \qquad \text{Equation 4}$$

Since the lifetime values of a new and old relationship are equivalent, this inequality holds as long as the cost of courtship (c_c) is higher than the cost of helping (c_S). In this situation, two individuals who bear the cost of start-up, who help when the cost of helping is less than c_S, and who end a relationship when the partner doesn't follow this rule have an advantage over con artists.

One important element of such courting gifts is that they are costly to the giver but cannot be used, resold, or regifted by the recipient. Otherwise, con artists could exploit the situation in two ways. First, they could simply take the gifts and run with their value intact. Second, they could exchange gifts, exploit the partner, and then regift what they received from their last victim to start a relationship with another potential victim. In a sense, such gifts must be worthless to anyone else. There are a number of ways to exchange such costly but worthless courtship gifts. These include spending time hanging out and giving gifts that lose much of their value in the act of giving or soon after giving (Bergstrom, Kerr, and Lachmann 2008; Sozou and Seymour 2005). Gifts with such characteristics include flowers (which wilt), perishable foods (which perish), and costly gift wrapping (which is destroyed upon opening). Arbitrary choosiness in one's partners also raises the cost of finding a partner.

Another important property of such gifts is that participants must give gifts prior to any communication between the two about the nature of the gift. Otherwise, two partners could agree to give start-up gifts that cost less than c_c, and they would do better than those who require c_c. This captures another important aspect of good gifts, that they must be surprises, and that the giver should decide in secret on what to give and how much to spend (Carmichael and MacLeod 1997).

Expanding the Scope of Help

Costly but worthless gifts provide an important safeguard against con artists. However, they also limit the scope of helping. When the cost of helping at a point in time exceeds the original courtship cost, then there is no incentive to help, because it would actually cost less to simply start a new relationship.

For this reason, partners are only able to achieve the benefits of helping up to some maximum cost, which is set by the original costs of courtship. Consider a situation where people can afford courtship gifts that are larger than the cost of small favors, but courtship gifts that would permit large favors are too expensive. In this case, it would never pay to help a friend who needs a large favor. Rather, partners would simply agree to maintain relationships where small favors, but not large favors, were granted. Let's call this the "courtship limited" strategy.

There is a way around this problem based on the way that friends frequently start with small favors and gradually raise the stakes of favors. It involves a "raise-the-stakes" strategy that judiciously increases the scope of helping over time, in such a way that the acceptable cost of helping can surpass the cost of courtship (Roberts and Sherratt 1998).

Consider two partners, Sally and Ann, who bear the cost of courtship and help each other as long as favors are small. At a point in the relationship, Ann needs a big favor. As discussed before, Sally has no incentive to help, because she could easily move to a new relationship for a smaller cost rather than bear the cost of helping Ann to maintain their relationship. However, there is a strategy that if followed by both Ann and Sally would create better outcomes than if they followed the "courtship limited" strategy.

According to this raise-the-stakes strategy, Sally does provide the large favor. If Ann follows the raise-the-stakes strategy, then she feels a great deal of gratitude and will provide a large favor the next time there is an opportunity. They then continue to provide big favors until one of them fails to; at which point, they return to the level of helping with only small favors. The expected future value of the relationship will be:

$$LV_L = \frac{D}{1-D}(p_s b_s + p_L b_L - p_s c_s - p_L c_L)$$

Equation 5

If, on the other hand, Ann follows the "courtship limited" strategy, then the next time that Sally needs a large favor, Ann doesn't help. The relationship continues, but Sally also returns to helping with only small favors. In this case, the expected value of the relationship will be:

$$LV_S = \frac{D}{1-D}(p_s b_s + p_L b_s - p_s c_s - p_L c_s)$$

Equation 6

As long as the following inequality holds, then Ann will do better by following the raise-the-stakes strategy. In short, she will have an incentive to reciprocate the large favor and thus take the friendship to a new level of favors.

$$LV_L - c_L > LV_S - c_S$$

Equation 7

Therefore, when this inequality holds, Sally in her original dilemma to raise the stakes also has an incentive to give the large favor, because she knows Ann will have an incentive to reciprocate in the future. If we make a simple assumption that the benefit-to-cost ratio (b/c) is constant for both large and small favors, this reduces to a very simple inequality:

$$\frac{b}{c} > \frac{1-D}{Dp_L} + 1$$

Equation 8

This equation means that the more people care about the future (larger D) and the more frequent opportunities for large favors become (larger p_L), the smaller the benefit-to-cost ratio must be for the raise-the-stakes strategy to be better than the courtship limited strategy. For the sake of simplicity, I have presented results when there are only two kinds of favor—big and small. This only permits raising the stakes once, from small to large gifts. However, similar results can be derived when favors can take on any number of costs and benefits, with different probability distributions (Watson 2002).

Loose Monitoring and Forgiveness

Until now, I have assumed that individuals will follow something like a tit-for-tat strategy to enforce favors. Specifically, if Sally fails to help with a small favor at the early stage of a relationship, then Ann ends it. Also, if Sally fails to provide a large favor, then Ann returns to helping only with small favors.

However, once Sally and Ann have built up the friendship to providing large favors, they can monitor each other at quite low levels, be very forgiving, and still maintain incentives to help the other. Suppose that Ann fails to help when Sally needs a favor; then Sally will return to only providing small favors with some probability, p_e. How high does this probability have to be for Ann to have an incentive to give help for a small favor?

If Ann helps, then the relationship continues with larger favors. If she doesn't, then Sally notices with probability p_e and returns to a relationship based on small favors. Otherwise the relationship continues as such. As long as the following inequality holds, then Ann will do better if she helps.

$$LV_L - c_S > p_e LV_S + (1 - p_e)LV_L \hspace{3cm} \text{Equation 9}$$

This simplifies to an inequality that describes the lowest value for Sally's vigilance in the relationship such that Ann has an incentive to help with small favors.

$$p_e > \frac{1}{p_L(\frac{b}{c} - 1)(\frac{c_L}{c_S} - 1)\frac{D}{1-D}} \hspace{3cm} \text{Equation 10}$$

Suppose that time steps occur in days and that we expect that exogenous forces would make the relationship end with a 1 percent chance on any given day ($D = 0.99$). Also suppose that large favors are needed every ten days, the benefit of help is twice its cost, and that the cost of helping with large favors is three times the cost for small favors. In that case, Sally would only need to respond to Ann's failure to provide small favors 5 percent of the time to give Ann an incentive to help with small favors. For large favors, Sally would still only need to respond 15 percent of the time to give Ann an incentive to provide big favors. Therefore, when individuals can cultivate a relationship at different levels of helping, even very low levels of monitoring and retaliation can provide an incentive to help a partner. If monitoring has any cost, then it would make sense to monitor at such low levels.

THE SHIFT FROM KNEE-JERK DEFENSE TO KNEE-JERK SUPPORT

Suppose that one can think of helping someone, Paul, in two ways:

1. Always help Paul, unless there is some reason to believe he is not a friend.
2. Don't help Paul, unless there is some reason to believe he is a friend.

The first rule describes a habitual tendency to help Paul, unless upon further deliberation there is a reason not to help him. The second rule describes a habitual tendency to not help, unless upon further deliberation there is a reason to.

According to the first rule, there will be very few cases where one fails to help Paul if he is a friend, but there will be more cases where one helps him even he is not really a friend. Therefore, there is a bias toward helping Paul. Conversely, according to the second rule, there is a bias against helping Paul.

Suppose that according to rule one, I accidentally don't help Paul with some probability (e_1), whereas I accidentally help a non-friend with some higher probability $(e_2 > e_1)$. Conversely, according to rule two, I accidentally don't help Paul with the higher probability (e_2), whereas I accidentally help a non-friend with the lower probability $(e_1 < e_2)$.

There are two kinds of costs that can arise from making bad decisions. First, one can lose whatever it costs to help Paul if he is actually not a friend (C_h). Second, by not helping Paul, it may end the relationship with some probability and at the cost of ending a mutually beneficial relationship. We'll call this the expected cost of not helping (C_n). Thus, we can write the expected payoffs of following rules one and two with Paul as:

$$EV(1) = -p(\mathit{friend})e_1 C_n - p(\sim \mathit{friend})e_2 C_h - p(\mathit{friend})(1 - e_1)C_h \quad \text{Equation 11}$$

$$EV(2) = -p(\mathit{friend})e_2 C_n - p(\sim \mathit{friend})e_1 C_h - p(\mathit{friend})(1 - e_2)C_h \quad \text{Equation 12}$$

When does the first decision rule give a better outcome? This occurs when $EV(1) - EV(2)$ is greater than 0.

$$EV(1) - EV(2) = [p(\mathit{friend})(C_n - C_h) - p(\sim \mathit{friend})C_h](e_2 - e_1) \quad \text{Equation 13}$$

Since $e_1 < e_2$, for rule one to give a better outcome, the following equivalent inequalities would need to hold.

$$p(\mathit{friend})(C_n - C_h) > (1 - p(\mathit{friend}))C_h \quad \text{Equation 14}$$

$$p(\textit{friend}) > \frac{C_h}{C_n}$$

Equation 15

In other words, if the probability that Paul is a friend is sufficiently high and the cost of not helping is sufficiently large relative to the cost of helping, then it makes sense to follow rule number one. In short, in a sufficiently long and valuable relationship, it makes sense to treat Paul with the adage "innocent until proven guilty" rather than "guilty until proven innocent." Positive feelings toward Paul may reflect changes in decision making necessary to make this shift.[1]

An important implication of this kind of decision making is that if people are uncertain about the cost of not helping—which requires estimating the probability of the relationship ending from not helping as well as the cost of losing the relationship—they may focus more on judging whether their partner is a friend than on the costs and benefits of the relationship. Therefore, decision making can transform from a concern with the costs and benefits of helping to judgments about a person's intentions toward oneself, which changes the focus to such signals of friendship as the longevity of the relationship and costly signals about one's intention toward a partner.

D-Statistics for Studies Cited

This table summarizes results of quantitative studies cited in the text. It includes the number of and types of informants, estimated d-statistic for the reported effect, and 95 percent confidence intervals for the d-statistic when calculation is possible.

N and Conditions	Participants	Variables	Derived from[a]	D-stat	95 Percent Confidence Interval[b]	Citation	Outcome Measure
More Likely to Share with Friends (Non-anonymous)							
20 reciprocated friend pairs, 25 acquaintance pairs	4th graders	Crayon sharing	t-stat	0.88		Staub and Sherk 1970	Behavioral
96 friend vs. enemy	7- to 11-year-olds in U.K.	Money	t-stat	1.76		Vaughan, Tajfel, and Williams 1981	Behavioral
41 friend pairs, 43 acquaintance pairs	3rd graders	Dictator game with tokens	prop	0.72	(0.27, 1.21)	Pataki, Shapiro, and Clark 1994	Behavioral
30 friend pairs, 30 acquaintance pairs	1st and 3rd graders	Trinkets	mean	0.42	(-0.09, 0.93)	Floyd 1964	Behavioral
56 friend pairs, 53 acquaintance pairs	5- to 9-year-olds	Use of crayon	t-stat	0.51		Jones 1985	Behavioral
30 friend pairs, 39 acquaintance pairs	1st graders	Dictator game with tokens	prop	0.35	(-0.14, 0.85)	Pataki, Shapiro, and Clark 1994	Behavioral
13 best friends vs. stranger	U.S. college students	Money	t-stat	2.06		Aron, Aron, Tudor, and Nelson 1991	Behavioral
14 friends, 14 acquaintances	9-year-olds	Tokens	t-stat	-0.36		Buhrmester, Goldfarb, and Cantrell 1992	Behavioral

Sample	Subjects	Measure	Statistic	Value	CI	Source	Method
14 friends, 14 acquaintances	13-year-olds	Tokens	t-stat	0.78		Buhrmester, Goldfarb, and Cantrell 1992	Behavioral
206 friends vs. 99, with 4 places removed	U.S. college students	Dictator game	from paper	0.52		Leider et al. 2009	Behavioral
36 friends, 36 acquaintances	Chinese kindergarten	Sharing snacks	mean	-0.15	(-0.61, 0.31)	Rao and Stewart 1999	Behavioral
36 friends, 36 acquaintances	Indian kindergarten	Sharing snacks	mean	-0.11	(-0.57, 0.36)	Rao and Stewart 1999	Behavioral
16 friend pairs, 16 acquaintance pairs	4- to 5-year-olds	Trinkets	mean	1.03	(0.28, 1.76)	Floyd 1964	Behavioral
57 friends, 57 strangers	3- to 5-year-olds	Number of snacks shared	mean	0.57	(0.19, 0.94)	Birch and Billman 1986	Behavioral
14 friends, 14 acquaintances	5-year-olds	Tokens	t-stat	-0.29		Buhrmester, Goldfarb, and Cantrell 1992	Behavioral

Cooperate More with Friends

Sample	Subjects	Measure	Statistic	Value	CI	Source	Method
19 friend pairs at varying degrees of friendship	4-year-olds	Cooperating in prisoner's dilemma	rho	1.01	(-0.01, 2.27)	Matsumoto et al. 1986	Behavioral
20 friend pairs, 20 stranger pairs	U.K. college	Donation to continuous prisoner's dilemma game	t-stat	0.96		Majolo et al. 2006	Behavioral

(continued)

N and Conditions	Participants	Variables	Derived from[a]	D-stat	95 Percent Confidence Interval[b]	Citation	Outcome Measure
Successfully Divide Surplus with Friends							
44 friend pairs, 43 acquaintance pairs	U.S. college students	Agreeing on how to share surplus from sale of lamp	rho	0.93	(0.47, 1.42)	McGinn and Keros 2002	Behavioral
50 friends, 50 strangers	U.S. college students	Successfully playing ultimatum game	rho	0.30	(-0.10, 0.71)	Polzer et al. 1993	Behavioral
10 groups	3- to 5-year-olds	Amount of time successfully viewing movie	rho	1.35	(-0.22, 3.70)	La Freniere and Charlesworth 1987	Behavioral
Helping (Non-anonymous)							
32 friend pairs, 32 strangers	College students	Buying lottery tickets	mean	2.03	(1.17, 2.89)	Boster et al. 1995	Behavioral
625 friends vs. 132, four places removed	U.S. college students	Helping game with tokens	mean	0.71	(0.42, 0.96)	Leider et al. 2009	Behavioral
12 friends, 12 acquaintances	W. German 17-year-olds	Money to needier (equal effort)	mean	1.81	(0.84, 2.76)	Schwinger and Lamm 1981	Hypothetical
12 friends, 12 acquaintances	W. German 17-year-olds	Money to needier (unequal effort)	mean	0.32	(-0.50, 1.11)	Schwinger and Lamm 1981	Hypothetical

32 friends, 32 acquaintances	W. German 17-year-olds	Money to needier (non-culpable)	mean	0.83	(0.32, 1.34)	Lamm and Schwinger 1980	Hypothetical
32 friends, 32 acquaintances	W. German 17-year-olds	Money to needier (culpable)	mean	0.92	(0.40, 1.44)	Lamm and Schwinger 1980	Hypothetical
32 friends, 32 acquaintances	6th graders	Dollars	mean	0.71	(0.20, 1.21)	McGillicuddy-De Lisi 1994	Hypothetical
32 friends, 32 acquaintances	3rd graders	Dollars	mean	0.31	(-0.19, 0.80)	McGillicuddy-De Lisi, Watkins, and Vinchur 1994	Hypothetical
32 friends, 32 acquaintances	Kindergarten	Dollars	mean	-0.22	(-0.71, 0.27)	McGillicuddy-De Lisi, Watkins, and Vinchur 1994	Hypothetical

Effect of Gift on Helping

32 friend pairs (16 gifts vs. 16 no gifts)	U.S. college students	Pre-giving vs. buying lottery tickets	mean	-0.19	(-0.89, 0.50)	Boster et al. 1995	Behavioral
32 stranger pairs (16 gifts vs. 16 no gifts)	U.S. college Students	Pre-giving vs. buying lottery tickets	mean	1.03	(0.29, 1.77)	Boster et al. 1995	Behavioral

(continued)

N and Conditions	Participants	Variables	Derived from[a]	D-stat.	95 Percent Confidence Interval[b]	Citation	Outcome Measure
Norm of Reciprocity—Sensitivity to Prior Gift							
32 friend pairs, 32 stranger pairs	U.S. college Students	Pre-giving vs. buying lottery tickets	mean	0.54	(0.04, 1.14)	Boster et al. 1995	Behavioral
20 friend pairs, 25 acquaintance pairs	U.S. 4th graders	Candy shared-eaten vs. time sharing crayon	rho	0.51		Staub and Sherk 1970	Behavioral
16 friend pairs, 16 acquaintance pairs	U.S. 4- to 5-year-olds	Trinkets vs. trinkets	mean	1.33	(0.55, 2.09)	Floyd 1964	Behavioral
20 friend pairs, 20 acquaintance pairs	U.S. 6- to 8-year-olds	Trinkets vs. trinkets	mean	1.04	(0.37, 1.70)	Floyd 1964	Behavioral
25 primed as friendship game, 25 non-primed	Maasai men	Shillings vs. percent returned	rho	0.60		Cronk 2007	Behavioral
Less Concern about Inputs							
13 friend pairs, 9 stranger pairs	U.S. college students	Glances at light indicating partner's input	mean	1.06	(0.16, 1.98)	Clark, Mills, and Corcoran 1989	Behavioral
More Concern about Needs							
11 friend pairs, 9 stranger pairs	U.S. college students	Glances at light indicating partner's needs	mean	0.83	(-0.13, 1.70)	Clark, Mills, and Corcoran 1989	Behavioral

Equality vs. Group Total

14 friend pairs, 14 acquaintance pairs	U.S. 10- to 12-year-olds	Group total	mean	0.53	(-0.23, 1.28)	Morgan and Sawyer 1967	Behavioral
12 friends, 12 acquaintances	W. German 17-year-olds	Money to needier (equal effort)	mean	1.81	(0.84, 2.76)	Schwinger and Lamm 1981	Hypothetical

Absolute Equity and Satisfaction

109 equal, 23 unequal	U.S. college students	Happy	mean	0.92	(0.57, 1.51)	Winn, Crawford, and Fischer 1991	Self-report
94 friendship pairs	U.S. college students	Positive feelings	rho	0.16	(-0.25, 0.57)	Mendelson and Kay 2003	Self-report[c]
52 on best friends	Average 69 years, U.S.	Satisfaction	rho	0.41	(-0.15, 1.00)	Jones and Vaughn 1990	Self-report
116 friends	Over 64 years	Satisfaction	rho	0.20	(-0.17, 0.58)	Roberto and Scott 1986	Self-report
94 on close friends	50 years or older	Happy/content	mean	0.35	(-0.10, 0.75)	Roberto 1996	Self-report
48 unequal best friends, 134 equal best friends	Dutch college students	Not lonely	t-stat	0.55		Prins and Buunk 1998	Self-report
120 on friends	60 years or older, U.S. women	Satisfaction	rho	< 0.32		Rook 1987	Self-report

Equity and Satisfaction—"Least Best Friends"

116 "least best friends"	Over 64 years	Satisfaction	rho	1.34	(0.92, 1.81)	Roberto and Scott 1986	Self-report

(continued)

N and Conditions	Participants	Variables	Derived from[a]	D-stat	95 Percent Confidence Interval[b]	Citation	Outcome Measure
Underbenefit and Anger							
14 under, 9 over		Anger	mean	1.02	(0.11, 1.89)	Winn, Crawford, and Fischer 1991	Self-report
35 over, 35 under		Anger	mean	-0.34	(-0.84, 0.16)	Roberto 1996	Self-report
Continuation							
3805	55 years or older, the Netherlands	Continuation vs. overbenefited instrumental	rho	0.08	(0.10, 0.22)	Ikkink and van Tilburg 1999	Self-report
3805	55 years or older, the Netherlands	Continuation vs. overbenefited emotional	rho	-0.08	(-0.10, -0.22)	Ikkink and van Tilburg 1999	Self-report
Input-Output and Satisfaction							
52 on best friends	Average 69 years, U.S.	Self-disclosure	rho	0.18	(-0.38, 0.76)	Jones and Vaughn 1990	Self-report
52 on best friends	Average 69 years, U.S.	Emotional support	rho	-0.54	(-1.15, 0.03)	Jones and Vaughn 1990	Self-report
52 on best friends	Average 69 years, U.S.	Tangible assistance	rho	0.26	(-0.30, 0.84)	Jones and Vaughn 1990	Self-report
52 on best friends	Average 69 years, U.S.	Socializing initiatives	rho	-0.02	(-0.59, 0.55)	Jones and Vaughn 1990	Self-report

116 on best friends	64 to 91 years	Global inequity	rho	0.20	(-0.17, 0.58)	Roberto and Scott 1986	Self-report
115 on friends	Average 72 years	Companionship, emotional, and instrumental support	rho	<0.32		Rook 1987	Self-report
94 on close friends	50 or older	Underbenefited vs. overbenefited	Mean	-0.49	(-0.99, 0.01)	Roberto 1996	Self-report
Anonymity							
Approximately 80 friends, 80 acquaintances	5- to 13-year-olds	Tokens	t-stat	0.45		Buhrmester, Goldfarb, and Cantrell 1992	Behavioral
206 friends vs. 97 acquaintances, 4 places removed	U.S. college students	Sharing game	from paper	0.52[d]	NA	Leider et al. 2007	Behavioral
876 friends vs. 181, 4 places removed	U.S. college students	Helping game with tokens	from paper	1.15[d]	NA	Leider et al. 2007	Behavioral
80 friends vs. 80, 4 places removed	10- to 12-year-olds	Dictator giving		2.39	(1.98, 2.80)	Goeree et al. 2007	Behavioral
22 friends vs. 27 not friends	Economics students, Spain	Dictator giving		0.70	(0.12, 1.28)	Branas-Garza, Duran, and Espinosa 2005	Behavioral
55 friends vs. strangers	U.S. college students	Social dilemma	t-stat	1.44		Aron, Aron, Tudor, and Nelson 1991	Behavioral
13 friends vs. strangers	U.S. college students	Social dilemma	t-stat	2.06		Aron, Aron, Tudor, and Nelson 1991	Behavioral

(continued)

N and Conditions	Participants	Variables	Derived from[a]	D-stat	95 Percent Confidence Interval[b]	Citation	Outcome Measure
Self-Serving Bias							
64 friend pairs	U.S. college students	Attribution of failure vs. success	mean	0.13	(-0.22, 0.47)	Campbell et al. 2000	Self-report
64 stranger pairs	U.S. college students	Attribution of failure vs. success	mean	0.83	(0.47, 1.19)	Campbell et al. 2000	Self-report
40 close pairs	U.S. college students	Attribution of failure vs. success	t-stat	-0.12		Sedikides et al. 1998	Self-report
40 non-close pairs	U.S. college students	Attribution of failure vs. success	t-stat	0.64		Sedikides et al. 1998	Self-report
52 close pairs	U.S. college students	Attribution of failure vs. success	t-stat	0.29		Sedikides et al. 1998	Self-report
52 non-close pairs	U.S. college students	Attribution of failure vs. success	t-stat	0.71		Sedikides et al. 1998	Self-report
Closeness and Helping							
91 close vs. non-close	Taiwan college students	Helping in academic task	t-stat	0.71		Han, Li, and Hwang 2005	Hypothetical
91 close vs. non-close	Taiwan college students	Helping in housekeeping	t-stat	0.64		Han, Li, and Hwang 2005	Hypothetical

77	U.S. college students	Helping partner	rho	1.81	(1.23, 2.47)	Korchmaros and Kenny 2006	Hypothetical
46	U.S. college students	Helping partner	rho	2.33	(1.51, 3.38)	Cialdini et al. 1997	Hypothetical
36	U.S. college students	Helping partner	rho	0.61	(-0.09, 1.37)	Cialdini et al. 1997	Hypothetical
29	U.S. college students	Helping partner	rho	1.62	(0.73, 2.76)	Korchmaros and Kenny 2001	Hypothetical

Closeness and Sacrifice

310 most close vs. least close	U.S. college students	Sacrificing for partner's gain	mean	3.53	(3.11, 3.96)	Jones and Rachlin 2006	Hypothetical
242 most close vs. least close	U.S. college students	Sacrificing for partner's gain	mean	3.43	(3.01, 3.85)	Rachlin and Jones 2008	Hypothetical
206 most close vs. least close	U.S. college students	Sacrificing for partner's gain	mean	1.43	(5.03, 6.21)	Rachlin and Jones 2008	Hypothetical

Induce Closeness and Helping

25 closeness condition, 25 control	U.S. college students	Amount of time offered to help	mean	0.98	(0.39, 1.57)	Maner et al. 2002	Behavioral
25 closeness condition, 25 control	U.S. college students	Amount of time offered to help	mean	0.61	(0.04, 1.17)	Maner et al. 2002	Behavioral

(continued)

N and Conditions	Participants	Variables	Derived from[a]	D-stat	95 Percent Confidence Interval[b]	Citation	Outcome Measure
Confusion on Memory Task							
14 best friend vs. celebrity	U.S. college students	Mistakenly recalling other for self	mean	2.13	(1.18, 3.05)	Mashek, Aron, and Boncimino 2003	Behavioral
14 best friend vs. parent	U.S. college students	Mistakenly recalling other for self	mean	1.66	(0.78, 2.51)	Mashek, Aron, and Boncimino 2003	Behavioral
Closeness and Similarity between Other-Self Activation							
16 women		Similarity of brain activity	rho	2.20	(0.83, 4.23)	Aron, Whitfield, and Lichty 2007	Biological
Oxytocin and Social Risk							
29 oxytocin, 29 non-oxytocin group		Investing 100 percent	rho	0.55	(0.02, 1.13)	Kosfeld et al. 2005	Behavioral
25 oxytocin, 24 non-oxytocin group		Investment after betrayal	mean	0.43	(-0.14, 1.00)	Baumgartner et al. 2008	Behavioral
Expressions of Liking							
33 opposite-sex friend pairs		Leaning	rho	1.12	(0.36, 2.04)	Gonzaga et al. 2001	Behavioral
33 opposite-sex friend pairs		Gesticulating	rho	0.87	(0.13, 1.73)	Gonzaga et al. 2001	Behavioral

Contribution of Adoptive Parents

Sample	Location	Outcome	Statistic	Value	CI	Reference	Method
161 2-parent adoptive families, 9661 2-parent biological families	U.S.	Involvement in school	rho	-.02		Hamilton, Cheng, and Powell 2007	Self-report
161 2-parent adoptive families, 9661 2-parent biological families	U.S.	Number of children's books	rho	-.01		Hamilton, Cheng, and Powell 2007	Self-report

Kin vs. Friends

Sample	Location	Outcome	Statistic	Value	CI	Reference	Method
1108 respondents	Israel (friends vs. kin)	Calling after an attack	prop	0.98	(0.85, 1.11)	Shavit, Fischer, and Koresh 1994	Self-report
1108 respondents	Israel (friends vs. kin)	Calling for social support	prop	0.06	(-0.06, 0.18)	Shavit, Fischer, and Koresh 1994	Self-report

Hormones and Romantic Love

Sample	Location	Outcome	Statistic	Value	CI	Reference	Method
48 in love, 48 single	Italian college students	Nerve growth factor	mean	0.79	(0.41, 1.16)	Emanuele et al. 2006	Biological
24 in love, 24 single	Italian college students	Cortisol	mean	0.41	(-0.16, 0.98)	Marazziti and Canale 2004	Biological

(continued)

N and Conditions	Participants	Variables	Derived from[a]	D-stat	95 Percent Confidence Interval[b]	Citation	Outcome Measure
Time Spent with Friends							
204 U.S. 11th graders, 222 Taiwanese 11th graders		Hours per week	mean	0.89	(0.60, 1.18)	Fuligni and Stevenson 1995	Self-report
204 U.S. 11th graders, 152 Japanese 11th graders		Hours per week	mean	0.49	(0.21, 0.77)	Fuligni and Stevenson 1995	Self-report
Attachment Anxiety							
48 adult women		Cortisol reactivity	rho	0.84	(0.24, 1.53)	Quirin, Pruessner, and Kuhl 2008	Survey and biological
Closeness after Cooperating							
22 pairs that cooperated in trust game (before and after)	U.S. college students	Perceived closeness	mean[e]	Approx. 0.83	(0.21, 1.44)	Krueger et al. 2007	Self-report
Tipping							
248 couples: touched by waiter vs. not touched	U.S. diners	Percent who tipped	mean	0.48	(0.21, 0.75)	Hornik 1992	Behavioral

Mood

Sample	Population	Task	Measure	Value	CI	Citation	Type
13 positive mood induced, 13 negative mood induced	U.S. college students	Helpful dictator game	rho	0.76	(-0.04, 1.77)	Capra 2004	Behavioral

Friendship Priming

Sample	Population	Task	Measure	Value	CI	Citation	Type
17 friendship prime, 16 coworker prime	U.S. college students	Willingness to help with next task	prop	0.76	(0.03, 1.59)	Fitzsimons and Bargh 2003	Behavioral
30 primary attachment prime, 30 acquaintance prime	U.S. college students	Agree to help	mean	0.68	(0.16, 1.20)	Mikulincer et al. 2005	Behavioral
30 primary attachment prime, 30 acquaintance prime	Israeli college students	Agree to help	mean	0.78	(0.25, 1.30)	Mikulincer et al. 2005	Behavioral
30 primary attachment prime, 30 acquaintance prime	U.S. college students	Agree to help	mean	0.55	(0.03, 1.06)	Mikulincer et al. 2005	Behavioral
30 primary attachment prime, 30 acquaintance prime	Israeli college students	Agree to help	mean	0.56	(0.04, 1.07)	Mikulincer et al. 2005	Behavioral

Empathic Accuracy

Sample	Population	Task	Measure	Value	CI	Citation	Type
24 friend pairs, 24 stranger pairs	U.S. college students	Content accuracy	t-stat	0.73		Stinson and Ickes 1992	Behavioral

(continued)

N and Conditions	Participants	Variables	Derived from[a]	D-stat	95 Percent Confidence Interval[b]	Citation	Outcome Measure
Blushing Study							
24 friends, 24 strangers	U.S. college students	Blushing	mean	0.67	(0.09, 1.25)	Shearn et al. 1999	Biological
24 friends	U.S. college students	Correlation of sympathetic response	rho	1.04	(0.14, 2.13)	Shearn et al. 1999	Biological
24 strangers	U.S. college students	Correlation of sympathetic response	rho	-0.30	(-1.20, 0.58)	Shearn et al. 1999	Biological
Not Asking Friend for Help							
119 asked for help vs. no help (computer)	Japanese college students	closeness	mean	1.89	(1.58, 2.20)	Niiya, Ellsworth, and Yamaguchi 2006	Hypothetical
119 asked for help vs. no help (dog)	Japanese college students	closeness	mean	1.44	(1.15, 1.73)	Niiya, Ellsworth, and Yamaguchi 2006	Hypothetical
119 asked for help vs. no help (hotel)	Japanese college students	closeness	mean	1.44	(1.15, 1.73)	Niiya, Ellsworth, and Yamaguchi 2006	Hypothetical
119 asked for help vs. no help (computer)	Japanese college students	sad	mean	-1.39	(-1.68, -1.10)	Niiya, Ellsworth, and Yamaguchi 2006	Hypothetical
119 asked for help vs. no help (dog)	Japanese college students	sad	mean	-1.40	(-1.69, -1.11)	Niiya, Ellsworth, and Yamaguchi 2006	Hypothetical
119 asked for help vs. no help (hotel)	Japanese college students	sad	mean	-2.05	(-2.37, -1.73)	Niiya, Ellsworth, and Yamaguchi 2006	Hypothetical

Forgiveness Studies

67 primed with close vs. non-close	Dutch college students	Willingness to forgive offenses	mean	0.59	(0.24, 0.93)	Karremans and Aarts 2007	Hypothetical
78 primed with close vs. non-close	Dutch college students	Willingness to forgive offenses	mean	1.02	(0.68, 1.35)	Karremans and Aarts 2007	Hypothetical
58 primed with close vs. non-close	Dutch college students	Willingness to forgive offenses	mean	0.69	(0.31, 1.06)	Karremans and Aarts 2007	Hypothetical
120 primed with close vs. non-close	Dutch college students	Willingness to forgive offenses	mean	0.63	(0.36, 0.90)	Karremans and Aarts 2007	Hypothetical

Social Support

14 explicit vs. control	Asian/Asian-American college students in U.S.	Reported stress	mean	0.83	(0.05, 1.60)	Taylor et al. 2007	Self-report
14 explicit vs. implicit	Asian/Asian-American college students in U.S.	Reported stress	mean	0.71	(-0.06, 1.47)	Taylor et al. 2007	Self-report
13 explicit vs. control	Majority white college students in U.S.	Reported stress	mean	-0.56	(-1.34, 0.24)	Taylor et al. 2007	Self-report
13 explicit vs. control	Majority white college students in U.S.	Reported stress	mean	-0.52	(-1.30, 0.26)	Taylor et al. 2007	Self-report

(continued)

N and Conditions	Participants	Variables	Derived from[a]	D-stat	95 Percent Confidence Interval[b]	Citation	Outcome Measure
Social Support (continued)							
14 explicit vs. 14 control	Asian/Asian-American college students in U.S.	Cortisol response	mean	0.96	(0.17, 1.74)	Taylor et al. 2007	Biological
14 explicit vs. 14 control	Asian/Asian-American college students in U.S.	Cortisol response	mean	0.86	(0.08, 1.63)	Taylor et al. 2007	Biological
Mobility and Local Friendships							
10,905 U.K. residents	U.K. adults	"How many of your personal friends live in this area (within about 15-minute walk of here)?" by length of stay	rho	0.49	(0.46, 0.53)	Sampson 1988	Self-report
Approximately 4800	U.S. families	Mobility distance or frequency x financial help	rho	0.49	(0.44, 0.55)	Magdol and Bessel 2003	Self-report
Approximately 4800	U.S. families	Mobility distance or frequency x task-related help	rho	0.37	(0.33, 0.44)	Magdol and Bessel 2003	Self-report

Online vs. Offline Friends

162 online friends, 162 offline friends	Hong Kong Internet users	Commitment	mean	1.12	(0.82, 1.42)	Chan and Cheng 2004	Self-report
162 online friends, 162 offline friends	Hong Kong Internet users	Integration	mean	0.75	(0.46, 1.04)	Chan and Cheng 2004	Self-report

[a] t-stat = t-statistic; prop = proportions; mean = difference in group means; from paper = directly from the referenced paper; rho = Pearson's rho or standardized regression coefficient.

[b] Confidence intervals calculated using ESCI delta Cumming and Finch 2001. Confidence intervals not calculated for d-statistic when derived from t-statistic. Exception: To calculate difference in change for friends and strangers requires an assumption of o correlation between a person's response at pre- and post-treatment. If one assumes a positive correlation, then this will increase the d-statistic. Also calculating d-statistic from Pearson's rho in this case requires an assumption of equal variance across conditions. It is equivalent to dividing the difference between d-statistics for each condition by sqrt(2). In this case, confidence intervals were not calculated.

[c] After controlling for other factors.

[d] Controlling for individual differences in altruism.

[e] Derived from graph in supplementary material.

Notes

Epigraph: Darwin and Darwin 1887, p. 37.

1. On Darwin's natural history of his own life, see ibid. On Darwin's friendship with John Henslow, see Walters and Stow 2001.

2. I use *West*, *Western*, and *Westerner* to refer to majority societies in Western Europe, the United States, Canada, Australia, and New Zealand.

3. Lewis 1960, p. 63.

4. The definition of friendship I use in this book is stricter than common notions of casual friendship current in the English-speaking West and closer to "close friendship." Chapters 1 and 2 discuss the definition in more detail.

5. Brain 1976; Bell and Coleman 1999.

6. Here I use *ecology* in the original sense of the study of an organism's relationship with its environment (Haeckel 1866), which can include not only the "natural" environment but also the social and cultural environment.

7. For exceptions to this neglect, see Smuts 1985; Silk 2003, 2002; Cords 1997; Tooby and Cosmides 1996; Smaniotto 2004.

8. Pakaluk compiles and reviews recent philosophical discussions of the relationship (Pakaluk 1991; Aristotle 2002; Derrida 1997). Friendship in non-Western traditions: Griffith 1889; Confucius 1998; Walshe 1995. In Griffith 1889, see book 10, hymn 71.

9. For theoretical and empirical reviews of social ties among humans, see Fiske 1991, Hinde 1997; and for cooperation among animals, see Dugatkin 1997.

10. The most prominent hypotheses for the natural selection of altruistic behavior among humans are kin selection (Griffith and West 2002; Hamilton 1964), reciprocal altruism (Trivers 1971, 2006), pair-bonding (Chapais 2008), and group selection (Richerson, Boyd, and Henrich 2003).

11. Kenrick and Trost 2000; Alexander 1979; Silk 2003.

12. Bshary and Noe 2003; Trivers 1971; Wilkinson 1984, 1985; Dugatkin 1997. The Russian naturalist Peter Kropotkin provided one of the earliest

descriptions of mutual aid in the context of evolutionary theory (Kropotkin 1902).

13. Axelrod and Hamilton 1981.

14. The simplicity of quid pro quo exchange and strictly balanced justice was also enticing to early English and Scottish social theorists who felt that the rules of friendship and close personal relationships were too vague, indeterminate, loose, and inaccurate. As Adam Smith lamented, the rules "admit of many exceptions, and require so many modifications, that it is scarce possible to regulate our conduct entirely by a regard to them." Adam Smith saw justice as a much more precise mode of governing behavior and a predictable means of ensuring justice and reciprocal exchange between individuals. Indeed, an automaton could follow and regulate them: "the man who . . . adheres with the most obstinate stedfastness [*sic*] to the general rules themselves, is the most commendable, and the most to be depended upon." "The moment he thinks of departing from the most staunch and positive adherence to what those inviolable precepts prescribe to him, he is no longer to be trusted, and no man can say what degree of guilt he may not arrive at" (Smith 2002, pp. 202, 204).

15. Some of these complications in isolation, such as the temporal spacing of favors, do not dramatically affect analytical solutions to the game (Boyd 1988).

16. Silk 2003.

17. Tomasello 2001.

18. Each of these forms of exchange may occur individually within a single non-human species, but the fact that all of these forms co-occur in a single species is unprecedented.

19. Watts 2002; de Waal and Brosnan 2006.

20. Chagnon and Bugos 1979; Wiessner 1982, 2002a.

1. AN OUTLINE OF FRIENDSHIP

1. Wandeki incident: Aufenanger 1966. "I eat you" among Fore (Lindenbaum 1979) and "internal excrement to be with" among Busama of New Guinea (Hogbin 1939) are other expressions for affection. Threatening to eat someone's liver as an insult that can cause war: Bensa and Goromido 1997.

2. Offensive behavior among friends: Brain 1976; Culwick, Culwick, and Kiwanga 1935; Griaule 1948; Radcliffe-Brown 1940, 1949. Prosocial (as opposed to antisocial) teasing and practical jokes in the establishment and maintenance of friendships: Barnett et al. 2004; Jones, Newman, and Bautista 2005; Keltner et al. 2001; Sanford and Eder 1984.

3. Kula partners among Trobriand islanders: Malinowski 1922; Baka Pygmies: Joiris 2003; U.S. high school students: Hruschka 2009.

4. More likely to share with close friends: fifteen behavioral experiments in ten articles, average d = 0.66 (Aron, Aron, Tudor, and Nelson, 1991; Birch and Billman 1986; Buhrmester, Goldfarb, and Cantrell 1992; Floyd 1964; Jones 1985; Leider et al. 2007; Pataki, Shapiro, and Clark 1994; Rao and Stewart 1999; Staub and Sherk 1970; Vaughan, Tajfel, and Williams 1981). To cooperate: two

behavioral experiments, average d = 0.98 (Majolo et al. 2006; Matsumoto et al. 1986). To successfully divide a surplus: three behavioral studies, average d = 0.63 (La Freniere and Charlesworth 1987; McGinn and Keros 2002; Polzer, Neale, and Glenn 1993). To help in times of need: two behavioral experiments, average d = 0.77 (Boster et al. 1995; Leider et al. 2007). Seven vignette experiments in three papers, average d = 0.58 (McGillicuddy-De Lisi, Watkins, and Vinchur 1994; Schwinger and Lamm 1981; Lamm and Schwinger 1980).

5. Axelrod 1984.

6. Cialdini 1998; Gouldner 1960, p. 174; Mauss 1954.

7. Individuals follow a norm of reciprocity with strangers and acquaintances (Boster et al. 1995; Burger et al. 2006), and this appears to be an internalized norm (Burger et al. 2009). Negative emotions if unable to repay a stranger who has helped them: Castro 1974; Gross and Latane 1974; Shumaker and Jackson 1979. Avoid asking strangers for favors they won't be able to reciprocate: Castro 1974; Greenberg and Shapiro 1971; Morris and Rosen 1973.

8. Boster et al. 1995.

9. Floyd 1964; Staub and Sherk 1970.

10. Chuuk: Marshall 1977. Tzeltal: Nash 1970. Shluh: Hatt 1974. Arapesh: Mead 1937a.

11. Walster, Walster, and Berscheid 1978.

12. Koryak: Jochelson 1908. Thai: Hanks 1962. Guarani: Watson 1952.

13. Less concern about inputs (d = -1.06) and more concern about needs (d = 0.83): Clark, Mills, and Corcoran 1989.

14. Result only held when students were allowed to bargain without stating their expectations prior to the experiment (Morgan and Sawyer 1967). It is supported by a study of hypothetical scenarios among West German high school students who were much more likely to meet the needs of friends than require an equal division of a surplus (d = 1.81) (Schwinger and Lamm 1981).

15. Friends in unbalanced relationships are not much lonelier or less satisfied in the friendship: Buunk and Prins 1998; Jones and Vaughan 1990; Mendelson and Kay 2003; Roberto 1996; Roberto and Scott 1986; Rook 1987; Winn, Crawford, and Fischer 1991. Underbenefited friends are no more angry or less satisfied than overbenefited friends: Roberto 1996; Roberto and Scott 1986; Rook 1987; Winn, Crawford, and Fischer 1991. Little effect on friendship duration: Ikkink and van Tilburg 1999, -0.10 < d < 0.10. Inequity with a "least best friend" was highly related to dissatisfaction in the relationship, d = 1.34: Roberto and Scott 1986.

16. Bo 2005; Groenenboom, Wilke, and Wit 2001; Gruder 1971; von Grumbkow et al. 1976; Heide and Miner 1992; Knippenberg and Steensma 2003; Mannix and Loewenstein 1993; Marlowe, Gergen, and Doob 1966; Sagan, Pondel, and Wittig 1981; Shapiro 1975.

17. Only one of these decisions was actually enacted by the experimenter, making it difficult for a decision maker to identify specifically which of their decisions came true.

18. The average is d = 0.52 for decisions to share and d = 1.15 for decisions

to help, after controlling for individual differences in altruism (Leider et al. 2009).

19. This finding has been replicated in seven behavioral experiments, average d = 1.14 (Aron, Aron, Tudor, Nelson 1991; Branas-Garza, Duran, and Espinosa 2005; Buhrmester, Goldfarb, and Cantrell 1992; Goeree et al. 2007).

20. Childs 1949; Papataxiarchis 1991; Shimony 1961.

21. The English and Russian examples are from Wierzbicka 1997. The Nepali example is taken from a personal communication with Brandon Kohrt (April 25, 2007). The Mongolian example is from my own fieldwork (1996–1998), and the French example from my experience in Belgium (1991–1992). For the Korean example, see Choi and Choi 2001.

22. Wharton 1934, p. 115.

23. Montaigne 1993, pp. 211–12.

24. Lakoff and Johnson 1999.

25. Aron, Aron, and Smollan 1992; Berscheid, Snyder, and Omoto 1989; Cialdini et al. 1997; Korchmaros and Kenny 2006; Kruger 2003; Maner and Gailliot 2006.

26. Maner et al. 2002.

27. Greater responsibility for failure and less responsibility for success: Campbell et al. 2000; Sedikides et al. 1998.

28. Closeness increases willingness to help: Cialdini et al. 1997; Han, Li, and Hwang 2005; Korchmaros and Kenny 2001, 2006. Closeness increases willingness to sacrifice for a partner's gain and to share goods: Jones and Rachlin 2006; Rachlin and Jones 2008.

29. Mashek, Aron, and Boncimino 2003.

30. Distress at losing a close other: Simpson 1987. Effect of partners' success on mood: Gardner, Gabriel, and Hochschild 2002; McFarland, Buehler, and MacKay 2001.

31. Moreover, these results could not be explained in terms of greater trait similarity with one's best friend, because in another study people rated themselves equally similar to their closest parent as to their best friend (Mashek, Aron, and Boncimino 2003).

32. Inclusion of other in self perspective: Agnew et al. 1998; Aron, Aron, and Smollan 1992; Aron, Aron, Tudor, and Nelson 1991; Mashek, Aron, and Boncimino 2003. Similar effects of closeness may also extend to groups: Coats et al. 2000; Otten and Epstude 2006; Schubert and Otten 2002; Tropp and Wright 2001.

33. Meta-analysis of these studies: Van Overwalle 2009.

34. Aron, Whitfield, and Lichty 2007.

35. Berndt 1985; McFarland, Buehler, and MacKay 2001; Pemberton and Sedikides 2001; Tesser and Smith 1980.

36. Brain 1976. The Nepali example is from a personal communication with Brandon Kohrt (April 25, 2007).

37. Soueif 2000, pp. 386–387.

38. One of the oldest lines of scientific research related to love is the study of

attachment, which grew out of the study of parent-child relationships (Ainsworth 1979; Bowlby 1958, 1982; Bretherton 1982; Cassidy and Shaver 1999) and has been extended to the study of attachment to a range of partners, including romantic partners (Hazan and Shaver 1987) and friends (Bartholomew and Horowitz 1991). The attachment perspective posits an attachment system, the goal of which is security, a state achieved by maintaining caregiver availability. I will review it in more detail in chapter 5.

39. Concept of love: Aron and Westbay 1996; Fehr and Russell 1991; Fisher 2004; Grote and Frieze 1994; Sprecher and Fehr 2006; Sternberg 1996.

40. Gonzaga et al. 2001.

41. Hatfield and Rapson 1996.

42. Maternal behaviors: Carter and Keverne 2002; Insel and Young 2001. Partner recognition in several mammalian species: Choleris et al. 2003; Winslow and Insel 2002.

43. Brain regions: Aron et al. 2005; Bartels and Zeki 2004. British clubgoers and Ecstasy: Wolff et al. 2006. Recent review on oxytocin: Donaldson and Young 2008.

44. Amygdala and oxytocin: Kirsch et al. 2005; Porges 2003.

45. Friend-like love and closeness (correlation = 0.64, d = 1.67): Aron, Aron, and Smollan 1992.

46. Kosfeld et al. 2005; Zak, Kurzban, and Matzner 2005.

47. Camerer 2003.

48. Kosfeld et al. 2005. The researchers also ruled out other possible interpretations. For example, oxytocin may simply make people more willing to hand over money to anonymous strangers. However, some trustees were also administered oxytocin, and they returned no more money than placebos. Another possibility is that oxytocin may make people less averse to risk in general, in this case the risk of being betrayed by a partner. However, the researchers also ran an experiment comparable to the trust game in terms of its riskiness, but one where the investor played with a machine rather than a person. In this case, oxytocin did not increase how much an investor sent. Rather, the results suggest that oxytocin made individuals less sensitive to a particular kind of risk—social risks.

49. Baumgartner et al. 2008; Heinrichs and Domes 2008; Huber, Veinante, and Stoop 2005; Kirsch et al. 2005; Phelps and LeDoux 2005; Rilling, King-Casas, and Sanfey 2008.

50. Eating together as a signal of friendship: Miller, Rozin, and Fiske 1998. Spending time together and exclusive interaction is greater among close friends than casual friends, d = 1.02 and d = 0.69: Hays 1989. Wasting time and extrinsically worthless gifts as a signal: Bergstrom, Kerr, and Lachmann 2008.

51. As with all signals, exclusive behaviors may not always appear as they are. A con artist, for example, may devote a great deal of time to winning over a mark. A status-hungry high school student may spend large amounts of time with a popular student because he wants to be seen as popular himself, not because he has any concern for that student's welfare. Nonetheless, the

absence of exclusive behaviors clearly signals a lack of personal interest and investment and raises questions about the partners' motivations toward each other.

52. Camerer 1988; Caplow 1984; Malinowski 1922; Mauss 1954; Weiner 1992.

53. Another interesting property of gifts is that they appear to violate key economic criteria of efficiency (Camerer 1988; Pieters and Robben 1999).

54. Gift givers prefer to give gift vouchers rather than their monetary equivalents as gifts, and compared to gift receivers, gift givers also prefer gifts that reflect the gift's exclusivity rather than its usefulness (Teigen, Olsen, and Solas 2005). Gift receivers prefer gifts that require time and effort on the part of givers (Robben and Verhallen 1994).

55. Ekman, Davidson, and Friesen 1990; Gonzaga et al. 2001. Duchenne quote cited in Ekman 2003, p. 212.

56. Gonzaga et al. 2001.

57. Newcomb and Bagwell 1995; Rose and Serafica 1986; Yamamoto and Suzuki 2006.

58. Dindia 1985; Kennedy 1986; Thibaut and Kelley 1959.

59. Barnett et al. 2004; Jones, Newman, and Bautista 2005; Keltner et al. 2001.

60. Burgoon 1991; Floyd 2006.

2. FRIENDSHIPS ACROSS CULTURES

1. Collected from Argentov's memoirs as well as from Ata'to himself (Bogoraz-Tan 1904–1909), p. 724.

2. Polygyny: Low 1988; household structures: Yanagisako 1979; love before marriage: Levine et al. 1995.

3. Cohen 1961; Paine 1969.

4. Comprehensive list of world cultures: Price 2004. HRAF dataset: http://www.yale.edu/hraf/.

5. Plato 1892, p. 39.

6. Of particular interest to the aims of this book is one subcategory in the HRAF coding system, *friendship,* which captures statements about how people conceive of friendship, how informal friendships develop, and how friends behave toward one another. To ensure that descriptions of friendship were not missed, I also independently searched the ethnographic accounts, using keywords such as *friendship* and *friend* and another code, *artificial kinship,* to identify descriptions that might have been missed by HRAF coders.

7. Peletz 1995, p. 345. Cross-cultural researchers have identified several ways that an ethnographer's prior mindset can bias the description of a culture—the tendency to notice and report behavior unlike that of one's own culture (bias of the "exotic") and to focus on norms and practices that fit one's theoretical biases (Naroll and Naroll 1963; Naroll 1962; Precourt 1979; Rohner, DeWalt, and Ness 1973; Divale 1976). Neglect of friendship: Marshall 1977; Paine 1969; Keesing 1972. Efforts to study friendship: Bell and Coleman 1999; Brain 1976; Leyton 1974b; Cohen 1961.

8. Rivers 1900; Bouquet 1996; Parkin 1997.

9. Durkheim and Mauss [1903] 1963; Morgan 1877; Rivers 1915; Tylor 1889; Murdock 1949; Radcliffe-Brown 1952.

10. Fortes 1970, p. 34.

11. Importance of kinship in ethnography: Barnes 1978, p. 121; Barnard and Good 1984.

12. *Notes and Queries on Anthropology* was published in six editions from 1874 to 1951 (Royal Anthropological Institute of Great Britain and Ireland [1874] 1951). History of its use in fieldwork: Urry 1972. Bott concludes anthropologists were too "bedazzled by kinship to take proper note of friendship" (Bott 1971, p. 234).

13. Early exceptions to this neglect of friendship: Radcliffe-Brown 1949, 1940; Eisenstadt 1956; Driberg 1935; Mandelbaum 1936; Herskovitz 1934. By contrasting the relative focus on kinship and friendship in anthropological literature, I do not mean to imply that these are two mutually exclusive constructs. I describe in chapter 3 how friendship and kinship overlap in diverse ways in different cultures.

14. Probability Sample File: Naroll 1967; Lagace 1979; Roe 2007.

15. C. J. Cherryh, *Foreigner* series.

16. Lepcha *ingzong:* Gorer 1938.

17. Female friendships in Hatzi: Kennedy 1986.

18. Ju/'hoansi *hxaro:* Wiessner 2002a, p. 421.

19. Barton 1930, p. 96. The societies for which ethnographers have stated that nothing like friendship exists are Iceland, Seri (Comcaac), Kogi (Cagaba), Klamath, Muria Gond, Greek herders, and Ifugao groups. Different perspectives on the importance of friendship in Icelandic society: Rich 1980, 1989; Pinson 1979, 1985. In Klamath society: Stern 1963; Spencer 1956. In Ifugao society: Barton 1930, pp. 97, 116; Barton 1919, pp. 79–80, 101; Beyer 1911, pp. 234, 244; Villaverde 1909, p. 241.

20. Kogi quotation: Reichel-Dolmatoff 1951, p. 239. Description of friendship among Muria Gond: Elwin 1947, p. 457; Greek herders: Campbell 1964, pp. 204–205, 231; rural Javanese farmers: Jay 1969; Cayapa farmers in Ecuador: Altschuler 1965.

21. Coser 1974.

22. Bigelow, Tesson, and Lewko 1996; Cole and Bradac 1996; Crawford 1977; Davis and Todd 1985; Hess 1972; Matthews 1986; Moffatt 1986; Naegele 1958; Rawlins 1992; Reisman 1979; Wiseman 1986; Wright 1974; Hays 1988; Kurth 1970.

23. Hinde 1997.

24. Western Tibetan: Peter 1963; Salish: Cline 1938; Mortlockese: Marshall 1977.

25. Mauss 1954; Camerer 1988; Komter 2004; Sherry 1983; Brown 1991. Here I define gift giving as the voluntary bestowing of goods without compensation.

26. Carmichael and MacLeod 1997; Sozou and Seymour 2005.

27. Importance of self-disclosure in friendship: Fehr 2004; Argyle and Henderson 1984; Davis and Todd 1985. Self-disclosure as producing collateral: Monsour 1992; La Gaipa 1977; Dindia 1985. Psychologists' use of self-disclosure of personal information to "foster" friendships between prior strangers in laboratory experiments: Campbell et al. 2000.

28. Bomana Prison inmates: Reed 2003, p. 160. Kanuri *ashirmanze*: Cohen 1960, p. 48.

29. Argyle and Henderson 1984; Li 2003.

30. Jacobson-Widding 1981; Reed-Danahay 1999.

31. Muria Gond: Elwin 1947, p. 457; Chinese: Levy 1949; Shona: Gelfand 1973; Thai peasants: Foster 1976.

32. Andaman Islanders: Radcliffe-Brown 1949; rituals as turning points: Lewis 1951.

33. Fischer 1992; Hartup 1975; Crawford 1977; Davis and Todd 1985; Weiss and Lowenthal 1975.

34. Gulliver 1955, 1951.

35. Use of far-off friends: Descola 1996; Gulliver 1955; Kiefer 1972; Shack 1963. In his analysis of Thai peasant friendships, Piker uses the fact that "friends to the death" live so far apart (60 percent live in another province) to argue that individuals rarely follow the ideals of friendship (Piker 1968). Good friends socializing less frequently in the U.S.: Jackson 1977; Hays 1985, 1988; Crawford 1977; Rose and Serafica 1986.

36. Burling 1963.

37. Behaviors toward friends and non-friends among high school students: Hruschka 2006. Students may act similarly in public to friends, acquaintances, and enemies but very differently in private (Buhrmester, Goldfarb, and Cantrell 1992).

38. Hechter 1987.

39. San: Wiessner 1994; Santal: Skrefsrud 1942, p. 112; Bena: Culwick, Culwick, and Kiwanga 1935, p. 147; Ashaninka: Santos-Granero 2007.

40. Wilkins and Gareis 2006. U.S. stereotypes of liking and love in friendship: Hess 1972; Kurth 1970; Reisman 1979. Use of the term *love* between friends in prior eras: Wierzbicka 1997. Societies where the equivalent for *love* is used to describe feelings for friends are the Himalayan Lepcha (Gorer 1938) and Swat Pathan (Lindholm 1982).

41. Jealousy among friends provides a theme in fiction and film. For studies of jealousy among friends: Parker et al. 2005; Roth and Parker 2001. Tarascan farmers: Friedrich 1958; Pashtun herders: Lindholm 1982; Copper Inuit youth: Condon 1987. Guatemalan village youth: Reina 1959.

42. As quoted in Slingerland 2003, p. 38.

43. Falk, Fehr, and Fischbacher 2008; Greenberg and Frisch 1972; Ames, Flynn, and Weber 2004; Tsang 2006; Tesser, Gatewood, and Driver 1968.

44. Lewis 1960, p. 71. Obligations underlying a norm of reciprocity: Gouldner 1960. Critique of their applicability in friendships: Clark and Mills 1979; Silk 2003. A norm of helping when a partner is in need closely resembles Fortes's description of *amity*: Fortes 1970, p. 110.

45. There are two exceptions from the broader literature. Abu-Sahra's description of friendship in a Tunisian village is that giving was strictly reciprocal and that violations of strict reciprocity would end the relationship (Abu-Zahra 1974). However, it is not clear whether "strict reciprocity" applied to gift giving or mutual aid. Jacobson-Widding describes a case among Buissi in Central Africa where friends could sue one another if an imbalance became too great (Jacobson-Widding 1981). However, the author uses this as an example of a friendship that is more like a contract relationship. Thai example: Benedict 1952, p. 30; De Young 1955; Foster 1976.

46. Nash 1970. Ortner describes how mutual aid groups between Tzeltal villager families within lineages may contribute labor and gifts and keep track of the balance loosely, whereas aid within the village (but outside the lineage) is common only in the form of gifts, and careful accounting is kept (Ortner 1978, pp. 22–23). Another example of the loose accounting of favors among friends can be found in Pitt-Rivers's description of life in an Andalusian village: "People assure one another that the favour they do is done with no afterthought. A pure favor which entails no obligation, an action which is done for the pleasure of doing it, prompted only by the desire to express esteem. . . . While a friend is entitled to expect a return of his feeling and favour he is not entitled to bestow them in that expectation" (Pitt-Rivers 1954, pp. 138–139).

47. Sahlins 1972. Later, Fiske expanded this scheme with four primary relationship types: communal sharing (Sahlins's generalized reciprocity), authority ranking and equality matching (balanced reciprocity), and market pricing (negative reciprocity) (Fiske 1991).

48. Wiessner 1977, 1982, 1994, 2002a. Reciprocating rounds of beer is another kind of gift giving where relatively strict reciprocity can be important and people attempt to restore balance if someone has either under-bought or over-bought (Karp 1980, p. 97; Moore 1990; Hunt and Satterlee 1986).

49. Kurth 1970; Fiebert and Fiebert 1969; Brain 1976; Moffatt 1986; Paine 1969; Suttles 1970.

50. Handy's description of Marquesans provides an example of avoiding low-status peers as friends: Handy 1923.

51. Brain 1976; Pitt-Rivers 1954, p. 140; Rezende 1999.

52. One reason for the difficulty in identifying equality in friend-like relationships stems from the problem of what counts as equal. For example, do partners have equal social assets? Do they have equal life chances? Do they lack an authority relation between themselves? Without a clearer understanding of what it means for members of a friend-like relationship to be social equals, it is difficult to argue that this constitutes a cross-culturally relevant aspect of friendship.

53. Hays 1988; Kurth 1970; Reisman 1979; Wright 1974; Hinde 1997; Krappmann 1996. Some scholars have claimed that freedom to choose one's friends is *the* defining feature of friendship (Eisenstadt 1974).

54. Bangwa example: Brain 1976, p. 11. Iroquois: Fenton 1953; Shimony 1961. Blood brotherhood example: Evans-Pritchard 1933; Burne 1902. U.S. example: Allan 1989; Feld and Carter 1998.

55. Wright 1974; Suttles 1970; Wiseman 1986.

56. Shack 1963, p. 205.

57. Silver contends that the privateness of friendship is a uniquely Western ideal (Silver 1990), and Cohen's cross-cultural study shows that friendships in many cultures are publicly recognized, contain a lengthy set of formal obligations, and are subject to third-party enforcement (Cohen 1961).

58. Tixier 1940, p. 216; Shack 1963.

59. Of course, there are cases in which these properties are not mentioned. However, given the incomplete nature of the ethnographic record, these absences in description cannot be considered as disconfirmations. The fact that these features are also common in the U.S. suggests that they do not represent an ethnographic bias toward "unfamiliar" and "exotic" features of non-U.S. cultures.

60. These expectations are beginning to change in particular regions, but this change has involved fervent debates between those who support and those who deny the moral and legal status of same-sex marriage.

61. David Livingstone, as referenced in Tegnaeus 1952, p. 149. Wood 1868.

62. Boswell 1994; Tulchin 2007, p. 635.

63. Review of blood brotherhood: Tegnaeus 1952; a critique of this review: Oschema 2006. Blood brotherhood remains common enough in parts of the world that some scholars have raised concern about its role in transmitting blood-borne diseases, such as HIV and hepatitis (Leblebicioglu et al. 2003; Kurcer and Pehlivan 2002). Friendship rites in Christian churches: Bray 2000; Boswell 1994.

64. Anderson 1911; Jacobson-Widding 1981; Kiefer 1972; Seligman and Seligman 1932.

65. Azande example: Evans-Pritchard 1933; other examples of supernatural sanctions: Kiefer 1972; Tegnaeus 1952. Role of fear of supernatural sanctions in moral behavior: Johnson and Kruger 2004.

66. Honigmann 1949, p. 191.

67. Taiwanese example: Diamond 1969. Kaska example: Honigmann 1949, p. 191. For other examples of touching among friends in Thailand: Embree 1950; Phillips 1966. Wogeo: Hogbin 1946. Aranda: Basedow 1925. Afghanistan and Pakistan: Lindholm 1982. Andaman Islands: Radcliffe-Brown 1922. Malaysia: Ahmad 1950. Going home drunk example: Norbeck and Norbeck 1956.

3. FRIENDSHIP AND KINSHIP

Epigraph: Leyton 1974a, p. 95.

1. Schwimmer 1974.

2. Fehr 1996; Silk 2002. The entire first definition of *friend* in the *Oxford English Dictionary* is as follows: "One joined to another in mutual benevolence and intimacy (J.). Not ordinarily applied to lovers or relatives" (1989).

3. Cohen 1961; Dubois 1974; Paine 1974.

4. Schneider 1968.

5. Tooby and Cosmides 1996; Vigil 2007.

6. Brown and Brown 2006; Korchmaros and Kenny 2001, 2006; Ackerman, Kenrick, and Schaller 2007; Park, Schaller, and Van Vugt 2008; Rushton 1989.

7. Hamilton 1964.

8. Described in Maynard Smith and Szathmary 1995.

9. Competition and limited kin dispersal can change the theory's predictions: Borgerhoff Mulder 2007; McElreath and Boyd 2007. Evidence consistent with theory: Berte 1988; Betzig and Turke 1986; Bowles and Posel 2005; Burnstein, Crandall, and Kitayama 1994; Cashdan 1985; Chagnon and Bugos 1979; Cialdini et al. 1997; Daly and Wilson 1988; Dunbar, Clark, and Hurst 1995; Essock-Vitale and McGuire 1985; Euler and Weitzel 1996; Grayson 1993; Gurven 2005; Gurven et al. 2001; Gurven et al. 2000; Hames 1987; Hurd 1983; Ivey 2000; Kruger 2001, 2003; Madsen et al. 2007; Morgan 1979; Rachlin and Jones 2008; Silk 1980; Webster 2003.

10. Sahlins 1976. Some of these debates are unproductive because critics assume that the theory of kin selection claims to account for all altruistic behavior, or all behavior, for that matter.

11. Mitani, Merriwether, and Zhang 2000; Mitani 2009.

12. Gaining access to material resources: Bourdieu 1972; avoiding incest taboos: Chagnon 2000; claiming a common ancestor: Virginia 2006.

13. Peng 2004; Virginia 2006.

14. Fischer 1992; Schneider 1968. In other societies, many relationships lie outside the formal system of kinship. Blood brothers among Azande in north-central Africa have mutual obligations that are directly opposed to those of kin, including burying their friends and conducting autopsies (Evans-Pritchard 1933, pp. 395–398). Therefore, it would be very difficult to become blood brothers with a biological brother or cousin. See Chun 1984 for Korean examples.

15. Eggan 1950, p. 27. For similar examples among the Aranda and the Maori, see Morton 1992, pp. 45–46, and Schwimmer 1974, respectively.

16. Schneider 1968.

17. Such non-genetic factors are important in Chuukese: Marshall 1977; Nukuorans: Carroll 1970; Navajo: Witherspoon 1975; Marshallese: Rynkiewich 1976; Fore: Glasse and Lindenbaum 1969, Lindenbaum and Glasse 1969; Palauans: Smith 1983; Telefol: Craig 1969; and Hawaiians: Linnekin 1985.

18. Glasse and Lindenbaum 1969, pp. 30–34.

19. Silk 1980, 1990.

20. All d-statistics: < 0.10 (Hamilton, Cheng, and Powell 2007).

21. Anderson, Kaplan, Lam, and Lancaster 1999; Anderson, Kaplan, and Lancaster 1999; Flinn 1988; Gibson 2004; Zvoch 1999.

22. Daly and Wilson 1996, 1999; Flinn, Leone, and Quinlan 1999.

23. Hawkes 1983.

24. Alvard 2003.

25. Distinction between communal relationships with close friends and family versus exchange relationships: Bugental 2000; Fiske 1991; Sahlins 1972.

26. In the Korean version of Facebook, Cyworld, special onsite friends are referred to as *Cy-Ilchons,* which draws from the Korean term *ilchon,* for the first circle of kinship between parents and children (Kim and Yun 2007).

27. Argyle et al. 1986.

28. This Yahgan sentiment (Gusinde 1937, p. 927) is echoed among native Hawaiians (Handy and Pukui 1958); Chuukese (Fischer 1950); Akan Ashanti (Rattray 1927); central Thai (Sharp and Rapaport 1966; Sharp 1978); Bengali (Klass 1978); Mentawaians (Schefold 1980); Shona (Gelfand 1973); Chukchee (Bogoraz-Tan 1904–1909); Marshallese (Wedgewood 1942); Kapauka (Pospisil 1978); Wogeo (Hogbin 1939, 1946); Ganda (Southwold 1965); Bahia (Borges 1992); Iroquois (Shimony 1961); Tukano (Goldman 1963); and Blackfoot (Goldfrank 1966).

29. Motivations that guide behavior toward extended and especially distant kin are different from those among immediate kin (Jones 2000). As opposed to acting out of goodwill, distant kin are frequently described as acting so as to improve their reputation or avoid sanctions or punishment. These qualitative findings from diverse cultures have more recently been backed up by systematic studies of motivations for helping in industrialized countries. In a survey in the 1990s, Dutch families reported giving comparable levels of help with childcare, meal preparation, and lodging to friends, children, and extended family (Komter and Vollebergh 2002). However, their motivations for helping differed across these relationship categories. Help given to friends was only occasionally reported as based on "obligation" (25 percent) rather than "love," a finding similar for helping one's own children (23 percent). However, among extended kin nearly half of all acts (46 percent) were reported as due to "obligations." For these reasons, I restrict comparisons to immediate kin.

30. Kruger 2003; Pulakos 1989; Stewart-Williams 2007.

31. Van Overwalle 2009.

32. Aron et al. 2005; Bartels and Zeki 2000, 2004.

33. Browne [1643] 1955, p. 85.

34. Zande: Evans-Pritchard 1933, p. 5; Kwoma: Whiting 1941. Expectations of helping for the Shluh: Hatt 1974; Wolof: Irvine 1974; Western Apache: Basso 1979; Slovene: Kremensek 1983; Turkana: Dyson-Hudson and McCabe 1985. For claims of comparable helping for the Lozi: Gluckman 1967, p. 71; Teda: Kramer 1978; Blackfoot: Goldfrank 1966; Seminole: Spoehr 1941; Montenegrin: Boehm 1983; Iroquois: Wallace 1969; Orokaiva: Reay 1953; Pashtun: Anderson 1982; Garo: Burling 1963.

35. Malays: Wilkinson 1925, p. 10.

36. Britain: Argyle and Furnham 1983; Australia: Miller and Darlington 2002; Canada: Wellman and Wortley 1990. The Netherlands: Komter and Vollebergh 2002. Australia: D'Abbs 1991. U.S. Hispanics, blacks, and whites: Muir and Weinstein 1962; Taylor, Chatters, and Mays 1988. Poor families in urban Chile: Espinoza 1999. Middle-class residents of Los Angeles: Essock-Vitale and McGuire 1985.

37. Madsen et al. 2007.

38. Jones 2000; Sahlins 1972.

39. O'Connell 1984. These findings fit with those of Stewart-Williams (2007) that favors were highly balanced among acquaintances but much less balanced among kin and friends.

40. Agnew et al. 1998; Brown and Brown 2006; Korchmaros and Kenny 2001, 2006.

41. Argument based on confusing friends with kin: Bailey 1989. Reeve argues that when the frequency of encountering kin is relatively high, then a bias in favor of falsely assuming kinship (rather than falsely assuming non-kinship) may be adaptive. Behavioral implication, in these contexts, is that individuals may be more likely to treat non-kin as kin rather than the reverse (Reeve 1998).

42. If the partner needed to make a phone call, had just been evicted from his or her apartment, had died in an accident leaving his or her two children without parents, or needed rescue from a burning house: Cialdini et al. 1997; needed help with an everyday errand: Korchmaros and Kenny 2001, 2006; or needed an organ or blood donation: Trief 2004.

43. Ackerman, Kenrick, and Schaller 2007; DeBruine 2002; Park and Schaller 2005; Park, Schaller, and Van Vugt 2008.

44. Cialdini et al. 1997; Korchmaros and Kenny 2001, 2006; Kruger 2003.

45. Rachlin and Jones 2008.

46. Essock-Vitale and McGuire 1985; Wellman and Wortley 1990.

47. Stewart-Williams 2007, 2008.

48. Sime 1983.

49. Shavit, Fischer, and Koresh 1993.

50. Most studies of helping behavior have generally examined either biological relatedness or psychological closeness in isolation. Studies in the tradition of behavioral ecology, for example, have carefully measured biological relatedness. However, they also have a methodological and theoretical bias against psychological measurement, and so psychological closeness is rarely assessed (Gurven 2005). This has the unfortunate side effect of lumping all non-kin, from close friends to strangers, into a single category. Meanwhile, studies in social psychology have examined the effects of closeness on actual helping behavior but have rarely gone beyond vignette methods to examine how people really differ in behavior toward kin of differing biological relatedness.

51. Such aid can lead to conflicts, such as the Garifuna fishermen in the mid-1900s who chided their kin for distributing food or clothing to friends over blood relations (Wright 1986). In some accounts, people make clear exhortations against friendship. Consider the warning given by an old San Ildefonso Tewa woman to her granddaughter, "Never have a friend. . . . You can never know whether she may or may not be a witch. And never, never say 'yes' to a friend if she should ask you to do something. She might be trying to get you into her power" (Whitman and Whitman 1947, p. 124).

52. Wallman 1974, p. 110.

53. Pitt-Rivers 1954, 1973.

54. Harding 1994.

55. Schwimmer 1974, p. 69.

56. Chagnon and Bugos 1979. Similar non-kin coalitions among Tausug horticulturalists counted in the hundreds: Kiefer 1972.

57. Alvard 2003; Chagnon and Bugos 1979; Hawkes 1983; Kiefer 1972; Schwartz 1974.

58. After marriage and the move to a new village: Lindenbaum 1979; migrations: Hammel 1968. If one's kin have died: Leyton 1974a, p. 96. In many of these cases, there is still a hierarchy of resort, where one turns first to kin and only to friends if kin are unable to help (Young and Willmott 1962). Among the Hare in the Canadian arctic, for example, in times of need an individual seeks aid first from his primary relatives, then from his close and more distant relatives, and finally from his friends: Hare (Hara 1980); Garifuna (Palacio 1982); Guarani (Reed 1995); Japanese (Cornell and Smith 1956); Hungarians and Americans (Litwak and Szelenyi 1969); Serbs (Hammel 1968). Nonetheless, friendship may play a role in "patching" holes and covering gaps when kin are not enough (Burridge 1957).

59. Anderson 2001.

60. Milicic 1998.

61. Bangwa: Brain 1976, pp. 27–28, 30–39; Chuuk: Gladwin and Sarason 1953. For more examples: Jivaro (Harner 1973); Bengali (Islam 1974, p. 131); Andalusian (Pitt-Rivers 1968).

62. Malinowski 1929, p. 412.

63. Nearly one in five British adults listed a relative as their closest friend, and over half listed their spouse or partner as their closest friend (Pahl and Pevalin 2005, p. 501). Between 5 and 15 percent of Russian and U.S. close friendships are with relatives (Sheets and Lugar 2005a, 2005b).

64. Gurdin 1996, p. 162.

65. Durrenberger and Palsson 1999, p. 67.

4. SEX, ROMANCE, AND FRIENDSHIP

1. Regan 2004.

2. Bush and Harbage 1961. This question of whether men and women can be friends is not limited to the U.S.: "From the majority point of view in Monteros, adult female friendships are anomalous. They are reasonably explained only as the result of homosexuality" (Brandes 1985, p. 116).

3. Kogan 2006, pp. 188, 246.

4. *Kama* and *metta:* Okada 1973; Silva 1990. *Inangaro kino:* Harris 1995.

5. Fisher 2004; Brand et al. 2007; Jankowiak and Fischer 1992; Tennov 1979. Men and women in many societies see this mutual attraction as the first criterion for choosing a spouse (Buss 1989), but it is not the case in all societies (Levine et al. 1995). Passionate and companionate love: Berscheid and Walster 1978.

6. Fisher 2004; Hazan and Shaver 1987; Hazan and Zeifman 1994; Money 1980; Sprecher and Regan 1998.

7. Aron, Fisher, and Strong 2006; Berscheid and Ammazzalorso 2001; Davis and Todd 1982.

8. Fisher 2004; Gonzaga et al. 2001.

9. Shostak 1981, p. 268.

10. Cimbalo, Faling, and Mousaw 1976; Sprecher and Regan 1998; Traupmann and Hatfield 1981. In the Traupmann and Hatfield study, the researchers conflated feelings of romantic love with those related to sexual desire, the main distinction being between companionate love and the other two kinds of love bundled up as "passion."

11. Ahrons and Wallisch 1987; Juhasz 1979; Masheter 1991; Metts and Cupach 1995; Metts, Cupach, and Bejlovec 1989.

12. Emanuele et al. 2006; Marazziti and Canale 2004.

13. Brain activation when viewing long- compared to short-term romantic partners: Aron et al. 2005; when viewing one's own child versus a romantic partner: Bartels and Zeki 2004. For speculations regarding common mammalian systems for pair-bonding, see Fisher, Aron, and Brown 2006.

14. Regan 2004; Regan and Berscheid 1995.

15. Hendrick and Hendrick 1987, pp. 282 and 293.

16. Aron and Aron 1991; Berscheid 1985; Regan 1998; Sprecher and Regan 1998; Tennov 1979.

17. Tennov 1979, p. 74.

18. Diamond 2000; Faderman 1981; Firth 1936; Gay 1986; Reina 1959; Rotundo 1989; Sahli 1979.

19. Diamond 2000; Diamond, Savin-Williams, and Dube 1999.

20. Diamond, Savin-Williams, and Dube 1999, p. 195.

21. Buster et al. 2005; Fisher 2004; Herdt and McClintock 2000; McClintock and Herdt 1996.

22. Hatfield et al. 1988.

23. Kephart 1967.

24. Simpson, Campbell, and Berscheid 1986.

25. Levine et al. 1995; Sprecher et al. 1994.

26. Broude 1983, 2003.

27. China: Pimentel 2000, Xiaohe and Whyte 1990; India: Sprecher and Chandak 1992; and Japan: Sprecher et al. 1994.

28. Male athletes: Moss, Panzak, and Tarter 1993. Testosterone patch: Buster et al. 2005. Discussions of brain regions: Bancroft 2005; Fisher 2004, 2006.

29. Bartels and Zeki 2000, 2004.

30. Fisher and Thomson 2007.

31. Fisher 1998.

32. Diamond 2003.

33. Allan 1989; Hendrick and Hendrick 1994; Pahl and Pevalin 2005. However, this is not always the case (Komarovsky 1962). The Delaware Native Americans would say, "My wife is not my friend" (Zeisberger 1910, p. 99).

34. Sternberg 1996.

35. Getz and Carter 1996; Insel 2003; Insel and Fernald 2004.

36. Baumgartner et al. 2008; Heinrichs and Domes 2008; Kosfeld et al. 2005.

37. Schapera 1930, p. 243; Brain 1976.

38. Schultze 1907.

39. Schapera 1930, pp. 242–43; Falk [1925] 1998; Schultze 1907, p. 319.

40. Murray 2004, s.v. *"soregus."*

41. Bilchitz and Hofmeyr 2006, pp. 27–28; Haacke and Eiseb 1998; Rust 1969.

42. Reeder 2000; Sapadin 1988.

43. Afifi and Faulkner 2000; Messman, Canary, and Hause 2000.

44. Ackerman, Kenrick, and Schaller 2007; Afifi and Burgoon 1998; Bleske and Buss 2000; Rose 1985; Schneider and Kenny 2000.

45. Bogle 2008. New Haven study: Sternberg 1996. Michigan State University study: Bisson and Levine 2009.

46. Reeder 2000.

47. Michael et al. 1994.

48. Montaigne 1993. Greek example: Loizos 1991, Papataxiarchis 1991; Iroquois example: Shimony 1961; Igbo example: Ottenberg 1989.

Cross-sex relationships permitted: Alutiiq: Befu 1970; Copper Inuit: Condon 1987; Tukano: Goldman 1963; Ndyuka: Bilby 1990; Scots, rare but possible: Ennew 1980; Kurds, rare but possible: Masters 1953; Serbs: Lodge 1941; Saramaca: Price 1991; Iroquois: Shimony 1961; Kapauka: Pospisil 1978; Chuuk: Marshall 1977; Hawaiians: Handy and Pukui 1958; Kuna: Marshall 1950; Banyoro, can be made: Nyakatura 1970; Turkana, occasionally: Dyson-Hudson and McCabe 1985; Tongans, usually between same-sex, but possible: Colson 1967; Tanala, both possible: Linton 1933; Kanuri, usually same-sex but not necessarily: Cohen 1960; Zuni: Parsons 1917; Russians: Sheets and Lugar 2005a.

Not permitted or treated with suspicion: In Korea, friendships between boy and girl are regarded with suspicion (Beirnatzki 1968). Among Santal in India, men and women do not become friends except in marriage (Skrefsrud 1942). For the Akan, a relationship between opposite sexes implies a sexual tie (Warren 1975). For the Igbo, a relationship between opposite sexes is not acceptable until a woman reaches menopause (Ottenberg 1989). Friendships should also be between members of the same sex: Ngoni, exist as a rule between pairs of the same sex (Read 1956); Shona (Gelfand 1973); Wolof (Gamble 1957); Ona (Gusinde 1931). Greek example (Loizos 1991; Papataxiarchis 1991). Dahomeans (Herskovitz 1934).

49. Kovacs, Parker, and Hoffman 1996; Maccoby 1990; McDougall and Hymel 2007; Parker and de Vries 1993. However, this is due to more than simply sexual prohibitions. Across cultural settings, children in the range of four to ten spend most of their time with peers of the same gender (Maccoby 1990, 1998; Serbin et al. 1994; Whiting and Edwards 1988). In the United States and Europe, time with the other gender increases across adolescence (Alsaker and Flammer 1999; Larson et al. 1996; Richards et al. 1998).

50. Ayres 1983; Canary et al. 1993; Messman, Canary, and Hause 2000.

51. Booth and Hess 1974; Kalmijn 2003; Rose 1985; Weiss and Lowenthal 1975.

52. Bisson and Levine 2009; Sapadin 1988; Werking 1997.

53. Afifi and Burgoon 1998; Schneider and Kenny 2000.

54. Bell 1981.

5. FRIENDSHIP: CHILDHOOD TO ADULTHOOD

1. Crick and Nelson 2002; Hendry and Reid 2000.

2. Horowitz and French 1979.

3. Wierzbicka 1997.

4. Sullivan 1953; Youniss 1980; Youniss and Smollar 1985.

5. Selective behaviors for toddlers: Howes 1983, 1988: Ross and Lollis 1989; Vandell and Mueller 1980. Stability of toddler friendship bonds: Dunn 2004; Freud and Burlingham 1944; Howes 1983, 1988; Howes, Hamilton, and Philipsen 1998.

6. Freud and Burlingham 1944, p. 581.

7. Furman and Bierman 1983; Gleason and Hohmann 2006; Howes 1983; Lederberg 1987.

8. Corsaro and Rizzo 1988.

9. Birch and Billman 1986; Floyd 1964; Rao and Stewart 1999. Sharing: Jones 1985; Pataki, Shapiro, and Clark 1994; Staub and Sherk 1970. What children say: Furman and Bierman 1983.

10. Muscovite living in Israel: Markowitz 1991, p. 640. Timbira: Nimuendaju 1946.

11. China: Chang and Holt 1991, Goodwin and Lee 1994; United States and Poland: Rybak and McAndrew 2006; Russia: Kon and Losenkov 1978; United States: Way 1998; Canada: Claes 1994.

12. Corsaro 2003, pp. 43, 71.

13. Berndt 1982; Berndt and Hoyle 1985; Berndt and Savin-Williams 1993; Bigelow 1977; Bigelow and La Gaipa 1975; French, Pidada, and Victor 2005; Furman and Bierman 1984; La Gaipa 1979; Sharabany, Gershoni, and Hofman 1981.

14. Bigelow 1977; Bigelow and La Gaipa 1975.

15. Youniss and Smollar 1985.

16. Selman 1980.

17. Pellegrini 1986.

18. Corsaro 2003.

19. Kurdek and Krile 1982. Two studies controlling for age among reciprocal friends, average $d = 1.04$. No association was found between unreciprocated friends or non-friends.

20. Gummerum and Keller 2008.

21. Scribner and Cole 1973.

22. Aguilar 1999.

23. Ik: Turnbull 1972, pp. 136–139. Muria Gond: Elwin 1947.

24. Konner 1975.

25. Mounts and Kim 2007; Suizzo 2007.

26. Wiessner 1998, p. 140.

27. Bangwa: Brain 1976, p. 27; Kwoma: Whiting 1941. Other examples of parents encouraging friendships: Marshallese (Wedgewood 1942); Moka (Strathern 1971, pp. 195–200).

28. Cheah and Chirkov 2008; Way and Greene 2006.

29. Last longer: Barbu 2003; Berndt and Hoyle 1985; Cairns 1994; Cairns, Leung, and Cairns 1995; Horrocks and Thompson 1946; Neckerman 1996; Thompson and Horrocks 1947. Understanding: Gummerum and Keller 2008.

30. Meeting potential mates: Newfoundlanders (Schwartz 1974, p. 87); Kwakiutl: (Ford 1941, p. 124).

31. Cohen 1961; Fourier 1928.

32. Kwakiutl: Ford 1941, p. 124.

33. Claes 1998; Douvan and Adelson 1966.

34. Gummerum and Keller 2008; Keller 2004; Keller et al. 1998.

35. For information regarding teenage Newfoundlanders, see Schwartz 1974, p. 87; for teenage Irish, see Leyton 1974a, p. 97; for Inuit teenagers, see Condon 1987; and for U.S. teenagers, see Weiss and Lowenthal 1975.

36. Hartup and Stevens 1999.

37. Berndt 2004; Hartup and Stevens 1999; Larson et al. 1996; Sullivan 1953.

38. Examples from semiliterate, rural societies: Kenya (Larson and Verma 1999); India (Larson, Verma, and Dworkin 2003); other societies (Whiting and Edwards 1988).

39. Fuligni and Stevenson 1995. African Americans: Csikszentmihalyi and Larson 1984; Richards et al. 1998. Koreans and Indians: Larson and Verma 1999. Europeans: Alsaker and Flammer 1999.

40. Schlegel and Barry 1991, p. 67.

41. Weiss and Lowenthal 1975.

42. Waite and Harrison 1992.

43. Johnson and Leslie 1982; Milardo 1982; Surra 1985.

44. Doherty and Feeney 2004.

45. Nix 1999.

46. Social network ties increase: Babchuk and Bates 1963; Fischer and Oliker 1983; Hess 1972; Milardo 1982; Moore 1990; Surra 1985; Van der Poel 1993. Reduction in interactions: Fischer 1989; Hurlbert and Acock 1990; Johnson and Leslie 1982; Kalmijn 2002; Kim and Stiff 1991; Roberto 1999; Surra 1985; Weiss and Lowenthal 1975. Turnover: Fischer and Oliker 1983; Wellman et al. 1997. No dramatic decrease in visiting: Waite and Harrison 1992. Unmarried friends rely much less on spending time together as a maintenance strategy than do married friends: Rose 1985. Married friends use affection more as a maintenance strategy than do unmarried friends: Rose and Serafica 1986.

47. Child rearing: Munch, McPherson, and Smith-Lovin 1997; divorce:

Kalmijn 2002, 2003; widowhood: Booth and Hess 1974; retirement: Dugan and Kivett 1998, Fischer and Oliker 1983, Jerrome and Wenger 1999, MacRae 1996.

48. Field 1999; Matthews 1986; Patterson, Bettini, and Nussbaum 1993.

49. Murray et al. 2008.

50. Hruschka 2009.

51. Bowlby 1982.

52. Ainsworth et al. 1978; Connell and Goldsmith 1982.

53. Hazan and Shaver 1987.

54. Baldwin and Fehr 1995; Baldwin et al. 1996.

55. Baldwin and Fehr 1995; Baldwin et al. 1996; Kurdek 2002; Overall, Fletcher, and Friesen 2003.

56. Kojetin 1996.

57. Baldwin et al. 1996.

58. Overall, Fletcher, and Friesen 2003.

59. Lemche et al. 2006; Quirin, Pruessner, and Kuhl 2008; Rholes and Simpson 2004; Vrticka et al. 2008.

60. What is known, and more important what is not known, about the neuroscience and physiology of attachment: Coan 2009.

61. Bippus and Rollin 2003; Grabill and Kerns 2000; Klohnen et al. 2005; Kurdek 2002; Mayseless and Scharf 2007; Mikulincer and Selinger 2001; Priel, Mitrany, and Shahar 1998; Shulman, Elicker, and Sroufe 1994; Weimer, Kerns, and Oldenburg 2004.

62. Tannen 1990; Gray 1993.

63. Lewis 1960.

64. Wright 1998.

65. Quote from Calcutta: Moorhouse 1972, p. 92.

66. Bank and Hansford 2000; Wright 1998.

67. Liberman 2006.

68. Hyde 2005; James and Drakich 1993; Mehl et al. 2007.

69. Height: $d = 1.79$ (Gray and Wolfe 1980); voice pitch: $d = 3.32$ (Apicella, Feinberg, and Marlowe 2007).

70. Moral orientation: Jaffee and Hyde 2000; prosociality: Eisenberg, Fabes, and Spinrad 2006; helping behavior: Eagly and Crowley 1986.

71. Barth and Kinder 1988; Fehr 1996; Weiss and Lowenthal 1975; Wright 1998.

72. Bell 1981, p. 55.

73. Bernard 1981; Gilligan 1982; Maccoby 1990, 1998; Tannen 1990.

74. Taylor et al. 2000.

75. Geary et al. 2003; Vigil 2007. A popular evolutionary argument proposed in the 1960s made exactly the opposite predictions of some of the evolutionary arguments we see today (Tiger 1969), suggesting that men, rather than women, form stronger friendships.

76. Dubois 1974; Kennedy 1986; Leyton 1974a; Mead 1937a; Saraswathi and Dutta 1988; Uhl 1991. Uhl suggests that anthropologists and informants (at least in Andalusia) may be involved in an "unintentional conspiracy" to por-

tray friendship as public, male, and important while minimizing or denying the existence of adult female friendship (Uhl 1991; Oliker 1989).

77. Fischer and Oliker 1983; Moore 1990; Munch, McPherson, and Smith-Lovin 1997.

78. Hacker 1981.

79. Dindia and Allen 1992.

80. Caldwell and Peplau 1982.

81. Berman, Murphy-Berman, and Pachauri 1988; Wheeler, Reis, and Bond 1989; Wong and Bond 1999.

82. Dubois and Hirsch 1990.

83. Dindia and Allen 1992; Hyde 2005; MacGeorge et al. 2004.

84. Lempers and Clark-Lempers 1992.

85. Similar expectations: Aries 1996; Ashton 1980; Caldwell and Peplau 1982; Cole and Bradac 1996; Enomoto 1999; Keller and Wood 1989; Kunkel and Burleson 1998; Monsour 1992; Parker and de Vries 1993; Sapadin 1988; Selman 1980; Tesch and Martin 1983; Wall, Pickert, and Paradise 1984; Weiss and Lowenthal 1975; Wright 1998. Special-purpose versus multifaceted friendships: Barth and Kinder 1988; Gouldner and Strong 1987; Weiss and Lowenthal 1975; Wright 1982. Further work is necessary to determine if any of the following proposed gender differences are supported: (1) helping friends (Eagly and Crowley 1986); (2) frequency of dyadic interactions (Claes 1998; Richards et al. 1998); (3) caring about having close friends (Aries and Johnson 1983; Johnson and Aries 1983); (4) jealousy in friendship; (5) conflict between best friends; (6) levels of friendship stress; (7) satisfaction in friendships (Harrison et al. 1995; Reisman 1990); (8) the receipt of affection, nurturance, and trust.

86. Rose and Rudolph 2006; Way and Greene 2006.

87. Burleson 1997, 2003; Reis 1998; Wright 1998.

88. Burleson 2003.

6. THE DEVELOPMENT OF FRIENDSHIPS

1. Studies of friendship duration generally do not ask how long people have been friends, but rather how long they have known one another (Fischer 1992; Ikkink and van Tilburg 1999). Studies that do ask how long people have been friends generally ask about broad periods of time, such as one month or less or three or more years (Geers, Reilley, and Dember 1998).

2. Murray, Holmes, and Collins 2006.

3. As described in Dunne 2004, p. 156.

4. Kin-biased ties: Gulliver 1971. Sex-biased ties: Larson and Verma 1999.

5. Feld 1981; Feld and Carter 1998; Adams and Allan 1998; Grossetti 2005; Festinger, Schachter, and Back 1950.

6. Papua New Guinea prisoners: Reed 2003; Bangwa children: Brain 1976, p. 22; Korean students: Brandt 1971.

7. The following are examples of separation from existing networks. Muria Gond and Khasi dormitories: Elwin 1947; Nakane 1967. Bangwa women who

are forced into polygynous marriages form new friendships in the chief's compound: Brain 1976, p. 42. Friendship formation in a Papua New Guinea prison: Reed 2003. Freshman orientation: Devlin 1996. In a monastery: Peter 1963. Same regiment: Sharp and Rapaport 1966. Cattle herding together as youth: Skrefsrud 1942. Ik children turned out from home: Turnbull 1972, pp. 136–139. U.S. high schoolers: Ortner 2003.

8. Examples: Bantu Kavirondo (Wagner 1949); Chagga (Guttman 1926, 1932); Kwoma (Whiting 1941); Malekula (Deacon 1934); San (Fourier 1928, Wiessner 1998); Dahomey (Herskovitz 1934).

9. Eggan 1950; Fabian 1992.

10. Levine et al. 1995; Lindenbaum 1979.

11. Whiting 1941.

12. Reputation for helping and kindness: Friedrichs 1960, Li et al. 2008; high status: Hruschka 2006; interpersonal similarity in social class, ethnicity, and attitudes: McPherson, Smith-Lovin, and Cook 2001, Rushton 1989. Knowing that a person likes you being enough to increase your liking for him or her: Backman and Secord 1959; Byrne and Rhamey 1965; Lowe and Goldstein 1970; Shrauger and Jones 1968; Montaya and Insko 2008.

13. Evans-Pritchard 1933; Baxter 1953. Commercial beginnings to friendship are also common in the U.S., where business ties can over time develop into close friendships (Haytko 2004; Ingram and Zou 2008).

14. Dunne 2004; Pakaluk 1991; Confucius 1998.

15. Shapiro 1975; Bo 2005; Heide and Miner 1992; von Grumbkow et al. 1976; Sagan, Pondel, and Wittig 1981; Knippenberg and Steensma 2003; Groenenboom, Wilke, and Wit 2001.

16. Carnegie 1937, p. 1. "Dale Carnegie's *How to Win Friends and Influence People*, the gold standard of the self-help genre, has sold more than 15 million copies since it was first published in 1937" (*Financial Post*, April 24, 2008).

17. Carnegie 1937, p. 1.

18. Tipping: Azar 2007, Seiter 2007, Hornik 1992; smiling: Mehu, Grammer, and Dunbar 2007; reciprocity: Cialdini 1998; and mood: Dovidio 1984, Capra 2004 (d = 0.76).

19. Sedikides et al. 1998.

20. Aron et al. 1997. In a meta-analysis of dozens of studies, Collins and Miller found a strong relationship between liking and self-disclosure in correlational studies (d = 0.85) but not in experimental studies (Collins and Miller 1994). Ruminating together over a distressing topic has also been shown to increase perception of friendship quality of over time (Rose, Carlson, and Waller 2007). All of these studies have been conducted in the United States and Europe, and as I will discuss in chapter 7, it is not clear that talking about highly personal topics has the same effect in all cultures.

21. This is a common strategy used by communes and family firms to induce members to exert greater effort for organizational goals (Kanter 1972; Carnegie 1937).

22. Treating members of an experimentally constituted group more favor-

ably than a member of a different but equally arbitrary group: Yamagishi and Mifune 2008; gossiping: Bosson et al. 2006; manipulating similarity: Maner et al. 2002, Burger et al. 2004, Oates and Wilson 2002, Park, Schaller, and Van Vugt 2008, Burger et al. 2001. Birthday mates are more likely to cooperate in prisoner's dilemma: Miller, Downs, and Prentice 1998.

23. Oates and Wilson 2002; Gueguen 2003.

24. Livingstone and Livingstone 1865, p. 148.

25. Fitzsimons and Bargh 2003; Mikulincer et al. 2005. Such a procedure also makes people more likely to express willingness to forgive (Karremans and Aarts 2007).

26. October 7, 2008, U.S. presidential debate, Nashville, TN. The American Presidency Project, http://www.presidency.ucsb.edu.

27. International relations: Roshchin 2006.

28. Basso 1979; Gulliver 1955.

29. Turkana: Gulliver 1951, p. 105.

30. The question of *why* these specific techniques invoke friendly situations is a very interesting one. Various explanations have been provided for some of these, including similarity being a cue to kinship, liking as a signal of another person's intent, and sharing personal details as a way of putting oneself in a vulnerable position.

31. Levine 2006.

32. Dunne 2004, p. 156.

33. Levine 2006.

34. Dana and Loewenstein 2003; the Henry J. Kaiser Family Foundation 2006.

35. Other examples of exploitation of friendship include Maasai: Merker 1910; Tongans and Zambian trader: Colson 1971; Indians: Reddy 1973; Nuer: Howell 1954.

36. Brain 1976, p. 30.

37. Maale: Donham 1985. Patron-client relationships: Campbell 1964; Wolf 1966; Boissevain 1966.

38. Cole and Teboul 2004.

39. Colvin, Vogt, and Ickes 1997; Stinson and Ickes 1992; Gesn 1995.

40. Planalp and Garvin-Doxas 1994.

41. Blushing study: Shearn et al. 1999. Using mutual knowledge in everyday life: Clark, Mills, and Corcoran 1989; Clark, Mills, and Powell 1986; Colvin, Vogt, and Ickes 1997; Planalp and Garvin-Doxas 1994.

42. Hays 1984; Mesch and Talmud 2006; Verbrugge 1979; Hays 1989.

43. Korean example cited in Choi and Choi 2001. Russian example cited in Pahl 2000, p. 155.

44. Basso 1970, p. 218; Basso 1979.

45. Much of the early stages of forming a relationship are taken up with individuals displaying the "wares" that they have to offer, as the following examples show. Trobrianders: Malinowski 1922; New Guinea highlanders: Meggitt 1965, Salisbury 1962; Azande: Baxter 1953; Kanuri: Cohen 1960; Lozi: Turner 1952; Malagasy traders: Fafchamps and Minten 1999.

46. Longevity predicting longevity: Ledbetter, Griffin, and Sparks 2007; Ikkink and van Tilburg 1999; Weisz and Wood 2005. "Moreover, they need time to grow accustomed to each other; for, as the proverb says, they cannot know each other before they have shared the traditional [peck of] salt, and they cannot accept each other or be friends until each appears lovable to the other and gains the other's confidence" (Aristotle 2002, p. 213).

47. Examples of exclusivity among the Irish: Arensberg and Kimball 1940; Tangu: Burridge 1957; Azande: Evans-Pritchard 1933; Tikopia: Firth 1936; Zuni: Roberts 1965.

48. Nyakatura 1970.

49. Parker et al. 2005; Aune and Comstock 1991.

50. Melville 1847, p. 177.

51. Owens 2003; Baxter et al. 1997.

52. This tension between focusing on intention or outcome in judging an act underlies many of the debates about the definition and existence of altruism in the human and animal world. Although biological definitions of altruism focus on the objective outcomes of behaviors (i.e., costs and benefits), definitions in the social sciences often capture the moral quality of the term by requiring some kind of altruistic intention on the part of the giver.

53. Kelley 1986.

54. Clark, Dubash, and Mills 1998; Lydon, Jamieson, and Holmes 1997.

55. Callaghan et al. 2005.

56. Kurth 1970; Moffatt 1986; Laumann 1973; Vaquera and Kao 2007; Gleason and Hohmann 2006; Vaughn et al. 2000; Gershman and Donald 1983.

57. Wieselquist 2007; Schneider et al. 2008; Wong and Bond 1999; Rubin 1985; Tagiuri, Blake, and Bruner 1953; Curry and Emerson 1970; Secord and Backman 1964; Rubin and Shenker 1978; Oswald, Clark, and Kelly 2004; Shah, Dirk, and Chervany 2006; Rosenblatt and Greenberg 1991; Deci et al. 2006; Vaughn et al. 2000.

58. Original Michigan study: Newcomb 1961. Later analysis: Kenny and Nasby 1980; Kenny and LaVoie 1982. This greater similarity may arise from two distinct processes—dropping friends with whom one has incongruent liking or converging in liking over time. Among five- to six-year-olds, old friends had higher agreement on liking than did new friends (correlation = 0.41 versus correlation = 0.11) (Dunn, Cutting, and Fisher 2002). Among college students, best friends had high congruence (positivity, correlation = 0.60; interaction, correlation = 0.61; supportiveness, correlation = 0.66; and openness, correlation = 0.58 [Oswald, Clark, and Kelly 2004]; commitment level, correlation = 0.52; trust level, correlation = 0.44 [Wieselquist 2007]).

59. Piker 1968, p. 203; Baxter et al. 1997; Walker 1995; Argyle and Henderson 1984; Hays 1985.

60. Shackelford and Buss 1996; Walker 1995.

61. D = 1.39 to 2.05 (Niiya, Ellsworth, and Yamaguchi 2006).

62. Bergstrom, Kerr, and Lachmann 2008.

63. Baxter et al. 1997; Weinstock and Bond 2000.

64. Tooby and Cosmides 1996.

65. As quoted in Smith 1954, p. 200.

66. Baumeister, Stillwell, and Wotman 1990; Way 1998; Singer and Doornenball 2006. Studies of the neuroscience of betrayal have focused on strangers interacting in economic games and specifically the influence of unreciprocated cooperation (Rilling et al. 2008).

67. Stealing a friend's partner, demanding immediate reciprocity, and failure to defend reputation: Shackelford and Buss 1996; deceit: Seiter, Bruschke, and Bai 2002, Maier and Lavrakas 1976; failure to help: Bleske and Shackelford 2001, Swingle 1966, Swingle and Gillis 1968, Shackelford and Buss 1996, Bar-Tal et al. 1977.

68. McCullough et al. 1998; McCullough 2008.

69. Perceived truthfulness: Anderson, Ansfield, and DePaulo 1999; DePaulo and Kashy 1998; Anderson, DePaulo, and Ansfield 2002. Positive intentions: Flannagan, Marsh, and Fuhrman 2005; Whitesell and Harter 1996. Although most research on positive illusions and similarity bias in close relationships has focused on romantic or marital partners (Gagne and Lydon 2004; Murray et al. 2002), some recent studies have extended this to positive illusions (Endo, Heine, and Lehman 2000) and perceived similarity (Morry 2007, 2005; Mashek, Aron, and Boncimino 2003) among friends who are not involved in a romantic or sexual relationship.

70. Bretherton 1986, p. 540.

71. Murray et al. 2008.

72. McCullough et al. 1998; Wieselquist et al. 1999; Karremans and Aarts 2007; Karremans and Van Lange 2008.

73. Meta-analysis comparing how friends and acquaintances deal with conflict: Newcomb and Bagwell 1995. Avoidance among acquaintances in conflict: Whitesell and Harter 1996; Laursen, Finkelstein, and Betts 2001; Hartup et al. 1988. Constructive resolution of problem: Vespo and Caplan 1993; Fonzi et al. 1997. Productive and positive feedback: Newcomb and Bagwell 1995; Burleson and Samter 1994.

74. Zuni: Goldman 1937; Zande: Evans-Pritchard 1933.

75. Rose 1984; Rose and Serafica 1986; Cairns 1994.

76. Sias et al. 2004.

77. Reina 1959, p. 47.

78. Rose 1984.

7. FRIENDSHIP, CULTURE, AND ECOLOGY

1. Pashtun: Lindholm 1982, p. 240; Orokaiva: Reay 1953, Schwimmer 1974. Of Orokaiva, "Although a man who is *otohu* [good] never quarrels and brawls, if he is involved in brawls in defence of a *naname* [friend] he is paradoxically judged to more *otohu* than ever" (Reay 1953, p. 117). Wolof: Gamble 1957; U.S. and European philosophers: Pakaluk 1991.

2. Trompenaars and Hampden-Turner 1998. Individual consistency: Cullen, Parboteeah, and Hoegl 2004; Stouffer and Toby 1951.

3. Eisenstadt 1974, p. 138; Parsons and Shils 1951.

4. The results of these choices were observed among a rarified set of individuals (i.e., corporate managers). Despite similar training and experience, they resolved the dilemma in very different ways depending on their cultural background. One can only imagine the differences one might observe with a more inclusive set of populations—including, for example, pastoralists, foragers, and other small-scale societies, many of which have traditionally been described at the far end of particularism.

5. Despite claims in some of these studies about cross-cultural difference, the findings rarely pass the d = 0.50 threshold used here. Perceived quality and emotional support: Berman, Murphy-Berman, and Pachauri 1988; Harrison et al. 1997, 1995; Penning and Chappell 1987; Rybak and McAndrew 2006; Schneider, Dixon, and Udvari 2007; Van Horn and Marques 2000; Verkuyten and Masson 1996; Way and Chen 2000; You and Malley-Morrison 2000. Expectations of a good friend: Cole and Bradac 1996; Gudykunst and Nishida 2001; You and Malley-Morrison 2000. Kinds of communication and influence: Cooper et al. 1993. Feelings of closeness: Li 2002. What counts as betrayal: Bradley, Flannagan, and Fuhrman 2001; Sheets and Lugar 2005b.

6. McPherson, Smith-Lovin, and Brashears 2006.

7. General social survey from 2002 (Davis and Smith 2007). In 1990, for example, the Gallup Poll found that only 3 percent of their sample reported no close friends; only 16 percent had fewer than three friends. Another recent telephone survey by Pew also found much larger numbers of core or close friends when it asked about a combination of types of contact (Boase et al. 2006).

8. Adams, Blieszner, and de Vries 2000; Fehr 1996; Oliker 1998; Wright 1978.

9. In his study of friendship in sixty cultures, Cohen found that talking and emotional support were only described as important in 25 percent of his sample societies (1961).

10. The papers did not contain sufficient detail to estimate a d-statistic (Won-Doornink 1991, 1985). In a replication of the General Social Survey in Tianjin, China, people were much less likely to talk about personal issues with family and friends and more likely to discuss such matters with coworkers than in the U.S. version of the survey (Ruan 1998).

11. Barnlund 1975; Chen 1995; French et al. 2006; French et al. 2001; Gudykunst and Nishida 1983; Gummerum and Keller 2008; Kito 2005; Li 2003; Oetzel et al. 2000; Rubin, Yang, and Porte 2000; Seki, Matsumoto, and Imahori 2002; Yum and Hara 2005. Daily activity study: Larson and Verma 1999.

12. Gao 1996, 1998; Ye 2006, p. 153.

13. Kim et al. 2006; Taylor et al. 2007.

14. Goodwin and Lee 1994; Gudykunst and Nishida 1983. Ting Jiang pointed out the possibility that the public nature of experiments could confound the results, and that very close friends in China do indeed share personal details. For example, one Chinese phrase for friends is "no words not said" (*wu hua bu shuo*; personal communication, 2008).

15. Gudykunst and Matsumoto 1996. Critique of the dimension of collectivism and individualism: Fiske 2002b.

16. Goodwin et al. 2001; Rybak and McAndrew 2006; Sheets and Lugar 2005a.

17. Adams and Plaut 2003; Berman, Murphy-Berman, and Pachauri 1988.

18. Hays 1988, p. 393. Antonio's advice to Shylock in *The Merchant of Venice* not to lend money to friends (Shakespeare 1.3).

19. In a cross-cultural survey of sixty societies, Cohen estimates the support provided by friendship is commonly economic (81 percent of his cases) and less frequently "sociopolitical and emotional" (25 percent; Cohen 1961, p. 373).

20. Castren and Lonkila 2004; Lonkila 1997.

21. Adams and Plaut 2003; Gans 1962; Gummerum and Keller 2008; Li 2003; Stack 1974.

22. Trompenaars and Hampden-Turner 1998.

23. McElreath 2004; Edgerton 1971.

24. Elbedour, Shulman, and Kedem 1997.

25. Support in school during conflicts: Grammer 1992; Strayer and Noel 1986. Schoolwork help: Rawlins 1992; Rizzo and Corsaro 1988. Thai friend: Piker 1968. Tausug: Kiefer 1972, 1968.

26. Granovetter 1995; Gulliver 1955; Heider 1969; MacDougall 1981; Shack 1963; Smith 1954.

27. Turnbull 1972, p. 104. Among the Gwembe Tonga, friendship is used both locally and more distally. With local friends one exchanges oxen and plows, implements, clothing, beer, labor, and so forth. With distant friends (those with access to the tobacco market or with herders on the plateau), one trades tobacco for goods or money or for safe refuge in times of hunger or cattle shortage (Colson 1967, 1971).

28. Another benefit provided by friends is that a third party singing one's praises can improve one's likeability (Pfeffer et al. 2006).

29. Lindenbaum 1979, p. 49.

30. Chagnon 1992, p. 159. Chagnon also describes the self-interest involved in such aid, as the hosts often demand that the guests provide some form of compensation after a lengthy stay.

31. Handy 1923, pp. 162–163.

32. Agta: Headland 1987; other examples of arranged friendships upon marriage include Bemba: Richards 1939; Khasi: Nakane 1967. Indian: Mayer 1966.

33. Ju/'hoansi: Wiessner 1982, 2002a, 1977, 1994. East African example: Gulliver 1955.

34. Burridge 1957; Shack 1963.

35. Chagnon and Bugos 1979.

36. Lonkila 1997; Salmi 2000; Shlapentokh 1989; Wierzbicka 1997.

37. Markowitz 1991, pp. 638–639.

38. Castren and Lonkila 2004; Ledeneva 1998; Manning 1995; Morton 1980.

39. Rose 1995.

40. Trompenaars and Hampden-Turner 1998.

41. Markowitz 1991.

42. Markowitz 1991, p. 645.

43. Food scarcity due to material unavailability: Espinoza 1999, Sik and Wellman 1999, Walker 1995; due to violence and lack of legal sanctions: Anderson 1969, Boissevain 1966, 1974, Fafchamps and Minten 1999, Fitch 1998, Foster 1961, Gans 1962, Geertz 1978, Granovetter 1985, Greif 2006, Kranton 1996, Stack 1974, Szanton 1972.

44. Lynd and Lynd 1956, p. 274.

45. Thus, Geertz's Moroccan merchants also maintained a social distance with their clients and suppliers, because too close, diffuse, and unconditional ties could be used "exploitatively—to delay delivery, pass off shoddy goods, pressure terms, etc." (Geertz 1978, p. 260; Khuri 1968).

46. Azande: Baxter 1953. Another example of blood brothers is the Kaguru: Beidelman 1928.

47. Long-distance trade ceases to be necessary or profitable: Gorer 1938, Grotanelli 1988, Pitt-Rivers 1968, p. 413; agriculture becomes less risky: Linton 1933; and welfare systems reduce individual uncertainty: McMichael and Manderson 2004.

48. Allan 1989; Giddens 1992, p. 58.

49. Allan 1989, pp. 56–57.

50. Working for money: Hirsch 1976, pp. 77–80; asocial pursuits such as watching TV: Putnam 2000. Such pursuits whittling away the time spent with friends: Bellah 1985; Brain 1976.

51. Anderson 2001; Coleman 1988; DeAngelis 1995; Park 1928.

52. D-statistic = 0.49 (Kasarda and Janowitz 1974; Sampson 1988).

53. South and Haynie 2004. In a recent study of the maintenance of long-distance friendships, Johnson (2001) found long-distance and geographically close friends did not vary in reported closeness, satisfaction, or expectation that the relationship would continue.

54. Financial support: d = 0.49; help with tasks: d = 0.37 (Magdol and Bessel 2003).

55. Boisjoly, Duncan, and Hofferth 1995; Briggs 1998.

56. McCollum 1990; Tilly and Brown 1967.

57. Butler, McAllister, and Kaiser 1973; Litwak 1960; Pettit 2004; Wellman et al. 1997. Nor is distance detrimental to mental health, as some have proposed (Dong et al. 2005).

58. Boisjoly, Duncan, and Hofferth 1995; Hofferth, Boisjoly, and Duncan 1999. Long-distant migrants have less contact with kin than do those who stay (Hendrix 1976), and less social support is exchanged among kin when they are geographically separated (Boisjoly, Duncan, and Hofferth 1995; Logan and Spitze 1996; Roschelle 1997).

59. Jones 1973; McCollum 1990; Wellman 1979.

60. Packard 1972, p. 149; Fischer 1992b; Jacobson 1973.

61. Serbia: Filipovic 1982, p. 28, Hammel 1968, p. 88; other examples of

using friends to "plug in" include Bena: Culwick, Culwick, and Kiwanga 1935; Copper Inuit: Jenness 1922.

62. Wolf 1966, pp. 11–12. Linton cited in Dubois 1974. Cohen puts forward a theory that the kinds of friendship observed in a society are associated with the social structure of the society. However, his definitions of social structure (maximally solidary, solidary-fissile, nonnucleated, and individual) as well as friendship (inalienable, close, casual, expedient) are so poorly operationalized that it is difficult to precisely articulate the predictions (Cohen 1961).

63. Examples of avoiding kin by making friends include Irish: Leyton 1974a, pp. 95–96; Eisenstadt 1974, p. 91; urban Uganda: Jacobson 1973; Chinese: Lang 1946.

64. Babchuk and Edwards showed that adults in Lincoln, Nebraska, who visited kin more actually visited friends more as well (1965).

65. Donne [1597] 1912, p. 180.

66. Most studies of telephone use suggest it didn't (Fischer 1992a).

67. Bargh and McKenna 2004; Gross, Juvonen, and Gable 2002.

68. Sproull and Kiesler 1986; Stoll 1995; Turkle 1996.

69. Brockman 1996, p. 279.

70. Katz and Aspden 1997; Rheingold 1994.

71. Bargh and McKenna 2004.

72. Bryant, Sanders-Jackson, and Smallwood 2006; Haythornthwaite and Wellman 2002; Katz and Rice 2002; Valkenberg and Peter 2007.

73. Parks and Floyd 1996.

74. McKenna, Green, and Gleason 2002.

75. Bargh and McKenna 2004; Bargh, McKenna, and Fitzsimons 2002; McKenna, Green, and Gleason 2002.

76. Henderson and Gilding 2004.

77. Booth and Hess 1974; Chan and Cheng 2004; Lenhart, Madden, and Hitlin 2005; Parks and Roberts 1998; Wolak, Mitchell, and Finkelhor 2002.

78. Chan and Cheng 2004. See also Parks and Roberts 1998; Ye 2007.

79. When friends and strangers are asked to split a surplus, friends are consistently more successful, regardless of the mode of communication, whereas strangers do increasingly better when more channels of communication are used (McGinn and Keros 2002; Valley, Moag, and Bazerman 1998).

80. Fischer 1992b; Hill and Dunbar 2003; Wellman and Wortley 1990.

81. Walther et al. 2008.

82. Johnson 2007.

83. Black 1904, pp. 39–40.

84. Fehr 1996; Rubin 1985; Wierzbicka 1997.

85. Kim and Yun 2007; Xie 2008.

8. PLAYING WITH FRIENDS

1. Freddie C. Velez, *Tempo*, May 29, 2008; Sibusiso Ngalwa, *Sunday Tribune*, April 4, 2004. In each of these cases, it is possible that the individual

would have done the same thing, even if the victim had been a stranger. More systematic study of such incidents could identify the relative rates at which individuals are likely to take risky steps to save friends versus strangers or acquaintances. At least one recent example suggests that friendship is a motivating factor. When asked why he saved his friend from drowning, a six-year-old from Fayetteville, Georgia, replied, "He's my friend and that's what friends do" (Kathy Jefcoats, *Atlanta Journal Constitution*, June 4, 2008).

2. Tooby and Cosmides 1996.

3. Luce and Raiffa 1957.

4. Trivers 1971, p. 38.

5. Axelrod 1984.

6. Axelrod and Dion 1988; Bendor and Swistak 1997.

7. Google Scholar search. Compare this to the number of articles and books published about other common games in game theory, including stag hunt, (2,000), the hawk-dove game (2,000), the battle of the sexes game (3,000), the coordination game (5,000), the ultimatum game (5,000), the minority game (1,000), and the public goods game (1,500; all numbers are approximate).

8. Axelrod and Hamilton 1981. If both players follow this hair trigger strategy, then neither has an incentive to defect, and they continue to cooperate until the relationship ends. In this case, it is crucial that neither player knows exactly when the relationship will end. Otherwise, each would have an incentive to defect on the last round, which would then give them an incentive to defect on the second-to-last-round, and so forth.

9. Approximately 18,000: Google Scholar search.

10. Bendor and Swistak 1997; Dugatkin 1997.

11. Mistakes: Trivers 1971, p. 38. Partner choice: Hruschka and Henrich 2006. See Henrich and Henrich 2007 for more detailed descriptions of complications to these models. Contrite forms of tit-for-tat have some model of intentions, their own intentions, but they do not model a partner's intentions.

12. It's important to point out here that tit-for-tat is only one mechanism that can maintain reciprocal altruism, even though these terms are occasionally conflated. The kinds of mechanisms underlying help among friends may be fundamentally different than tit-for-tat but nonetheless can also be mechanisms for supporting reciprocal altruism.

13. Another common criticism of the repeated prisoner's dilemma game is that opportunities to help may not occur simultaneously, they may become very imbalanced, and they may be separated by very long periods of time. It turns out that this detail does not dramatically change the predictions from the model (Boyd 1988).

14. Carmichael and MacLeod 1997; Mailath and Samuelson 2006; Neilson 1999; Watson 2002.

15. Axelrod and Wu 1995; Nowak and Sigmund 1992.

16. This becomes especially difficult if there are long intervals between possible exchanges. Trivers resolves this problem by suggesting that humans might expect "interest" to be added to a long overdue altruistic debt. However,

imbalances often arise that can last for years, and the friends are not expected to pay interest. Moreover, such calculations would have been very difficult when exchanges could occur in so many currencies.

17. Baumeister, Vohs, and Tice 2007; Loewenstein and Small 2007; Rick and Loewenstein 2008.

18. Baumeister, Vohs, and Tice 2007; Loewenstein and Small 2007. Loss of willpower when energy depleted, specifically related to altruism toward strangers: Gailliot and Baumeister 2007; Gailliot et al. 2007.

19. Some authors have proposed that positive affect is a mechanism that commits people to behaviors that yield long-term payoffs, despite the possibility for short-term gains (Fessler and Haley 2003; Fiske 2002a). Emotions may convert the costs and benefits of diverse interactions into a "common currency" to make sure one is not sacrificing too much for an exploitative partner; emotions may also provide a simple, imperfect, but good enough assessment of our history and current status with a partner (Aureli and Schaffner 2002).

20. Haselton and Nettle (2006) proposed that such biases were selected by natural or sexual selection, whereas others (Yamagishi et al. 2007) remain agnostic about the process, suggesting it could be either cultural selection or learning. In the case of friendship, the biases would be learned over the course of a relationship.

21. A similar argument about the golden rule: Wilkins and Thurner nd. Emotions as an automatic, involuntary, and rapid response that helps humans regulate, maintain, and use different social relationships, usually (though not always) for their own benefit: Frank 1988. The argument that positive feelings for a partner mediate the kind of knee-jerk altruism found among close friends is different than an argument that people do things due to what Goldschmidt has called "affect hunger": Goldschmidt 2006.

22. This use of error management theory extends the traditional approach to thinking of such biases as "built in" by natural selection. In the case of friendship, one's approach to error management shifts over the course of the relationship and at some point flips from being knee-jerk defensive to being knee-jerk supportive.

23. Tooby and Cosmides 1996.

24. Niv, Joel, and Dayan 2006. Thye, Yoon, and Lawler 2002.

25. Odling-Smee, Laland, and Feldman 2003; Flack et al. 2006.

26. Blake [1790] 1994. There is extensive evidence that most people are intrinsically motivated to cultivate friendships, and when such relationships are cultivated, they change their decision making toward maintenance. Moreover, when people lose friendships, they are motivated to build new ones (Aron et al. 2004; Baumeister and Leary 1995). Rejection by strangers increases people's tendency to behave in ways that might elicit positive social responses from novel others, such as a willingness to solicit friendships: Maner et al. 2007; to conform: Williams, Cheung, and Choi 2000; and to work harder on a collective task: Williams and Sommer 1997.

27. Bergstrom, Kerr, and Lachmann 2008. Although game theory traditionally focuses on how the rules of particular games motivate self-reinforcing behavior, it is also possible to take an engineering approach to games by asking how to change the rules of a game so that individuals will behave in a certain way.

CONCLUSION

Epigraph: Forster 1924, p. 247.

1. Falk, Fehr, and Fischbacher 2008; Greenberg and Frisch 1972; Ames, Flynn, and Weber 2004; Tsang 2006; Tesser, Gatewood, and Driver 1968.

2. Pair-bonding hypothesis: Chapais 2008; caregiver-infant hypothesis: Bowlby 1982; psychological kinship hypothesis: Bailey 1989. Tentative evidence indicates that love expressed toward family members is psychologically different from love expressed toward close friends and romantic partners. Specifically, in one of the earliest quantitative studies of closeness, people who described feeling closer to their friends also expressed greater feelings of loving and liking for their friends (correlation = 0.43) and also for romantic partners (correlation = 0.57). However, this association was attenuated among close family members (Berscheid, Snyder, and Omoto 1989; Aron, Aron, and Smollan 1992). These findings suggest that you can love kin without feeling close, and you can feel close to kin without feeling love. However, in the case of friendships (and romantic relationships), love and closeness somehow occur together.

3. Although there is much research on resource transfers and helping in human behavioral ecology, it is difficult to use them for an understanding of friendship because in most cases they rely exclusively on behavioral measurement (Gurven 2005). Such a methodological bias is a useful antidote to an overreliance on psychological measures, but it also lumps all non-kin together, since people are not asked to whom they feel close or who they identify as a close friend.

4. Dunbar and Shultze 2007; Herrmann et al. 2008.

5. Dutch, for example, does not appear to use a spatial metaphor for friendship, focusing on other qualities—*goede vriende* (good friend), *dikke maatjes* (thick mate), *beste vriende* (good friend). *Warm* is another common metaphor in interpersonal contexts (Alberts and Decsy 1990).

6. Guroglu et al. 2008; Aron, Whitfield, and Lichty 2007.

7. Rilling, King-Casas, and Sanfey 2008; Bartels and Zeki 2004, 2000; Ortigue and Bianchi-Demicheli 2008; Aron et al. 2005; Fisher, Aron, and Brown 2006; Guroglu et al. 2008; Rilling 2008.

8. Ross 1995.

9. Are people also less likely to feel gratitude for help from a friend than from a stranger? Are there other feelings and psychological states that are less sensitive to the behaviors of friends?

10. Shlapentokh, cited in Pahl 2000, p. 154.

11. Economists have developed mathematical models showing that heavy reliance on personalized exchange often makes it difficult for people to engage in market interactions with strangers (Kranton 1996).

12. Campbell describes the conflicts that arise in a Greek herding village between the altruistic ideals of friendship and the limits placed on it by family loyalties and the fear that one will be bested or exploited by an "outsider" (Campbell 1964, p. 194).

13. Friends as job finders: Korpi 2001; Eve 2002; Yakubovich and Kozina 2000; as conduits for gossip: Festinger, Schachter, and Back 1950. In early-twentieth-century Serbia, diverse ethnic groups, such as Vlachs, Serbs, and Gypsies, and Christian and Muslim religious communities were knit together through cross-ethnic god-sibling ties (Hammel 1968). In a recent study of changing international relations during the European Reformation, friendship among princes and kings was often the prime method for arranging peaceful inter-state relations, once the Catholic church's canon law began to break down (Roshchin 2006). Cross-group friendships also appear to improve the exchange of valuable information across competing groups, such as corporations and firms. In one study of the hotel industry in Sydney, Australia, managers who had friends among managers of competing hotels had higher annual hotel yields by hundreds of thousands of Australian dollars, possibly due to tacit collusion through norms against price cutting (Ingram and Roberts 2000).

14. Farrell 2001. Stark and Bainbridge 1980; della Porta 1988; Diani and McAdam 2003; Ibrahim 1980; Gould 1991. For example, 23 percent of jointly owned U.S. start-ups have at least one pair of friends as co-owners (Leider et al. 2007).

15. Song and Olshfski 2008.

16. Tenbrunsel et al. 1999. Hays and Oxley found that the most adaptive social networks for first-term university students were those which were permeable, in that they brought in new friends who were also college students rather than holding on completely to one's old high school or neighborhood friends who did not also enter college (Hays and Oxley 1986).

17. Friends as informal caregivers: National Alliance for Caregiving and AARP 2004; informal exchange: Dimaggio and Louch 1998. In Britain the cards exchanged at Christmas alone were worth 40 million pounds in 1964. Over 4 percent of consumer expenditures—1,401 million in 1968 and about a third of what is spent on housing—was spent on gifts (Davis 1972). The economist Joel Waldfogel estimated that holiday gift expenses in the United States totaled about 40 billion dollars in 1992 (Waldfogel 1993). Even if some portion of these expenditures was on friends, it would constitute a non-trivial portion of the economy.

18. Stark and Bainbridge 1980; della Porta 1988; Diani and McAdam 2003; Ibrahim 1980; Gould 1991.

19. Biggart 1989. Other companies use the same strategy to sell kitchen supplies, scrap-booking materials, party supplies, dietary supplements, rubber stamps, jewelry, cosmetics, soy candles, essential oils, wine accessories, spa

products, home decor, and children's books. In such cases, the host receives a 10 to 50 percent commission.

20. Martin and Smith 2008, p. 48.

21. Saudi Arabia, 55 percent; Israel, 72 percent; United Kingdom, 81 percent (Elbedour, Shulman, and Kedem 1997; Larson and Bradney 1988).

22. Less reactive with friends: Heinrichs et al. 2003; O'Donovan and Hughes 2008. Monetary value of increased interaction with friends: Powdthavee 2008.

23. McCorkle et al. 2008.

24. Kohlberg 1969.

25. Hopkins 2002.

26. Goldschmidt 2006; Fry 2007.

27. Loewenstein and Small 2007.

APPENDIX B

1. For a similar argument about the golden rule, see Wilkins and Thurner nd. Emotions as an automatic, involuntary, and rapid response that help humans regulate, maintain, and use different social relationships, usually (though not always) for their own benefit: Frank 1988.

References

Aberle, D. F. 1951. *The Psychosocial Analysis of a Hopi Life-History.* Berkeley: University of California Press.

Abraham, R. C. 1933. *The Tiv People.* Lagos: Government Printing Office.

Abu-Lughod, L. 1986. *Veiled Sentiments: Honor and Poetry in a Bedouin Society.* Berkeley: University of California Press.

Abu-Zahra, N. 1974. Material power, honour, friendship and etiquette of visiting. *Anthropological Quarterly* 47 (1): 120–138.

Ackerman, J. M., D. T. Kenrick, and M. Schaller. 2007. Is friendship akin to kinship? *Evolution and Human Behavior* 28:365–374.

Adams, G., and V. C. Plaut. 2003. The cultural grounding of personal relationships: Friendship in North American and West African worlds. *Personal Relationships* 10 (3): 333–348.

Adams, R. G., and G. Allan, eds. 1998. *Placing Friendship in Context.* Cambridge: Cambridge University Press.

Adams, R. G., R. Blieszner, and B. de Vries. 2000. Definitions of friendship in the third age: Age, gender, and study location effects. *Journal of Aging Studies* 14 (1): 117–133.

Adriani, N., and A. C. Kruijt. 1950. *The Bare'e-speaking Toradja of Central Celebes (the East Toradja).* Amsterdam: Noord-Hollandsche Uitgevers Maatschappij.

Afifi, W. A., and J. K. Burgoon. 1998. "We never talk about it": A comparison of cross-sex friendships and dating relationships on uncertainty and topic avoidance. *Personal Relationships* 5 (3): 255–272.

Afifi, W. A., and S. L. Faulkner. 2000. On being "just friends": The frequency of sexual activity in cross-sex friendships. *Journal of Social and Personal Relationships* 17 (2): 205–222.

Agnew, C. R., P. A. M. Van Lange, C. E. Rusbult, and C. A. Langston. 1998. Cognitive interdependence: Commitment and the mental representation of close relationships. *Journal of Personality and Social Psychology* 74 (4): 939–954.

Aguilar, M.I. 1999. Localized kin and globalized friends: Religious modernity and the "educated self" in East Africa. In *The Anthropology of Friendship*, ed. S.M. Bell and S. Coleman. Oxford: Berg.

Ahmad, A. 1950. Malay manners and etiquette. *Royal Asiatic Society* 23 (3): 43–73.

Ahrons, C.R., and L.S. Wallisch. 1987. The relationship between former spouses. In *Intimate Relationships: Development, Dynamics, and Deterioration*, ed. D. Perlman and S.W. Duck. Los Angeles: Sage.

Ainsworth, M.D.S. 1979. Infant-mother attachment. *American Psychologist* 34 (10): 932–937.

Ainsworth, M.D.S., M.C. Blehar, E. Waters, and S. Wall. 1978. *Patterns of Attachment: Assessed in the Strange Situation and at Home*. Hillsdale, NJ: Lawrence Erlbaum.

Ajzen, I., T.C. Brown, and F. Carvajal. 2004. Explaining the discrepancy between intentions and actions: The case of hypothetical bias in contingent valuation. *Personality and Social Psychology Bulletin* 30 (9): 1108–1121.

Alberts, J.R., and G.J. Decsy. 1990. Terms of endearment. *Developmental Psychobiology* 23 (7): 569–584.

Alexander, C.S., and H.J. Becker. 1978. Use of vignettes in survey research. *Public Opinion Quarterly* 42:93–104.

Alexander, R.D. 1979. Natural selection and social exchange. In *Social Exchange in Developing Relationships*, ed. R.L. Burgess and T.L. Huston. New York: Academic Press.

Allan, G. 1989. *Friendship: Developing a Sociological Perspective*. Boulder, CO: Westview Press.

Alsaker, F.D., and A. Flammer. 1999. *The Adolescent Experience: European and American Adolescents in the 1990s*. Mahwah, NJ: Lawrence Erlbaum.

Altschuler, M. 1965. The Cayapa: A study in legal behavior. PhD diss., University of Minnesota, Minneapolis.

Alvard, M.S. 2003. Kinship, lineage and an evolutionary perspective on cooperative hunting groups in Indonesia. *Human Nature* 14 (2): 129–163.

Alvarsson, J. 1988. *The Mataco of the Gran Chaco: An Ethnographic Account of Change and Continuity in Mataco Socio-economic Organization*. Uppsala, Sweden: Academiae Upsaliensis.

Ames, D.R., F.J. Flynn, and E.U. Weber. 2004. It's the thought that counts: On perceiving how helpers lend a hand. *Personality and Social Psychology Bulletin* 30 (4): 461–474.

Anderson, D.C. 2001. *Losing Friends*. London: Social Affairs Unit.

Anderson, D.E., M.E. Ansfield, and B.M. DePaulo. 1999. Love's best habit: Deception in the context of relationships. In *The Social Context of Nonverbal Behavior*, ed P. Philippot, R.S. Feldman, and E.S. Coats. Cambridge: Cambridge University Press.

Anderson, D.E., B.M. DePaulo, and M.E. Ansfield. 2002. The development of deception detection skill: A longitudinal study of same-sex friends. *Personality and Social Psychology Bulletin* 28 (4): 536–545.

Anderson, J. 1982. Social structure and the veil: The comportment and the composition of interaction in Afghanistan. *Anthropos* 77:397–420.

Anderson, J. N. 1969. Buy-and-sell and economic personalism: Foundations for Philippine entrepreneurship. *Asian Survey* 9 (9): 641–668.

Anderson, K. G., H. Kaplan, D. Lam, and J. Lancaster. 1999. Paternal care by genetic fathers and stepfathers: Reports by Xhosa high school students. *Evolution and Human Behavior* 20 (6): 433–451.

Anderson, K. G., H. Kaplan, and J. Lancaster. 1999. Paternal care by genetic fathers and stepfathers: Reports from Albuquerque men. *Evolution and Human Behavior* 20 (6): 405–431.

Anderson, M. 1978. Saami ethnoecology: Resource management in Norwegian Lapland. PhD diss., Yale University, New Haven, CT.

Anderson, R. G. 1911. *Some Tribal Customs in Their Relation to Medicine and Morals of the Nyam-Nyam and Gour People Inhabiting the Eastern Bahr-El-Ghazal.* Khartoum: Wellcome Tropical Research Laboratories at the Gordon Memorial College.

Apicella, C. L., D. R. Feinberg, and F. W. Marlowe. 2007. Voice pitch predicts reproductive success in male hunter-gatherers. *Biology Letters* 3 (6): 682–684.

Archer, W. G. 1984. *Tribal Law and Justice: A Report on the Santal.* New Delhi: Concept.

Arensberg, C., and S. Kimball. 1940. *Family and Community in Ireland.* Cambridge: Harvard University Press.

Argyle, M., and A. Furnham. 1983. Sources of satisfaction and conflict in long-term relationships. *Journal of Marriage and Family* 45 (3): 481–493.

Argyle, M., and M. Henderson. 1984. The rules of friendship. *Journal of Social and Personal Relationships* 1 (2): 211–237.

Argyle, M., M. Henderson, M. Bond, Y. Jizuka, and A. Contarello. 1986. Cross-cultural variations in relationship rules. *International Journal of Psychology* 21:287–315.

Aries, E. 1996. *Men and Women in Interaction: Reconsidering Differences.* Oxford: Oxford University Press.

Aries, E., and F. L. Johnson. 1983. Close friendship in adulthood: Conversational content between same-sex friends. *Sex Roles* 9 (12): 1183–1196.

Aristotle. 2002. *Nicomachean Ethics.* Oxford: Oxford University Press.

Aron, A., and E. N. Aron. 1991. Love and sexuality. In *Sexuality in Close Relationships,* ed. K. McKinney and S. Sprecher. Mahwah, NJ: Lawrence Erlbaum.

Aron, A., E. N. Aron, and D. Smollan. 1992. Inclusion of other in the self scale and the structure of interpersonal closeness. *Journal of Personality and Social Psychology* 63 (4): 596–612.

Aron, A., E. N. Aron, M. Tudor, and G. Nelson. 1991. Close relationships as including other in the self. *Journal of Personality and Social Psychology* 60 (2): 241–253.

Aron, A., H. Fisher, D. J. Mashek, G. Strong, H. Li, and L. L. Brown. 2005.

Reward, motivation, and emotion systems associated with early-stage intense romantic love. *Journal of Neurophysiology* 94:327–337.

Aron, A., H. E. Fisher, and G. Strong. 2006. Romantic love. In *The Cambridge Handbook of Personal Relationships,* ed. A. L. Vangelisti and D. Perlman. Cambridge: Cambridge University Press.

Aron, A., T. McLaughlin-Volpe, D. Mashek, G. Lewandowski, S. C. Wright, and E. N. Aron. 2004. Including others in self. *European Review of Social Psychology* 15:101–132.

Aron, A., E. Melinat, E. N. Aron, R. D. Vallone, and R. J. Bator. 1997. The experimental generation of interpersonal closeness: A procedure and some preliminary findings. *Personality and Social Psychology Bulletin* 23 (4): 363–377.

Aron, A., and L. Westbay. 1996. Dimensions of the prototype of love. *Journal of Personality and Social Psychology* 70 (3): 535–551.

Aron, A., S. Whitfield, and W. Lichty. 2007. Whole brain correlations: Examining similarity across conditions of overall patterns of neural activation in fMRI. In *Real Data Analysis,* ed. S. S. Sawilowsky. Charlotte, NC: Information Age Publishing.

Ashton, N. L. 1980. Exploratory investigation of perceptions of influences on best-friend relationships. *Perceptual and Motor Skills* 50:379–386.

Assam, Department of Economics and Statistics. 1963. *Report on Rural Economic Survey in United K. and J. Hills.* Shillong, India: Government Press.

Aufenanger, H. 1966. Friendship in the Highlands of New Guinea. *Anthropos* 61:305–306.

Aune, K. S., and J. Comstock. 1991. Experience and expression of jealousy: Comparison between friends and romantics. *Psychological Reports* 69:315–319.

Aureli, F., and C. M. Schaffner. 2002. Relationship assessment through emotional mediation. *Behaviour* 139 (2–3): 393–420.

Axelrod, R. 1984. *The Evolution of Cooperation.* New York: Basic Books.

Axelrod, R., and D. Dion. 1988. The further evolution of cooperation. *Science* 242:1385–1390.

Axelrod, R., and W. D. Hamilton. 1981. The evolution of cooperation. *Science* 211:1390–1396.

Axelrod, R., and J. Wu. 1995. How to cope with noise in the iterated prisoner's dilemma. *Journal of Conflict Resolution* 39 (1): 183–189.

Ayres, J. 1983. Strategies to maintain relationships: Their identification and perceived usage. *Communication Quarterly* 3 (1): 62–67.

Azar, O. H. 2007. The social norm of tipping: A review. *Journal of Applied Social Psychology* 37 (2): 380–402.

Azevedo, T. 1969. *The Colored Elite in a Brazilian City.* New Haven, CT: HRAF.

Babchuk, N., and A. P. Bates. 1963. The primary relations of middle-class couples: A study of male dominance. *American Sociological Review* 28:377–384.

Babchuk, N., and J. N. Edwards. 1965. Voluntary associations and the integration hypothesis. *Sociological Inquiry* 35 (2): 149–162.

Backman, C. W., and P. F. Secord. 1959. The effect of perceived liking in interpersonal attraction. *Human Relations* 12:379–384.

Bailey, K. G. 1989. Psychological kinship, love, and liking: Preliminary validity data. *Journal of Clinical Psychology* 45 (4): 587–594.

Baldwin, M. W., and B. Fehr. 1995. On the instability of attachment style ratings. *Personal Relationships* 2:247–261.

Baldwin, M. W., J. P. H. Keelan, B. Fehr, V. Enns, and E. Koh-Rangarajoo. 1996. Social-cognitive conceptualization of attachment working models: Availability and accessibility effects. *Journal of Personality and Social Psychology* 71:94–109.

Bancroft, J. 2005. The endocrinology of sexual arousal. *Journal of Endocrinology* 186:411–427.

Bank, B., and S. L. Hansford. 2000. Gender and friendship: Why are men's best same-sex friendships less intimate and supportive. *Personal Relationships* 7:63–78.

Barbu, S. 2003. Stability and flexibility in preschoolers' social networks: A dynamic analysis of socially directed behavior allocation. *Journal of Comparative Psychology* 117 (4): 429–439.

Bargh, J. A. 2006. What have we been priming all these years? On the development, mechanisms, and ecology of nonconscious social behavior. *European Journal of Social Psychology* 36:147–168.

Bargh, J. A., and K. Y. A. McKenna. 2004. The Internet and social life. *Annual Review of Psychology* 55:573–590.

Bargh, J. A., K. Y. A. McKenna, and G. M. Fitzsimons. 2002. Can you see the real me? Activation and expression of the "true self" on the Internet. *Journal of Social Issues* 58 (1): 33–48.

Barker, J. 1953. Memoir on the culture of the Waica. *Boletin Indigenista Venezolano* 1:433–489.

Barnard, A., and A. Good. 1984. *Research Practices in the Study of Kinship.* London: Academic Press.

Barnes, J. A. 1978. Genealogies. In *The Craft of Social Anthropology*, ed. A. L. Epstein. New Brunswick, NJ: Transaction Publishers.

Barnett, M. A., S. R. Burns, F. W. Sanborn, J. S. Bartel, and S. J. Wilds. 2004. Antisocial and prosocial teasing among children: Perceptions and individual differences. *Social Development* 13 (2): 292–310.

Barnett, W. K. 1971. An ethnographic description of Sanlei Ts'un, Taiwan, with emphasis on women's roles: Overcoming research problems caused by the presence of a great tradition. PhD diss., Michigan State University, East Lansing.

Barnlund, D. C. 1975. *Public and Private Self in Japan and the United States: Communicative Styles of Two Cultures.* Tokyo: Simul Press.

Bar-Tal, D., Y. Bar-Zohar, M. S. Greenberg, and M. Hermon. 1977. Reciprocity behavior in the relationship between donor and recipient and between harm-doer and victim. *Sociometry* 40 (3): 293–298.

Bartels, A., and S. Zeki. 2000. The neural basis of romantic love. *Neuroreport* 11 (17): 3829–3834.

———. 2004. The neural correlates of maternal and romantic love. *Neuroimage* 21:1155–1166.

Barth, R. J., and B. N. Kinder. 1988. A theoretical analysis of sex differences in same-sex friendships. *Sex Roles* 19 (5–6): 349–363.

Bartholomew, K., and L. M. Horowitz. 1991. Attachment styles among young adults: A test of a four-category model. *Journal of Personality and Social Psychology* 61 (2): 226–244.

Barton, R. F. 1919. *Ifugao Law*. Berkeley: University of California Press.

———. 1930. *The Half-way Sun: Life among the Headhunters of the Philippines*. New York: Brewer & Warren, Inc.

Basedow, H. 1925. *The Australian Aboriginal*. Adelaide: F. W. Preece and Sons.

Basso, K. H. 1970. "To give up on words": Silence in western Apache culture. *Southwestern Journal of Anthropology* 26: 213–230.

———. 1979. *Portraits of "the Whiteman": Linguistic Play and Cultural Symbols among the Western Apache*. Cambridge: Cambridge University Press.

Baumeister, R. F., and M. R. Leary. 1995. The need to belong: Desire for interpersonal attachments as a fundamental human motivation. *Psychological Bulletin* 117 (3): 497–529.

Baumeister, R. F., A. Stillwell, and S. R. Wotman. 1990. Victim and perpetrator accounts of intepersonal conflict: Autobiographical narratives about anger. *Journal of Personality and Social Psychology* 59 (5): 994–1005.

Baumeister, R. F., K. D. Vohs, and D. M. Tice. 2007. The strength model of self-control. *Current Directions in Psychological Science* 16 (6): 351–355.

Baumgartner, T., M. Heinrichs, A. Vonlanthen, U. Fischbacher, and E. Fehr. 2008. Oxytocin shapes the neural circuitry of trust and trust adaptation in humans. *Neuron* 58 (4): 639–650.

Baxter, L. A., M. Mazanec, J. Nicholson, G. Pittman, K. Smith, and L. West. 1997. Everyday loyalties and betrayals in personal relationships. *Journal of Social and Personal Relationships* 14:655–678.

Baxter, P. T. W. 1953. *The Azande and Related Peoples of the Anglo-Egyptian Sudan and Belgian Congo*. London: International African Institute.

Beaglehole, E. 1935. *Hopi of the Second Mesa*. Menasha, WI: American Anthropological Association.

Beaglehole, E., and P. Beaglehole. 1937. *Notes on Hopi Economic Life*. New Haven, CT: Yale University Press.

Befu, H. 1970. An ethnographic sketch of Old Harbor, Kodiak: An Eskimo village. *Arctic Anthropology* 6 (2): 29–42.

Behnke, R. H. 1980. *The Herders of Cyrenaica: Ecology, Economy and Kinship among the Bedouin of Eastern Libya*. Chicago: University of Illinois Press.

Beidelman, T. O. 1928. The blood-covenant and the concept of blood in Ukaguru. *Africa* 33:321–42.

Beirnatzki, W. E. 1968. Varieties of Korean lineage structure. PhD diss., St. Louis University, St. Louis.

Bell, R. 1981. *Worlds of Friendship*. Newbury Park, CA: Sage.

Bell, S. M., and S. Coleman. 1999. *The Anthropology of Friendship*. Oxford: Berg.

Bellah, R. N. 1985. *Habits of the Heart: Individualism and Commitment in American Life*. Berkeley: University of California Press.

Bendor, J., and P. Swistak. 1997. The evolutionary stability of cooperation. *American Political Science Review* 91 (2): 290–307.

Benedict, R. 1952. *Thai Culture and Behavior: An Unpublished War Time Study, Dated September 1943*. Ithaca, NY: Department of Far Eastern Studies, Cornell University.

Bennett, W. C. 1935. *The Tarahumara: An Indian Tribe of Northern Mexico*. Chicago: University of Chicago Press.

Bensa, A., and A. Goromido. 1997. The political order and corporal coercion in Kanak societies of the past (New Caledonia). *Oceania* 68 (2): 84–106.

Bergstrom, C. T., B. Kerr, and M. Lachmann. 2008. Building trust by wasting time. In *Moral Markets*, ed. P. J. Zak. Princeton, NJ: Princeton University Press.

Berman, J. J., V. Murphy-Berman, and A. Pachauri. 1988. Sex differences in friendship patterns in India and in the United States. *Basic and Applied Social Psychology* 9 (1): 61–71.

Bernard, J. 1981. *The Female World*. New York: Simon and Schuster.

Bernardi, B. 1955. *The Age System of the Masai*. New Haven, CT: HRAF.

Berndt, T. J. 1982. The features and effects of friendship in early adolescence. *Child Development* 53 (6): 1447–1460.

———. 1985. Effects of friendship on prosocial intentions and behavior. *Child Development* 52 (2): 636–643.

———. 2004. Children's friendships: Shifts over a half-century in perspectives on their development and their effects. *Merrill-Palmer Quarterly* 50 (3): 206–233.

Berndt, T. J., and S. J. Hoyle. 1985. Stability and change in childhood and adolescent friendships. *Developmental Psychology* 21 (6): 1007–1015.

Berndt, T. J., and R. C. Savin-Williams. 1993. Peer relations and friendships. In *Handbook of Clinical Research and Practice with Adolescents*, ed. P. H. Tolan and B. J. Coler. New York: Wiley.

Berscheid, E. 1985. Interpersonal attraction. In *The Handbook of Social Psychology*, ed. G. Lindzey and E. Aronson. New York: Random House.

Berscheid, E., and H. Ammazzalorso. 2001. Emotional experience in close relationships. In *Blackwell Handbook of Social Psychology*, ed. G. J. O. Fletcher and M. S. Clark. London: Blackwell.

Berscheid, E., M. Snyder, and A. M. Omoto. 1989. The relationships closeness inventory: Assessing the closeness of interpersonal relationships. *Journal of Personality and Social Psychology* 57 (5): 792–807.

Berscheid, E., and E. Walster. 1978. *Interpersonal Attraction*. Reading, MA: Addison-Wesley.

Berte, N. A. 1988. K'ekchi' horticultural labor exchange: Productive and reproductive implications. In *Human Reproductive Behavior: A Darwinian Perspective*, ed. L. L. Betzig, M. Borgerhoff Mulder, and P. W. Turke. Cambridge: Cambridge University Press.

Bertrand, M., and S. Mullainathan. 2001. Do people mean what they say? Implications for subjective survey data. *American Economic Review* 91 (2): 67–72.

Bettelheim, B. 1969. *The Children of the Dream*. New York: MacMillan Co.

Betzig, L. L., and P. W. Turke. 1986. Food sharing on Ifaluk. *Current Anthropology* 27 (4): 397–400.

Beyer, H. O. 1911. An Ifugao burial ceremony. *Philippine Journal of Science* 6:227–252.

Biella, P., N. A. Chagnon, and G. Seaman. 1997. *Yanomamo Interactive*. New York: Harcourt Brace & Company.

Bigelow, B. J. 1977. Children's friendship expectations: A cognitive-developmental study. *Child Development* 48 (1): 246–253.

Bigelow, B. J., and J. J. La Gaipa. 1975. Children's written descriptions of friendships: A multidimensional analysis. *Developmental Psychology* 11 (6): 857–858.

Bigelow, B. J., G. Tesson, and J. H. Lewko. 1996. *Learning the Rules: The Anatomy of Children's Relationships*. New York: Guilford Press.

Biggart, N. W. 1989. *Charismatic Capitalism*. Chicago: University of Chicago Press.

Bilby, K. M. 1990. The remaking of the Aluku: Culture, politics and Maroon ethnicity in French South America. PhD diss., Johns Hopkins University, Baltimore.

Bilchitz, D., and K. Hofmeyr. 2006. Parliamentary submission, civil union bill. September 29.

Bippus, A. M., and E. Rollin. 2003. Attachment style differences in relational maintenance and conflict behaviors: Friends' perceptions. *Communication Reports* 16:113–124.

Birch, L. L., and J. Billman. 1986. Preschool children's food sharing with friends and acquaintances. *Child Development* 57 (2): 387–395.

Bishop, C. A. 1974. *The Northern Ojibwa and the Fur Trade: An Historical and Ecological Study*. Toronto: Holt, Rinehart and Winston of Canada.

Bishop, I. 1898. *Korea and Her Neighbors*. New York: Fleming H. Revell.

Bisson, M. A., and T. R. Levine. 2009. Negotiating a friends with benefits relationship. *Archives of Sexual Behavior* 38 (1): 66–72.

Black, H. 1904. *Friendship*. London: Hodder & Stoughton.

Blake, W. [1790] 1994. *The Marriage of Heaven and Hell*. Mineola, NY: Courier Dover Publications.

Bleske, A. L., and D. M. Buss. 2000. Can men and women be just friends? *Personal Relationships* 7 (2): 131–151.

Bleske, A. L., and T. K. Shackelford. 2001. Poaching, promiscuity, and deceit: Combatting mating rivalry in same-sex friendships. *Personal Relationships* 8 (4): 407–424.

Bloch, M. 1973. The long-term and the short-term: The economic and political significance of the morality of kinship. In *The Character of Kinship*, ed. J. Goody and M. Fortes. Cambridge: Cambridge University Press.

Bo, P. D. 2005. Cooperation under the shadow of the future: Experimental evidence from infinitely repeated games. *American Economic Review* 95 (5): 1591–1604.

Board of Governors of the Federal Reserve System. 2008. *Survey of Consumer Finances*. Washington, D.C.: Board of Governors of the Federal Reserve System.

Boase, J., J. B. Horrigan, B. Wellman, and L. Rainie. 2006. *The Strength of Internet Ties*. Washington, D.C.: Pew Internet and American Life Project.

Boehm, C. 1983. *Montenegrin Social Organization and Values: Political Ethnography of a Refuge Area Tribal Adaptation*. New York: AMS Press.

Bogle, K. A. 2008. *Hooking Up: Sex, Dating, and Relationships on Campus*. New York: New York University.

Bogoraz-Tan, V. G. 1904–1909. *The Chukchee*. Leiden: E. J. Brill.

Bohannan, P. 1954. Circumcision among the Tiv. *Man* 54:2–6.

———. 1957. *Tiv Farm and Settlement*. London: Her Majesty's Stationery Office.

Bohannan, P., and L. Bohannan. 1968. *Tiv Economy*. Evanston, IL: Northwestern University Press.

Boisjoly, J., G. J. Duncan, and S. Hofferth. 1995. Access to social capital. *Journal of Family Issues* 16 (5): 609–631.

Boissevain, J. 1966. Patronage in Sicily. *Man* 1 (1): 18–33.

———. 1974. *Friends of Friends: Networks, Manipulators and Coalitions*. Oxford: Blackwell.

Bollig, L. 1927. *The Inhabitants of the Truk Islands: Religion, Life and a Short Grammar of a Micronesian People*. Munster: Aschendorff.

Booth, A., and E. Hess. 1974. Cross-sex friendships. *Journal of Marriage and Family* 36 (1): 38–46.

Borgerhoff Mulder, M. 2007. Hamilton's rule and kin competition: The Kipsigis case. *Evolution and Human Behavior* 28:299–312.

Borges, D. E. 1992. *The Family in Bahia, Brazil, 1870–1945*. Stanford, CA: Stanford University Press.

Bosson, J. K., A. B. Johnson, K. Niederhoffer, and W. B. Swann. 2006. Interpersonal chemistry through negativity: Bonding by sharing negative attitudes about others. *Personal Relationships* 13 (2): 135–150.

Boster, F. J., J. I. Rodriguez, M. G. Cruz, and L. Marshall. 1995. The relative effectiveness of a direct request message and a pregiving message on friends and strangers. *Communication Research* 22 (4): 475–484.

Boswell, J. 1994. *Same-Sex Unions in Premodern Europe*. New York: Villard.

Bott, E. 1971. *Family and Social Network: Roles, Norms and External Relationships in Ordinary Urban Families*. New York: Free Press.

Bouquet, M. 1996. Family trees and their affinities: The visual imperative of the genealogical diagram. *Journal of the Royal Anthropological Institute* 2 (1): 43–66.

Bourdieu, P. 1972. Les strategies matrimoniales dans le systeme de reproduction. *Annales. Economies, Societes, Civilisations* 4–5:1105–1127.

Bowlby, J. 1958. The nature of the child's tie to his mother. *International Journal of Psycho-analysis* 39:350–373.

———. 1982. *Attachment*. 2d ed. New York: Basic Books.

Bowles, S., and D. Posel. 2005. Genetic relatedness predicts South African migrant workers' remittances to their families. *Nature* 434 (7031): 380–383.

Boyd, D. 2006. Friends, friendsters, top 8: Writing community into being on social network sites. *First Monday* 11:12.

Boyd, R. 1988. Is the repeated prisoner's dilemma a good model of reciprocal altruism? *Ethology and Sociobiology* 9 (2–4): 211–221.

Bradfield, M. 1973. *A Natural History of Associations: A Study in the Meaning of Community*. New York: International Universities Press.

Bradley, L. A., D. Flannagan, and R. Fuhrman. 2001. Judgment biases and characteristics of friendships of Mexican American and Anglo-American girls and boys. *Journal of Early Adolescence* 21 (4): 405–424.

Brain, R. 1976. *Friends and Lovers*. New York: Basic Books.

Branas-Garza, P., P. M. Duran, and M. P. Espinosa. 2005. Do experimental subjects favor their friends? Papers 05/14, University of Granada.

Brand, S., M. Luethi, A. von Planta, M. Hatzinger, and E. Holsboer-Trachsler. 2007. Romantic love, hypomania, and sleep patterns in adolescents. *Journal of Adolescent Health* 41 (1): 69–76.

Brandes, S. 1985. Women of southern Spain. *Anthropology* 9 (1–2): 111–128.

Brandt, V. S. R. 1971. *A Korean Village between Farm and Sea*. Cambridge: Harvard University Press.

Bray, A. 2000. Friendship, the family and liturgy: A rite for blessing friendship in traditional Christianity. *Theology and Sexuality* 7:15–33.

Bretherton, I. 1982. The origins of attachment theory: John Bowlby and Mary Ainsworth. *Developmental Psychology* 28 (5): 759–775.

———. 1986. Learning to talk about emotions: A functionalist perspective. *Child Development* 57:529–548.

Briggs, X. 1998. Brown kids in white suburbs: Housing mobility and the many faces of social capital. *Housing Policy Debate* 9 (1): 177–221.

Brockman, J. 1996. *Digerati: Encounters with the Cyber Elite*. San Francisco: Hardwired.

Broude, G. J. 1983. Male-female relationships in cross-cultural perspective: A study of sex and intimacy. *Cross-cultural Research* 18 (2): 154–181.

———. 2003. Husband-wife interactions and aloofness. In *Encyclopedia of Sex and Gender*, ed. C. R. Ember and M. Ember. New York: Springer US.

Brown, D. 1991. *Human Universals*. New York: McGraw-Hill.

Brown, S., and R.M. Brown. 2006. Selective investment theory: Recasting the functional significance of close relationships. *Psychological Inquiry* 17 (1): 1–29.

Browne, T. [1643] 1955. *Religio Medico*. Cambridge: Cambridge University Press.

Bryant, J.A., A. Sanders-Jackson, and A.M.K. Smallwood. 2006. IMing, text messaging, and adolescent social networks. *Journal of Computer Mediated Communication* 11:577–592.

Bshary, R., and R. Noe. 2003. Biological markets: The ubiquitous influence of partner choice on the dynamics of cleaner fish–client reef fish interactions. In *Genetic and Cultural Evolution of Cooperation*, ed. P Hammerstein. Cambridge: MIT Press.

Buechler, H.C. 1971. *The Bolivian Aymara*. New York: Holt, Rinehart and Winston.

Bugental, D.B. 2000. Acquisition of the algorithms of social life: A domain-based approach. *Psychological Bulletin* 126 (2): 187–219.

Buhrmester, D., J. Goldfarb, and D. Cantrell. 1992. Self-presentation when sharing with friends and nonfriends. *Journal of Early Adolescence* 12 (1): 61–79.

Bunnag, J. 1973. *Buddhist Monk, Buddhist Layman: A Study of Urban Monastic Organization in Central Thailand*. Cambridge: Cambridge University Press.

Burger, J.M., A.M. Ehrlichman, N.C. Raymond, J.M. Ishikawa, and J. Sandoval. 2006. Reciprocal favor exchange and compliance. *Social Influence* 1 (3): 169–184.

Burger, J.M., N. Messian, D. Patel, A. del Prado, and C. Anderson. 2004. What a coincidence! The effects of incidental similarity on compliance. *Personality and Social Psychology Bulletin* 30 (1): 35–43.

Burger, J.M., J. Sanchez, J.E. Imberi, and L.R. Grande. 2009. The norm of reciprocity as an internalized social norm: Returning favors even when no one finds out. *Social Influence* 4 (1): 169–184.

Burger, J.M., S. Soroka, K. Gonzago, E. Murphy, and E. Somervall. 2001. The effect of fleeting attraction on compliance to requests. *Personality and Social Psychology Bulletin* 27 (12): 1578–1586.

Burgoon, J.K. 1991. Relational message interpretations of touch, conversational distance, and posture. *Journal of Nonverbal Behavior* 15 (4): 233–259.

Burleson, B.R. 1997. A different voice on different cultures: Illusion and reality in the study of sex differences in personal relationships. *Personal Relationships* 4 (3): 229–241.

———. 2003. The experience of emotional support: What the study of cultural and gender differences can tell us about close relationships, emotion, and interpersonal communication. *Personal Relationships* 10 (1): 1–23.

Burleson, B.R., and W. Samter. 1994. A social skills approach to relationships maintenance: How individual differences in communication affect the achievement of relationship functions. In *Communication and Relational Maintenance*, ed. D.J. Canary and L. Stafford. San Diego, CA: Academic Press.

Burling, R. 1963. *Rengsanggri: Family and Kinship in a Garo Village*. Philadelphia: University of Pennsylvania Press.

Burne, C.S. 1902. How to annul "blood-brotherhood." *Folklore* 13 (4): 430.

Burnstein, E., C. Crandall, and S. Kitayama. 1994. Some neo-Darwinian decision rules for altruism: Weighing cues for inclusive fitness as a function of the biological importance of the decision. *Journal of Personality and Social Psychology* 67 (5): 773–789.

Burridge, K.O.L. 1957. Friendship in Tangu. *Oceania* 27:177–189.

Bush, D., and A. Harbage. 1961. *Pelican Shakespeare*. Baltimore: Pelican Books.

Buss, D.M. 1989. Sex differences in human mate preferences: Evolutionary hypotheses tested in 37 cultures. *Behavioral and Brain Sciences* 12 (1): 1–49.

Buster, J.E., S.A. Kingsberg, O. Aguirre, C. Brown, J. Breaux, A. Buch, C. Rodenberg, K. Wekselman, and P. Casson. 2005. Testosterone patch for low sexual desire in surgically menopausal women: A randomized trial. *Obstetrics and Gynecology* 105 (5): 944–952.

Butler, E.W., R.J. McAllister, and E.J. Kaiser. 1973. The effects of voluntary and involuntary residential mobility on females and males. *Journal of Marriage and Family* 35:219–227.

Buunk, B.P., and K.S. Prins. 1998. Loneliness, exchange orientation, and reciprocity in friendships. *Personal Relationships* 5:1–14.

Byrne, D., and R. Rhamey. 1965. Magnitude of positive and negative reinforcements as a determination of attraction. *Journal of Personality and Social Psychology* 2:884–889.

Cairns, R.B. 1994. *Lifelines and Risks: Pathways of Youth in Our Time*. Cambridge: Cambridge University Press.

Cairns, R.B., M.C. Leung, and B.D. Cairns. 1995. Social networks over time and space in adolescence. In *Pathways through Adolescence*, ed. L.J. Crockett and A.C. Crouter. Mahwah, NJ: Lawrence Erlbaum.

Caldwell, M.A., and L.A. Peplau. 1982. Sex differences in same-sex friendship. *Sex Roles* 8 (7): 721–732.

Callaghan, T., P. Rochat, A. Lillard, M.L. Claux, H. Odden, S. Hakura, S. Tapauya, and S. Singh. 2005. Synchrony in the onset of mental-state reasoning: Evidence from five cultures. *Psychological Science* 16 (5): 378–384.

Calonne-Beaufaict, A. 1921. *Azande: Introduction a une ethnographie generale des bassins de l'Ubangi-Uele et d l'Aruwimi*. Bruxelles: M. Lamertin.

Camerer, C.F. 1988. Gifts as economic signals and social symbols. *American Journal of Sociology* 94:S180-S214.

———. 2003. *Behavioral Game Theory: Experiments in Strategic Interaction*. Princeton, NJ: Princeton University Press.

Campbell, J.K. 1964. *Honour, Family and Patronage: A Study of Institutions and Moral Values in a Greek Mountain Community.* Oxford: Clarendon Press.

Campbell, W.K., C. Sedikides, G.D. Reeder, and A.J. Elliott. 2000. Among friends? An examination of friendship and the self-serving bias. *British Journal of Social Psychology* 39:229–239.

Canary, D.J., L. Stafford, K.S. Hause, and L.A. Wallace. 1993. An inductive analysis of relational maintenance strategies: Comparisons among lovers, relatives, friends, and others. *Communication Research Reports* 10 (1): 5–14.

Caplow, T. 1984. Rule enforcement without visible means: Christmas gift giving in Middletown. *American Journal of Sociology* 89 (6): 1306–1323.

Capra, C.M. 2004. Mood-driven behavior in strategic interactions. *American Economic Review* 94 (2): 367–372.

Carmichael, H.L., and W.B. MacLeod. 1997. Gift giving and the evolution of cooperation. *International Economic Review* 38 (3): 485–509.

Carnegie, D. 1937. *How to Win Friends and Influence People.* New York: Simon and Schuster.

Carroll, V. 1970. Adoption on Nukuoro. In *Adoption in Eastern Oceania*, ed. V. Carroll. Honolulu: University of Hawaii Press.

Carter, C.S., and E.B. Keverne. 2002. The neurobiology of social affiliation and pair bonding. In *Hormones, Brain and Behavior*, ed. D. Pfaff. New York: Academic Press.

Casati, G. 1891. *Ten Years in Equatoria and the Return with Emin Pasha.* London: Frederick Warne and Company.

Cashdan, E.A. 1985. Coping with risk: Reciprocity among the Basarwa of Northern Botswana. *Man* 20 (3): 454–474.

Cassidy, J., and P.R. Shaver, eds. 1999. *Handbook of Attachment.* New York: Guilford Press.

Castren, A.-M., and M. Lonkila. 2004. Friendship in Finland and Russia from a micro perspective. In *Between Sociology and History: Essays on Micro-history, Collective Action, and Nation-Building*, ed. A.-M. Castren, M. Lonkila, and M. Peltonen. Helsinki: SKS/Finnish Literature Society.

Castro, M.A.C. 1974. Reactions to receiving aid as a function of cost to donor and opportunity to aid. *Journal of Applied Social Psychology* 4 (3): 194–209.

Chagnon, N.A. 1967. Yanomamo warfare, social organization and marriage alliances. PhD diss., University of Michigan, Ann Arbor.

———. 1992. *Yanomamo.* New York: Harcourt Brace Chicago Publishers.

———. 2000. Manipulating kinship rules: A form of male Yanomamo reproductive competition. In *Adaptation and Human Behavior*, ed. L. Cronk, N.A. Chagnon and W. Irons. New York: Aldine Transactions.

Chagnon, N.A., and P.E. Bugos. 1979. Kin selection and conflict: An analysis of a Yanomamo ax fight. In *Evolutionary Biology and Human Social Behavior: An Anthropological Perspective*, ed. N.A. Chagnon and W. Irons. North Scituate, MA: Duxbury Press.

Champion, J. R. 1963. A study in culture persistence: The Tarahumara of northwestern Mexico. PhD diss., Columbia University, New York.

Chan, D. K. S., and G. H. L. Cheng. 2004. A comparison of offline and online friendship qualities at different stages of relationship development. *Journal of Social and Personal Relationships* 21 (3): 305–320.

Chang, H., and G. R. Holt. 1991. More than relationship: Chinese interaction and the principle of kuan-hsi. *Communication Quarterly* 39 (3): 251–271.

Chapais, B. 2008. *Primeval Kinship: How Pair-Bonding Gave Birth to Human Society.* Cambridge: Harvard University Press.

Cheah, C. S. L., and V. Chirkov. 2008. Parents' personal and cultural beliefs regarding young children: A cross-cultural study of aboriginal and Euro-Canadian mothers. *Journal of Cross-cultural Psychology* 39 (4): 402–423.

Chen, G. M. 1995. Differences in self-disclosure patterns among Americans versus Chinese. *Journal of Cross-cultural Psychology* 26 (1): 84.

Cherryh, C. J. 1994. *Foreigner.* New York: DAW.

Chewings, C. 1936. *Back in the Stone Age: The Natives of Central Australia.* Sydney, Australia: Angus & Robertson.

Childs, G. M. 1949. *Umbundu Kinship and Character: Being a Description of the Ovimbundu of Angola.* London: Oxford University Press.

Choi, S. C., and S. H. Choi. 2001. Cheong: The socioemotional grammar of Koreans. *International Journal of Group Tensions* 30 (1): 69–80.

Choleris, E., J. Gustafsson, K. S. Korach, L. J. Muglia, D. W. Pfaff, and S. Ogawa. 2003. An estrogen-dependent four-gene micronet regulating social recognition: A study with oxytocin and estrogen receptor-alpha and -beta knockout mice. *Proceedings of the National Academy of Sciences* 100 (10): 6192–6197.

Chun, K.-S. 1984. *Reciprocity and Korean Society: An Ethnography of Hasami.* Seoul: Seoul National University Press.

Cialdini, R. B. 1998. *Influence: The Psychology of Persuasion.* New York: Collins.

Cialdini, R. B., S. L. Brown, B. P. Lewis, C. Luce, and S. L. Neuberg. 1997. Reinterpreting the empathy-altruism relationship: One into one equals oneness. *Journal of Personality and Social Psychology* 73 (3): 481–494.

Cimbalo, R. S., V. Faling, and P. Mousaw. 1976. The course of love: A cross-sectional design. *Psychological Reports* 38:1292–1294.

Cipriani, L. 1966. *The Andaman Islanders.* New York: F. A. Praeger.

Claes, M. 1994. Friendship characteristics of adolescents referred for psychiatric treatment. *Journal of Adolescent Research* 9 (2): 180–192.

———. 1998. Adolescents' closeness with parents, siblings, and friends in three countries: Canada, Belgium and Italy. *Journal of Youth and Adolescence* 27 (2): 165–184.

Clark, M. S. 1981. Noncomparability of benefits given and received: A cue to the existence of friendship. *Social Psychology Quarterly* 44 (4): 375–381.

———. 1984. Record keeping in two types of relationship. *Journal of Personality and Social Psychology* 47 (3): 549–557.

Clark, M. S., P. Dubash, and J. Mills. 1998. Interest in another's consideration of one's needs in communal and exchange relationships. *Journal of Experimental Social Psychology* 34 (3): 246–264.

Clark, M. S., and J. Mills. 1979. Interpersonal attraction in exchange and communal relationships. *Journal of Personality and Social Psychology* 37 (1): 12–24.

Clark, M. S., J. Mills, and D. M. Corcoran. 1989. Keeping track of needs and inputs of friends and strangers. *Personality and Social Psychology Bulletin* 15 (4): 533–542.

Clark, M. S., J. Mills, and M. C. Powell. 1986. Keeping track of needs in communal and exchange relationships. *Journal of Personality and Social Psychology* 51 (2): 333–338.

Clemmer, R. O. 1995. *Roads in the Sky: The Hopi Indians in a Century of Change*. Boulder, CO: Westview Press.

Cline, W. B. 1938. *The Sinkaietk or Southern Okanagon of Washington.* Menasha, WI: George Banta Publishing Company.

Coan, J. A. 2009. Toward a neuroscience of attachment. In *Handbook of Attachment: Theory, Research and Applications*, ed. J. Cassidy and P. R. Shaver. New York: Guilford Publications.

Coats, S., E. R. Smith, H. M. Claypool, and M. J. Banner. 2000. Overlapping mental representations of self and in-group: Reaction time evidence and its relationship with explicit measures of group identification. *Journal of Experimental Social Psychology* 36 (3): 304–315.

Cohen, A. 1969. *Custom and Politics in Urban Africa: A Study of Hausa Migrants in Yoruba Towns*. Berkeley: University of California Press.

Cohen, R. 1960. The structure of Kanuri society. PhD diss., University of Wisconsin, Madison.

———. 1967. *The Kanuri of Bornu*. New York: Holt, Rinehart and Winston.

Cohen, Y. 1961. Patterns of friendship. In *Social Structure and Personality: A Casebook*, ed. Y. Cohen. New York: Holt, Rinehart and Winston.

Cole, D. P. 2003. Where have the Bedouin gone? *Anthropological Quarterly* 76 (2): 235–268.

Cole, J. T. 1969. The human soul in the Aymara culture of Pumasara: An ethnographic study in the light of George Herbert Mead and Martin Buber. PhD diss., University of Pennsylvania, Philadelphia.

Cole, T., and J. J. Bradac. 1996. A lay theory of relational satisfaction with best friends. *Journal of Social and Personal Relationships* 13 (1): 57–83.

Cole, T., and J. C. B. Teboul. 2004. Non-zero-sum collaboration, reciprocity, and the preference for similarity: Developing an adaptive model of close relational functioning. *Personal Relationships* 11 (2): 135–160.

Coleman, J. D. B. 1976. Language shift in a bilingual Hebridean crofting community. PhD diss., University of Massachusetts.

Coleman, J. S. 1988. Social capital in the creation of human capital. *American Journal of Sociology* 3:95–120.

Collins, N.L., and L.C. Miller. 1994. Self-disclosure and liking: A meta-analytic review. *Psychological Bulletin* 116 (3): 457–475.

Colson, E.F. 1967. *Marriage and the Family among the Plateau Tonga.* Manchester: Manchester University Press.

———. 1971. *The Social Consequences of Resettlement: The Impact of the Kariba Resettlement upon the Gwembe Tonga.* Manchester: Manchester University Press.

Colvin, C.R., D. Vogt, and W. Ickes. 1997. Why do friends understand each other better than strangers do? In *Empathic Accuracy,* ed. W. Ickes. New York: Guilford Publications.

Condon, R.G. 1987. *Inuit Youth: Growth and Change in the Canadian Arctic.* New Brunswick, NJ: Rutgers University Press.

Confucius. 1998. *The Original Analects: Sayings of Confucius and His Successors, Translations from the Asian Classics.* Ed. E.B. Brooks and A.T. Brooks. New York: Columbia University Press.

Connell, J.P., and H.H. Goldsmith. 1982. A structural modeling approach to the study of attachment and strange situation behaviors. In *The Development of Attachment and Affiliative Systems,* ed. R. Emde and R. Harmon. New York: Plenum Press.

Coontz, S. 2000. *The Way We Never Were.* New York: Basic Books.

Cooper, C.R., H. Baker, D. Polichar, and M. Welsh. 1993. Values and communication of Chinese, Filipino, European, Mexican, and Vietnamese American adolescents with their families and friends. *New Directions for Child Development* 62:73.

Cooper, J.M. 1917. *Analytical and Critical Bibliography of Tierra del Fuego and Adjacent Territory.* Washington, D.C.: Government Printing Office.

Cords, M. 1997. Friendships, alliances, reciprocity and repair. In *Machiavellian Intelligence 2,* ed. A. Whiten and R. Byrne. Cambridge: Cambridge University Press.

Cornell, J.B., and R.J. Smith. 1956. *Two Japanese Villages: Matsunagi, a Japanese Mountain Community.* Ann Arbor: University of Michigan Press.

Corsaro, W.A. 1981. Friendship in the nursery school: Social organization in a peer environment. In *The Development of Children's Friendships,* ed. S. Asher and J. Gottman. New York: Cambridge University Press.

———. 2003. *We're Friends, Right? Inside Kids' Cultures.* Washington, D.C.: Joseph Henry Press.

Corsaro, W.A., and T.A. Rizzo. 1988. "Discussione" and friendship: Socialization processes in the peer culture of Italian nursery school. *American Sociological Review* 53 (6): 879–894.

Coser, L.A. 1974. *Greedy Institutions: Patterns of Undivided Commitment.* New York: Free Press.

Cox, B.A. 1969. *Law and Conflict Management among the Hopi.* Berkeley: University of California Press.

Craig, R. 1969. Marriage among the Telefolmin. In *Pigs, Pearlshells, and*

Women: Marriage in the New Guinea Highlands, ed. V. Glasse. Englewood Cliffs, NJ: Prentice Hall.

Crawford, M. 1977. What is a friend? *New Society* 42:116–117.

Crick, N. R., and D. A. Nelson. 2002. Relational physical victimization within friendships: Nobody told me there'd be friends like these. *Journal of Abnormal Child Psychology* 30 (6): 599–607.

Cronk, L. 2007. The influence of cultural framing on play in the trust game: A Maasai example. *Evolution and Human Behavior* 28 (5): 352–358.

Cronk, L., and H. Wasielewski. 2008. An unfamiliar social norm rapidly produces framing effects in an economic game. *Journal of Evolutionary Psychology* 6 (4): 283–308.

Csikszentmihalyi, M., and R. Larson. 1984. *Being Adolescent: Conflict and Growth in the Teenage Years*. New York: Basic Books.

Cullen, J. B., K. P. Parboteeah, and M. Hoegl. 2004. Cross-national differences in managers' willingness to justify ethically suspect behaviors: A test of institutional anomie theory. *Academy of Management Journal* 47 (3): 411–421.

Culshaw, W. J. 1949. *Tribal Heritage: A Study of the Santals*. London: Lutterworth Press.

Culwick, A. T., G. M. Culwick, and T. Kiwanga. 1935. *Ubena of the Rivers*. London: G. Allen and Unwin.

Curry, T. J., and R. M. Emerson. 1970. Balance theory: A theory of interpersonal attraction. *Sociometry* 33 (2): 216–238.

Czekanowski, J. 1924. *Forschungen im Nil-Kongo-Zwischengebiet*. Leipzig: Klinkhardt & Biermann.

D'Abbs, P. 1991. *Who Helps? Support Networks and Social Policy in Australia*. Melbourne: Australian Institute of Family Studies.

Daly, M., and M. Wilson. 1988. *Homicide: Foundations of Human Behavior*. New York: Aldine Transaction.

———. 1996. Violence against stepchildren. *Current Directions in Psychological Science* 5 (3): 77–81.

———. 1999. *The Truth about Cinderella*. New Haven, CT: Yale University Press.

Damas, D. 1972. *The Copper Eskimo*. New York: Holt, Rinehart and Winston.

Damon, F. H. 1980. The Kula and generalised exchange: Considering some unconsidered aspects of the elementary structures of kinship. *Man* 15 (2): 267–292.

Dana, J., and G. F. Loewenstein. 2003. A social science perspective on gifts to physicians from industry. *Journal of the American Medical Association* 290:252–255.

Dante. 1892. *The Divine Comedy*. Translated by C. E. Norton. Boston: Houghton Mifflin and Co.

Darwin, C. 1871. *Descent of Man*. London: John Murray.

Darwin, C., and F. Darwin. 1887. *The Life and Letters of Charles Darwin:*

Including an Autobiographical Chapter. 2 vols. New York: D. Appleton and Company.

Davis, J. 1972. Gifts and the UK economy. *Man* 7:408–429.

———. 1988. *Libyan Politics: Tribe and Revolution; An Account of the Zuwaya and their Government.* Berkeley: University of California Press.

Davis, J. A., and T. W. Smith. 2007. *General Social Surveys, 1972–2006* [machine-readable data file]. Chicago: National Opinion Research Center.

Davis, K. E., and M. J. Todd. 1982. Friendship and love relationships. *Advances in Descriptive Psychology* 2:79–112.

———. 1985. Assessing friendship: Prototypes, paradigm cases and relationship description. In *Understanding Personal Relationships: An Interdisciplinary Approach.* Beverly Hills, CA: Sage.

De Laguna, F. 1972. *Under Mount Saint Elias: The History and Culture of the Yakutat Tlingit.* Washington, D.C.: Smithsonian Institution Press.

de Waal, F., and S. F. Brosnan. 2006. Simple and complex reciprocity in primates. In *Cooperation in Humans and Primates,* ed. P. M. Kappeler and C. van Schaik. Berlin: Springer.

De Young, J. E. 1955. *Village Life in Modern Thailand.* Berkeley: University of California Press.

Deacon, A. B. 1934. *Malekula: A Vanishing People in the New Hebrides.* London: Routledge.

DeAngelis, T. 1995. A nation of hermits: The loss of community. *APA Monitor* 26 (1):46.

DeBruine, L. M. 2002. Facial resemblance enhances trust. *Proceedings of the Royal Society B* 269:1307–1312.

Deci, E. L., J. G. La Guardia, A. C. Moller, M. J. Scheiner, and R. M. Ryan. 2006. On the benefits of giving as well as receiving autonomy support: Close friendships. *Personality and Social Psychology Bulletin* 32 (3): 313–327.

DeGlopper, D. R. 1974. City of the sands: Social structure in a nineteenth-century Chinese city. PhD diss., Cornell University, Ithaca, NY.

della Porta, D. 1988. Recruitment processes in clandestine political organizations: Italian leftwing terrorism. In *From Structure to Action,* ed. S. Tarrow, B. Klandermans, and H. Kriesi. New York: JAI Press.

Denich, B. 1977. Women, work, and power in modern Yugoslavia. In *Sexual Stratification: A Cross-cultural View,* ed. A. Schlegel. New York: Columbia University Press.

Dennis, W. 1940. *The Hopi Child.* New York: D. Appleton-Century Company.

DePaulo, B. M., and D. A. Kashy. 1998. Everyday lies in close and casual relationships. *Journal of Personality and Social Psychology* 71 (1): 63–79.

Derrida, J. 1997. *Politics of Friendship.* London; New York: Verso.

Descola, P. 1996. *The Spears of Twilight.* New York: Free Press.

Devlin, A. S. 1996. Survival skills training during freshman orientation: Its role in college adjustment. *Journal of College Student Development* 37 (3): 324–334.

Diamond, L. M. 2000. Passionate friendships among lesbian, bisexual and heterosexual women. *Journal of Research on Adolescence* 10 (2): 191–209.

———. 2003. What does sexual orientation orient? A biobehavioral model distinguishing romantic love and sexual desire. *Psychological Review* 110 (1): 173–192.

Diamond, L. M., R. C. Savin-Williams, and E. M. Dube. 1999. Intimate peer relations among lesbian, gay and bisexual adolescents. In *The Development of Romantic Relationships in Adolescence,* ed. W. Furman, B. B. Brown, and C. Feiring. Cambridge: Cambridge University Press.

Diamond, N. 1969. *K'un Shen: A Taiwan Village.* New York: Holt, Rinehart and Winston.

Diani, M., and D. McAdam, eds. 2003. *Social Movement Analysis: The Network Perspective.* Oxford: Oxford University Press.

Dimaggio, P., and H. Louch. 1998. Socially embedded consumer transactions: For what kinds of purchases do people most often use social networks? *American Sociological Review* 63 (5): 619–637.

Dindia, K. 1985. A functional approach to self-disclosure. In *Sequence and Pattern in Communicative Behavior,* ed. R. L. Street and J. N. Cappella. London: Edward Arnold.

Dindia, K., and M. Allen. 1992. Sex differences in self-disclosure: A meta-analysis. *Psychological Bulletin* 112:106–124.

Dinstein, I., C. Thomas, M. Behrmann, and D. J. Heeger. 2008. A mirror up to nature. *Current Biology* 18 (1): R13-R18.

Divale, W. T. 1976. Female status and cultural evolution: A study in ethnographer bias. *Cross-cultural Research* 11 (3): 169–211.

Doherty, N. A., and J. A. Feeney. 2004. The composition of attachment networks throughout the adult years. *Personal Relationships* 11 (4): 469–488.

Donaldson, Z. R., and L. J. Young. 2008. Oxytocin, vasopressin, and the neurogenetics of sociality. *Science* 322:900–904.

Dong, M., R. F. Anda, V. J. Felitti, D. F. Williamson, S. R. Dube, D. W. Brown, and W. H. Giles. 2005. Childhood residential mobility and multiple health risks during adolescence and adulthood. *Archives of Pediatric and Adolescent Medicine* 159:1104–1110.

Donham, D. 1985. History at one point in time: "Working together" in Maale, 1975. *American Ethnologist* 12 (2): 262–284.

Donne, J. [1597] 1912. To S[ir] Henry Wotton. In *Poems of John Donne,* ed. H. J. C. Grierson. Oxford: Oxford University Press.

Dorsey, G. A. 1940. *Notes on Skidi Pawnee Society.* Chicago: Field Museum Press.

Douvan, E., and J. Adelson. 1966. *The Adolescent Experience.* New York: Wiley.

Dovidio, J. F. 1984. Helping behavior and altruism. In *Advances in Experimental Social Psychology,* ed. L. Berkowitz. New York: Academic Press.

Driberg, J. H. 1935. The "best friend" among the Didinga. *Man* 35:101–102.

Dubois, C. 1974. The gratuitous act: An introduction to the comparative study of friendship patterns. In *The Compact: Selected Dimensions of Friendship*, ed. E. Leyton. St. John's, Canada: Memorial University of Newfoundland.

Dubois, D. L., and B. J. Hirsch. 1990. School and neighborhood friendship patterns of blacks and whites in early adolescence. *Child Development* 61 (2): 524–536.

Ducey, P. R. 1956. Cultural continuity and population change on the Isle of Skye. PhD diss., Columbia University, New York.

Duck, S. W., and P. H. Wright. 1993. Re-examining gender differences in same-gender relationships. *Sex Roles* 28:709–727.

Dugan, E., and V. R. Kivett. 1998. Implementing the Adams and Blieszner conceptual model: Predicting interactive friendship processes of older adults. *Journal of Social and Personal Relationships* 15 (5): 607–622.

Dugatkin, L. A. 1997. *Cooperation among Animals: An Evolutionary Perspective*. Oxford Series in Ecology and Evolution. New York: Oxford University Press.

Dunbar, R. I. M., A. Clark, and N. L. Hurst. 1995. Conflict and cooperation among the Vikings: Contingent behavioral decisions. *Ethology and Sociobiology* 16 (3): 233–246.

Dunbar, R. I. M., and S. Shultze. 2007. Evolution in the social brain. *Science* 317:1344–1347.

Dunn, J. 2004. *Children's Friendships: The Beginnings of Intimacy*. Malden, MA: Blackwell.

Dunn, J., A. L. Cutting, and N. Fisher. 2002. Old friends, new friends: Predictors of children's perspectives on their friends at school. *Child Development* 73 (2): 621–635.

Dunne, M. 2004. Good friends or bad friends? The *amicitia* of Boncompagno da Signa. In *Amor Amicitiae: On the Love That Is Friendship*, ed. J. J. McEvoy, T. A. F. Kelly and P. Rosemann. Leuven: Peeters Publishers.

Durkheim, E., and M. Mauss. [1903] 1963. *Primitive Classification*. Translated by R. Needham. Chicago: University of Chicago Press.

Durrenberger, E. P., and G. Palsson. 1999. The importance of friendship in the absence of states, according to the Icelandic sagas. In *The Anthropology of Friendship*, ed. S. M. Bell and S. Coleman. Oxford: Berg.

Dyson-Hudson, R., and J. T. McCabe. 1985. *South Turkana Nomadism: Coping with an Unpredictably Varying Environment*. New Haven, CT: HRAFlex Books.

Eagly, A. H., and M. Crowley. 1986. Gender and helping behavior: A meta-analytic review of the social psychological literature. *Psychological Bulletin* 100 (3): 283–308.

Edgerton, R. 1971. *The Individual in Cultural Adaptation*. Berkeley: University of California Press.

Eggan, F. R. 1950. *Social Organization of the Western Pueblos*. Chicago: University of Chicago.

Eisenberg, N., R. A. Fabes, and Tracy L. Spinrad. 2006. Prosocial development.

In *Handbook of Child Psychology,* ed. W. Damon, R. M. Lerner and N. Eisenberg. New York: Wiley.

Eisenstadt, S. N. 1956. Ritualized personal relations: Blood brotherhood, best friends, compadre, etc.: Some comparative hypotheses and suggestions. *Man* 56:90–95.

———. 1974. Friendship and the structure of trust and solidarity in society. In *The Compact: Selected Dimensions of Friendship,* ed. E. Leyton. Toronto: University of Toronto Press.

Ekman, P. 2003. Darwin, deception, and facial expression. *Annual of the New York Academy of Sciences* 1000:205–221.

Ekman, P., R. J. Davidson, and W. V. Friesen. 1990. The Duchenne smile: Emotional expression and brain physiology, part 2. *Journal of Personality and Social Psychology* 58 (2): 342–353.

Elbedour, S., S. Shulman, and P. Kedem. 1997. Adolescent intimacy. *Journal of Cross-cultural Psychology* 28 (1): 5–22.

Elwin, V. 1947. *The Muria and Their Ghotul.* Oxford: Oxford University Press.

Emanuele, E., P. Politi, M. Bianchi, P. Minoretti, M. Bertona, and D. Geroldi. 2006. Raised plasma nerve growth factor levels associated with early-stage romantic love. *Psychoneuroendocrinology* 31 (3): 288–294.

Embree, J. 1950. Thailand: A loosely structured social system. *American Anthropologist* 52:181–192.

Endo, Y., S. J. Heine, and D. R. Lehman. 2000. Culture and positive illusions in close relationships: How my relationships are better than yours. *Personality and Social Psychology Bulletin* 26 (12): 1571–1586.

Ennew, J. 1980. *The Western Isles Today.* Cambridge: Cambridge University Press.

Enomoto, J. 1999. Socio-emotional development of friendship adolescents: Activities with friends and the feelings of friends. *Japanese Journal of Educational Psychology* 47 (2): 180–190.

Espinoza, V. 1999. Social networks among the urban poor: Inequality and integration in a Latin American city. In *Networks in the Global Village: Life in Contemporary Communities,* ed. B. Wellman. Boulder, CO: Westview Press.

Essock-Vitale, S. M., and M. T. McGuire. 1985. Women's lives viewed from an evolutionary perspective. Two. Patterns of helping. *Ethology and Sociobiology* 6 (3): 155–173.

Estrada, P. 1995. Adolescents' self-report of prosocial responses to friends and acquaintances: The role of sympathy-related cognitive, affective, and motivational processes. *Journal of Research on Adolescence* 5 (2): 173–200.

Euler, H. A., and B. Weitzel. 1996. Discriminative grandparental solicitude as reproductive strategy. *Human Nature* 7 (1): 39–59.

Evans-Pritchard, E. E. 1933. Zande blood-brotherhood. *Africa* 6 (4): 369–401.

Eve, M. 2002. Is friendship a sociological topic? *Archives Europeennes de sociologie* 43 (3): 386–409.

Ewers, J. C. 1958. *The Blackfeet: Raiders of the Northwestern Plains.* Norman: University of Oklahoma Press.

Fabian, S. M. 1992. *Space-Time of the Bororo of Brazil.* Gainesville: University Press of Florida.

Faderman, L. 1981. *Surpassing the Love of Men: Romantic Love and Friendship between Women from the Renaissance to the Present.* New York: William Morrow & Co.

Fafchamps, M., and B. Minten. 1999. Relationships and traders in Madagascar. *Journal of Developmental Studies* 35 (6): 1–35.

Falk, A., E. Fehr, and U. Fischbacher. 2008. Testing theories of fairness—intentions matter. *Games and Economic Behavior* 62 (1): 287–303.

Falk, K. [1925] 1998. Homosexuality among the natives of southwest Africa (1925–1926). In *Boy-Wives and Female Husbands: Studies of African Homosexualities,* ed. S. O. Murray and W. Roscoe. New York: Macmillan.

Farrell, M. P. 2001. *Collaborative Circles: Friendship Dynamics and Creative Work.* Chicago: University of Chicago Press.

Fehr, B. 1996. *Friendship Processes.* London: Sage.

———. 2004. Intimacy expectations in same-sex friendships: A prototype interaction-pattern model. *Journal of Personality and Social Psychology* 86 (2): 265–284.

Fehr, B., and J. A. Russell. 1991. The concept of love viewed from a prototype perspective. *Journal of Personality and Social Psychology* 60 (3): 425–438.

Feld, S. L. 1981. The focused organization of social ties. *American Journal of Sociology* 86 (5): 1015–1035.

Feld, S. L., and W. C. Carter. 1998. Foci of activity as changing contexts for friendship. In *Placing Friendship in Context,* ed. R. G. Adams and G. A. Allan. Cambridge: Cambridge University Press.

Fenton, W. N. 1953. *The Iroquois Eagle Dance: An Offshoot of the Calumet Dance.* Washington, D.C.: Smithsonian Institution.

Fessler, D., and K. J. Haley. 2003. The strategy of affect: Emotions in cooperation. In *Genetic and Cultural Evolution of Cooperation,* ed. P. Hammerstein. Cambridge: MIT Press.

Festinger, L., S. Schachter, and K. Back. 1950. *Social Pressures in Informal Groups: A Study of a Housing Project.* Palo Alto, CA: Stanford University Press.

Fiebert, M. S., and P. B. Fiebert. 1969. A conceptual guide to friendship formation. *Perceptual Motor Skills* 28:383–390.

Field, D. 1999. Continuity and change in friendships in advanced old age: Findings from the Berkeley Older Generation Study. *International Journal of Aging and Human Development* 48 (4): 325–346.

Field, M. J. 1960. *Search for Security: An Ethno-psychiatric Study of Rural Ghana.* Evanston, IL: Northwestern University Press.

Filipovic, M. S. 1982. *Among the People, Native Yugoslav Ethnography: Selected Writing of Milenko S. Filipovic.* Ann Arbor: Michigan Slavic Publications.

Firth, R. 1936. Bond-friendship in Tikopia. In *Custom Is King: Essays Presented to R.R. Marett*, ed. L.H.D. Buxton. London: Hutchinson's Scientific and Technical Publications.

Fischer, A.M. 1950. The role of Trukese mother and its effect on child training. SIM Report No. 8. Washington, D.C.: Pacific Science Board.

Fischer, C.S. 1992a. *America Calling: A Social History of the Telephone to 1940*. Berkeley: University of California Press.

———. 1992b. *To Dwell among Friends*. Chicago: University of Chicago Press.

Fischer, C.S., and S.J. Oliker. 1983. Research notes on friendship, gender and the life cycle. *Social Forces* 62:124–133.

Fischer, J.L. 1989. Marital status and career stage influences on social networks of young adults. *Journal of Marriage and Family* 51 (2): 521–534.

Fisher, H.E. 1998. Lust, attraction and attachment in mammalian reproduction. *Human Nature* 9 (1): 23–52.

———. 2004. *Why We Love: The Nature and Chemistry of Romantic Love*. New York: Henry Holt and Company.

———. 2006. The drive to love: The neural mechanism for mate selection. In *The New Psychology of Love*, ed. R.J. Steinberg and K. Weis. New Haven, CT: Yale University Press.

Fisher, H.E., A. Aron, and L.L. Brown. 2006. Romantic love: A mammalian brain system for mate choice. *Philosophical Transactions of the Royal Society B: Biological Sciences* 361 (1476): 2173–2186.

Fisher, H.E., and J.A. Thomson. 2007. Lust, romance, attachment: Do the side effects of serotonin-enhancing antidepresssants jeopardize love, marriage and fertility? In *Evolutionary Cognitive Neuroscience*, ed. S.M. Platek, J.P. Keenan and T.K. Shackelford. Cambridge: MIT Press.

Fiske, A.P. 1991. *Structures of Social Life: The Four Elementary Forms of Human Relations: Communal Sharing, Authority Ranking, Equality Matching, Market Pricing*. New York: Free Press.

———. 2002a. Socio-moral emotions motivate action to sustain relationships. *Self and Identity* 1 (2): 169–175.

———. 2002b. Using individualism and collectivism to compare cultures: Critique of the validity and measurement of the constructs: Comment on Oyserman et al. (2002). *Psychological Bulletin* 128 (1): 78–88.

Fitch, K.L. 1998. *Speaking Relationally: Cuture, Communication and Interpersonal Connection*. New York: Guilford Press.

Fitzsimons, G.M., and J.A. Bargh. 2003. Thinking of you: Nonconscious pursuit of interpersonal goals associated with relationship partners. *Journal of Personality and Social Psychology* 84 (1): 148–164.

Flack, J.C., M. Girvan, F.B.M. De Waal, and D. Krakauer. 2006. Policing stabilizes construction of social niches in primates. *Nature* 439:426–429.

Flannagan, M., D.L. Marsh, and R. Fuhrman. 2005. Judgments about the hypothetical behaviors of friends and romantic partners. *Journal of Social and Personal Relationships* 22 (6): 797–815.

Flinn, M. V. 1988. Step- and genetic parent/offspring relationships in a Caribbean village. *Ethology and Sociobiology* 9:335–369.

Flinn, M. V., D. V. Leone, and R. J. Quinlan. 1999. Growth and fluctuating asymmetry of stepchildren. *Evolution and Human Behavior* 20:465–479.

Floyd, J. M. K. 1964. Effects of amount of reward and friendship status of the other on the frequency of sharing in children. PhD diss., University of Minnesota, Minneapolis.

Floyd, K. 2006. *Communicating Affection: Interpersonal Behavior and Social Context.* Cambridge: Cambridge University Press.

Fonzi, A., B. H. Schneider, F. Tani, and G. Tomada. 1997. Predicting children's friendship status from their dyadic interaction in structured situations of potential conflict. *Child Development* 68 (3): 496–506.

Ford, C. S. 1941. *Smoke from Their Fires: The Life of a Kwakiutl Chief.* New Haven, CT: Yale University Press.

Forster, E. M. 1924. *A Passage to India.* New York: Harcourt Brace.

Forster, J., N. Liberman, and R. S. Friedman. 2008. What do we prime? On distinguishing between semantic priming, procedural priming, and goal priming. In *Oxford Handbook of Human Action,* ed. E. Morsella, J. A. Bargh, and P. M. Gollwitzer. New York: Oxford University Press.

Fortes, M. 1950. *African Systems of Kinship and Marriage.* London: Oxford University Press.

———. 1970. *Time and Social Structure and Other Essays.* New York: Humanities Press.

Fosbrooke, H. A. 1948. An administrative survey of the Masai social system. *Tanganyika Notes and Records* 26:1–51.

Foster, B. L. 1976. Friendship in rural Thailand. *Ethnology* 15 (3): 251–267.

Foster, G. M. 1961. The dyadic contract: A model for the social structure of a Mexican peasant village. *American Anthropologist* 63:1173–1192.

Fourier, L. 1928. The bushmen in southwest Africa. In *The Native Tribes of Southwest Africa.* Cape Town: Cape Times Limited.

Frank, R. H. 1988. *Passions Within Reason: The Strategic Role of the Emotions.* New York: W. W. Norton & Company.

French, D. C., A. Bae, S. Pidada, and O. C. Lee. 2006. Friendships of Indonesian, South Korean, and US college students. *Personal Relationships* 13 (1): 69–81.

French, D. C., S. Pidada, and A. Victor. 2005. Friendships of Indonesian and United States youth. *International Journal of Behavioral Development* 29 (4): 304–313.

French, D. C., M. Rianasari, S. Pidada, P. Nelwan, and D. Buhrmester. 2001. Social support of Indonesian and U.S. children and adolescents by family members and friends. *Merrill-Palmer Quarterly* 47 (3): 377–394.

Freud, A., and D. Burlingham. 1944. *Infants without Families: The Case for and against Residential Nurseries.* New York: International Universities Press.

Fried, J. 1952. Ideal norms and social control in Tarahumara society. PhD diss., Yale University, New Haven, CT.

Friedrich, P. 1958. A Tarascan cacicazgo: Structure and function. In *Systems of Political Control and Bureaucracy in Human Societies*, ed. V. F. Ray. Seattle: American Ethnological Society.

Friedrichs, R. W. 1960. Alter versus ego: An exploratory assessment of altruism. *American Sociological Review* 25:496–508.

Fry, D. 2007. *The Human Potential for Peace: An Anthropological Challenge to Assumptions about War and Violence.* New York: Oxford University Press.

Fuligni, A. J., and H. W. Stevenson. 1995. Time use and mathematics achievement among American, Chinese, and Japanese high school students. *Child Development* 66 (3): 830–842.

Furman, W. 1999. Friends and lovers: The role of peer relationships in adolescent romantic relationships. In *Relationships as Developmental Contexts*, ed. W. A. Collins, B. P. Laursen, and W. W. Hartup. Mahwah, NJ: Lawrence Erlbaum.

Furman, W., and K. L. Bierman. 1983. Developmental changes in young children's conceptions of friendship. *Child Development* 54 (3): 549–556.

———. 1984. Children's conceptions of friendship: A multimethod study of developmental changes. *Developmental Psychology* 20 (5): 925–931.

Furman, W., and D. Buhrmeister. 1992. Age and sex differences in perceptions of networks of personal relationships. *Child Development* 63:103–115.

Gagne, F. M., and J. E. Lydon. 2004. Bias and accuracy in close relationships: An integrative review. *Personality and Social Psychology Bulletin* 8 (4): 322–338.

Gailliot, M. T., and R. F. Baumeister. 2007. The physiology of willpower: Linking blood glucose to self-control. *Personality and Social Psychology Review* 11 (4): 303–327.

Gailliot, M. T., R. F. Baumeister, C. N. DeWall, J. K. Maner, E. A. Plant, D. M. Tice, L. E. Brewer, and B. J. Schmeichel. 2007. Self-control relies on glucose as a limited energy source: Willpower is more than a metaphor. *Journal of Personality and Social Psychology* 92 (2): 325–336.

Gallin, B. 1966. *Hsin Hsing: A Chinese Village in Change.* Berkeley: University of California Press.

Gallin, B., and R. S. Gallin. 1974. The integration of village migrants in Taipei. In *The Chinese between Two Worlds*, ed. M. Elvin and G. W. Skinner. Stanford, CA: Stanford University Press.

Gamble, D. P. 1957. *The Wolof of Senegambia: Together with Notes on the Lebu and the Serer.* London: International African Institute.

Gans, H. J. 1962. *The Urban Villagers.* New York: Free Press.

Gao, G. 1996. Self and other: A Chinese perspective on interpersonal relationships. In *Communication in Personal Relationships across Cultures*, ed. W. B. Gudykunst, S. Ting-Toomey, and T. Nishida. Thousand Oaks, CA: Sage.

———. 1998. "Don't take my word for it." Understanding Chinese speaking practices. *International Journal of Intercultural Relations* 22 (2): 163–186.

Gardner, W. L., S. Gabriel, and L. Hochschild. 2002. When you and I are "we," you are not threatening: The role of self-expansion in social comparison. *Journal of Personality and Social Psychology* 82 (2): 239–251.

Gates, H. 1973. Prosperity settlement: The politics of *paipai* in Taipei, Taiwan. PhD diss., University of Michigan, Ann Arbor.

Gay, J. 1986. "Mummies and babies" and friends and lovers in Lesotho. In *Anthropology and Homosexual Behavior*, ed. E. Blackwood. Binghamton, NY: Haworth Press.

Geary, D. C., J. Byrd-Craven, M. K. Hoard, and J. M. Vigil. 2003. Evolution and development of boys' social behavior. *Developmental Review* 23 (4): 444–470.

Geers, A. L., S. P. Reilley, and W. N. Dember. 1998. Optimism, pessimism, and friendship. *Current Psychology* 17 (1): 3–19.

Geertz, C. 1978. Bazaar economy: Information and search in peasant marketing. *American Economic Review* 68:28–32.

———. 1979. Suq: The bazaar economy in Sefrou. In *Meaning and Order in Moroccan Society*, ed. C. Geertz, H. Geertz, and L. Roren, pp. 123–224. New York: Cambridge University Press.

Gelfand, M. 1973. *The Genuine Shona: Survival Values of an African Culture:* Mambo Press.

Gershman, E., and H. S. Donald. 1983. Differential stability of reciprocal friendships and unilateral relationships among preschool children. *Merrill-Palmer Quarterly* 29 (2): 169–177.

Gesn, P. R. 1995. Shared knowledge between same-sex friends. PhD diss., University of Texas, Arlington.

Getz, L., and S. Carter. 1996. Prairie-vole partnerships. *American Scientist* 84 (1): 56–62.

Gibson, K. R. 2004. Relatedness and investment in adoptive households. PhD diss., University of Kansas, Lincoln, Nebraska.

Giddens, A. 1992. *The Transformation of Intimacy: Sexuality, Love and Eroticism.* Palo Alto, CA: Stanford University Press.

Gilligan, C. 1982. *In a Different Voice: Women's Conceptions of the Self and Morality.* Cambridge: Harvard University Press.

Gillis, M. T., and P. L. Hettler. 2007. Hypothetical and real incentives in the ultimatum game and Andreoni's public goods game: An experimental study. *Eastern Economic Journal* 33 (4): 491–510.

Gladwin, T., and S. B. Sarason. 1953. *Truk: Man in Paradise.* New York: Viking Fund Publications in Anthropology.

Glaeser, E. L., D. Laibson, J. Scheinkman, and C. Soutter. 2000. Measuring Trust. *Quarterly Journal of Economics* 115 (3): 811–846.

Glasse, R., and S. Lindenbaum. 1969. Marriage in South Fore. In *Pigs, Pearlshells, and Women: Marriage in the New Guinea Highlands*, ed. R. Glasse and M. J. Meggitt. Englewood Cliffs, NJ: Prentice Hall.

Gleason, M. E. J. 2002. Social provisions of real and imaginary relationships in early childhood. *Developmental Psychology* 38:979–992.

Gleason, T. R., and L. M. Hohmann. 2006. Concepts of real and imaginary friendships in early childhood. *Social Development* 15 (1): 128–144.

Gluckman, M. 1967. *The Judicial Process among the Barotse of Northern Rhodesia.* Manchester: Manchester University Press.

Goeree, J. K., M. A. McConnell, T. Mitchell, T. Tromp, and L. Yariv. 2007. Linking and giving among teenage girls. Working paper, California Institute of Technology.

Goldfrank, E. S. 1966. *Changing Configurations in the Social Organization of a Blackfoot Tribe during the Reserve Period (the Blood of Alberta, Canada).* Seattle: University of Washington Press.

Goldman, I. 1937. The Zuni Indians of New Mexico. In *Cooperation and Competition among Primitive Peoples,* ed. M. Mead. New York.

———. 1963. *The Cubeo: Indians of the Northwest Amazon.* Urbana: University of Illinois Press.

Goldschmidt, W. 2006. *Bridge to Humanity: How Affect Hunger Trumps the Selfish Gene.* New York: Oxford University Press.

Gomes, E. H. 1911. *Seventeen Years among the Sea Dyaks of Borneo: A Record of Intimate Association with the Natives of the Bornean Jungles.* London: Seeley & Co., Ltd.

Gonzaga, G. C., D. Keltner, E. A. Londahl, and M. D. Smith. 2001. Love and the commitment problem in romantic relations and friendship. *Journal of Personality and Social Psychology* 81 (2): 247–262.

Goodwin, R., and I. Lee. 1994. Taboo topics among Chinese and English friends. *Journal of Cross-cultural Psychology* 25 (3): 325–338.

Goodwin, R., G. Nizharadze, L. A. N. Luu, E. Kosa, and T. Emelyanova. 2001. Social support in a changing Europe: An analysis of three post-communist nations. *European Journal of Social Psychology* 31 (4): 379–393.

Gorer, G. 1938. *Himalayan Village: An Account of the Lepchas of Sikkim.* London: Michael Joseph, Ltd.

Gould, Roger V. 1991. Multiple networks and mobilization in the Paris Commune. *American Sociological Review* 56 (6): 716–729.

Gould-Martin, K. 1976. Women asking women: An ethnography of health care in rural Taiwan. PhD diss., Rutgers University, New Brunswick, NJ.

Gouldner, A. W. 1960. The norm of reciprocity: A preliminary statement. *American Sociological Review* 25:161–178.

Gouldner, H., and M. S. Strong. 1987. *Speaking of Friendship: Middle-Class Women and Their Friends.* New York: Greenwood Press.

Grabill, C. M., and K. A. Kerns. 2000. Attachment styles and intimacy in friendship. *Personal Relationships* 7 (4): 363–378.

Grammer, K. 1992. Intervention in conflicts among children: Contexts and consequences. In *Coalitions and Alliances in Human and Other Animals,* ed. A. H. Harcourt and F. B. M. de Waal. Oxford: Oxford University Press.

Granovetter, M. 1985. Economic action and social structure: The problem of embeddedness. *American Journal of Sociology* 83 (1): 142–171.

———. 1995. *Getting a Job: A Study of Contacts and Careers.* Chicago: University of Chicago Press.

Gray, J. 1993. *Men Are from Mars, Women Are from Venus.* New York: HarperCollins.

Gray, J.P., and L.D. Wolfe. 1980. Height and sexual dimorphism of stature among human societies. *American Journal of Physical Anthropology* 53:441–456.

Grayson, D.K. 1993. Differential mortality and the Donner Party disaster. *Evolutionary Anthropology* 2 (5): 151–159.

Greenberg, M.S., and D.M. Frisch. 1972. Effect of intentionality on willingness to reciprocate a favor. *Journal of Experimental Social Psychology* 8:99–111.

Greenberg, M.S., and S. Shapiro. 1971. Indebtedness: An adverse aspect of asking for and receiving help. *Sociometry* 34 (2): 290–301.

Greif, A. 2006. *Institutions and the Path to the Modern Economy: Lessons from Medieval Trade.* Cambridge: Cambridge University Press.

Griaule, M. 1948. L'alliance cathartique. *Africa: Journal of the International African Institute* 18 (4): 242–358.

Griffith, A.S., and S.A. West. 2002. Kin selection: Fact and fiction. *Trends in Ecology and Evolution* 17 (1): 15–21.

Griffith, R.T.H. 1889. *The Hymns of the Rig Veda.* Kotagiri.

Groenenboom, A., H.A.M. Wilke, and A.P. Wit. 2001. Will we be working together again? The impact of future interdepedence on group members' task motivation. *European Journal of Social Psychology* 31:369–378.

Gross, A.E., and J.G. Latane. 1974. Receiving help, reciprocation, and interpersonal attraction. *Journal of Applied Social Psychology* 4 (3): 210–223.

Gross, E.F., J. Juvonen, and S.L. Gable. 2002. Internet use and well-being in adolescence. *Journal of Social Issues* 58 (1): 75–90.

Grossetti, M. 2005. Where do social relations come from? A study of personal networks in the Toulouse area of France. *Social Networks* 27 (4): 289–300.

Grotanelli, V.L. 1988. *The Python Killer.* Chicago: University of Chicago Press.

Grote, N.K., and I.H. Frieze. 1994. The measurement of friendship-based love in intimate relationships. *Personal Relationships* 1 (3): 275–300.

Gruder, C.L. 1971. Relationships with opponent and partner in mixed-motive bargaining. *Journal of Conflict Resolution* 15 (3): 403–416.

Gudykunst, W.B., and Y. Matsumoto. 1996. Cross-cultural variability of communication in personal relationships. In *Communication in Personal Relationships across Cultures,* ed. W.B. Gudykunst, S. Ting-Toomey, and T. Nishida. Newbury Park, CA: Sage.

Gudykunst, W.B., and T. Nishida. 1983. Social penetration in Japanese and American close friendships. *Communication Yearbook* 7:147–166.

———. 2001. Anxiety, uncertainty, and perceived effectiveness of communication across relationships and cultures. *International Journal of Intercultural Relations* 25 (1): 55–71.

Gueguen, N. 2003. Help on the web: The effect of the same first name between the sender and the receptor in a request made by email. *Psychological Record* 53 (3): 459–466.

Gulliver, P. 1951. *A Preliminary Survey of the Turkana.* Cape Town: Commonwealth School of African Studies.

———. 1955. *The Family Herds: A Study of Two Pastoral Tribes.* London: Routledge & Kegan Paul.

———. 1971. *Neighbors and Networks: The Idiom of Kinship in Social Action among the Ndendeuli of Tanzania.* Berkeley, CA: University of California Press.

Gummerum, M., and M. Keller. 2008. Affection, virtue, pleasure, and profit: Developing an understanding of friendship closeness and intimacy in western and Asian societies. *International Journal of Behavioral Development* 32 (3): 218–231.

Gurdin, J. B. 1996. *Amitie/Friendship: An Investigation into Cross-cultural Styles in Canada and the United States.* Bethesda, MD: Austin and Winfield.

Gurdon, P. R. T. 1907. *The Khasis.* London: David Nutt.

Guroglu, B., G. J. T. Haselager, C. van Lieshout, A. Takashima, M. Rijpkema, and G. Fernandez. 2008. Why are friends special? Implementing a social interaction simulation task to probe the neural correlates of friendship. *Neuroimage* 39 (2): 903–910.

Gurven, M. 2005. To give or to give not: The behavioral ecology of human food transfers. *Behavioral and Brain Sciences* 27:543–583.

Gurven, M., W. Allen-Arave, K. Hill, and A. M. Hurtado. 2001. Reservation food sharing among the Ache of Paraguay. *Human Nature* 12 (4): 273–297.

Gurven, M., K. Hill, H. Kaplan, A. M. Hurtado, and R. Lyles. 2000. Food transfers among Hiwi foragers of Venezuela: Test of reciprocity. *Human Ecology* 28 (2): 171–218.

Gurven, M., and J. Winking. 2008. Collective action in action: Prosocial behavior in and out of the laboratory. *American Anthropologist* 110 (2): 179–190.

Gusinde, M. 1931. *The Fireland Indians.* Vol. 1, *The Selk'nam on the Life and Thought of a Hunting People of the Great Island of Tierra del Fuego.* Modling: Verlag der Internationalen Zeitschrift.

———. 1937. *The Yahgan: The Life and Thought of the Water Nomads of Cape Horn.* Vienna: Verlag.

Guttman, Bruno. 1926. *Das recht der Dschagga.* Munich: Beck.

———. 1932. *Die stammeslehren der Chagga.* Munich: Beck.

Haacke, W. H. G., and E. Eiseb. 1998. *A Khoekhoegowab Dictionary with an English Khoekhoegowab Index.* Windhoek, Namibia: Gamsberg Macmillan.

Hacker, H. M. 1981. Blabbermouths and clams: Sex differences in self-disclosure in same-sex and cross-sex friendships. *Psychology of Women Quarterly* 5 (3): 385–401.

Haeckel, Ernst. 1866. *General Morphology of Organisms*. Berlin: Reimer.

Halpern, J. M. 1986. *A Serbian Village in Historical Perspective*. Prospect Heights, IL: Waveland Press, Inc.

Hames, R. 1987. Garden labor exchange among the Ye'kwana. *Ethology and Sociobiology* 8:354–392.

Hamilton, L., S. Cheng, and B. Powell. 2007. Adoptive parents, adaptive parents: Evaluating the importance of biological ties for parental investment. *American Sociological Review* 72:95–116.

Hamilton, W. D. 1964. The genetical evolution of social behaviour, parts 1 and 2. *Journal of Theoretical Biology* 7 (1): 1–52.

Hammel, E. 1968. *Alternative Social Structures and Ritual Relations in the Balkans*. Englewood Cliffs, NJ: Prentice-Hall.

Han, C. C. [1949] 1970. *Social Organization of Upper Han Hamlet in Korea*. Ann Arbor, MI: HRAF.

Han, K.-H., M.-C. Li, and K.-K. Hwang. 2005. Cognitive responses to favor requests from different social targets in a Confucian society. *Journal of Social and Personal Relationships* 22 (2): 283–294.

Handy, E. S. C. 1923. *The Native Culture in the Marquesas*. Honolulu: Bernice P. Bishop Museum.

Handy, E. S. C., and M. K. Pukui. 1958. *The Polynesian Family System in Ka-'u, Hawai'i*. Wellington: Polynesian Society.

Hanks, J. R., and L. M. Hanks. 1950. *Tribe under Trust: A Study of the Blackfoot Reserve of Alberta*. Toronto: University of Toronto Press.

Hanks, L. M. 1962. Merit and power in the Thai social order. *American Anthropologist* 64 (6): 1246–1261.

Hara, H. S. 1980. *The Hare Indians and Their World*. Ottawa: National Museum of Canada.

Harding, T. G. 1994. Precolonial New Guinea trade. *Ethnology* 33 (2): 101–125.

Harner, M. J. 1973. *The Jivaro*. New York: American Museum of Natural History.

Harrell, S. 1974. *Belief and Unbelief in a Taiwan Village*. Stanford, CA: Stanford University Press.

Harrington, J. P. 1912. Tewa relationship terms. *American Anthropologist* 14 (3): 472–498.

Harris, H. 1995. Rethinking Polynesian heterosexual relationships: A case study of Mangaia, Cook Islands. In *Romantic Passion: A Universal Experience?* ed. W. Jankowiak. New York: Columbia University Press.

Harrison, A. O., R. B. Stewart, K. Myambo, and C. Teveraishe. 1995. Perceptions of social networks among adolescents in Zimbabwe and the United States. *Journal of Black Psychology* 21 (4): 382–407.

———. 1997. Social networks among early adolescent Zimbabweans in extended families. *Journal of Research on Adolescence* 7 (2): 153–172.

Hartup, W. W. 1975. The origins of friendship. In *Friendship and Peer Relations*, ed. M. Lewis and L. A. Rosenblum. New York: Wiley.

Hartup, W. W., B. P. Laursen, M. I. Stewart, and A. Eastenson. 1988. Conflict and the friendship relations of young children. *Child Development* 59 (6): 1590–1600.

Hartup, W.W., and N. Stevens. 1999. Friendships and adaptation across the lifespan. *Current Directions in Psychological Science* 8 (3): 76–79.

Haselton, M.G., and D. Nettle. 2006. The paranoid optimist: An integrative evolutionary model of cognitive biases. *Personality and Social Psychology Review* 10 (1): 47–66.

Hassan, M., M. Shuaibu, and F.L. Heath. 1952. *A Chronicle of Abuja.* Ibadan: Ibadan University Press.

Hatfield, E., and R.L. Rapson. 1996. *Love and Sex: Cross-cultural Perspectives.* New York: Allyn & Bacon.

Hatfield, E., E. Schmitz, J. Cornelius, and R.L. Rapson. 1988. Passionate love: How early does it begin? *Journal of Psychology and Human Sexuality* 1 (1): 35–51.

Hatt, D.G. 1974. *Skullcaps and Turbans: Domestic Authority and Public Leadership among the Idaw Tanan of the Western High Atlas, Morocco.* Ann Arbor, MI: University Microfilms.

Hawkes, K. 1977. Co-operation in Binumarien: Evidence for Sahlin's model. *Man* 12 (3–4): 459–483.

———. 1983. Kin selection and culture. *American Ethnologist* 10 (2): 345–363.

Hays, R.B. 1984. The development and maintenance of friendship. *Journal of Social and Personal Relationships* 1 (1): 75–98.

———. 1985. A longitudinal study of friendship development. *Journal of Personality and Social Psychology* 48 (4): 909–924.

———. 1988. Friendship. In *Handbook of Personal Relationships,* ed. S.W. Duck. London: Wiley.

———. 1989. The day-to-day functioning of close versus casual friendships. *Journal of Social and Personal Relationships* 6:21–37.

Hays, R.B., and D. Oxley. 1986. Social network development and functioning during a life transition. *Journal of Personality and Social Psychology* 50 (2): 305–313.

Haythornwaite, C., and B. Wellman, eds. 2002. *The Internet in Everyday Life.* New York: Wiley-Blackwell.

Haytko, D.L. 2004. Firm-to-firm and interpersonal relationships: Perspectives from advertising agency account managers. *Journal of Academy of Marketing Science* 32 (3): 312–328.

Hazan, C., and P.R. Shaver. 1987. Romantic love conceptualized as an attachment process. *Journal of Personality and Social Psychology* 52 (3): 511–524.

Hazan, C., and D. Zeifman. 1994. Sex and the psychological tether. In *Advances in Personal Relationships,* ed. D. Perlman and K. Bartholomew. London: Jessica Kingsley.

Headland, T.N. 1987. Kinship and social behavior among Agta Negrito hunter-gatherers. *Ethnology* 26 (4): 261–280.

Hechter, M. 1987. *Principles of Group Solidarity.* Berkeley: University of California Press.

Heide, J. B., and A. S. Miner. 1992. The shadow of the future: Effects of antici-pated interaction and frequency of contact on buyer-seller cooperation. *Academy of Management Journal* 35 (2): 265–291.

Heider, K. G. 1969. Visiting trade institutions. *American Anthropologist* 71:462–471.

Heinrichs, M., T. Baumgartner, C. Kirschbaum, and U. Ehlert. 2003. Social support and oxytocin interact to suppress cortisol and subjective responses to psychosocial stress. *Biological Psychiatry* 54:1389–1398.

Heinrichs, M., and G. Domes. 2008. Neuropeptides and social behaviour: Effects of oxytocin and vasopressin in humans. *Progress in Brain Research* 170:337–350.

Hektner, J. M., J. A. Schmidt, and M. Csikszentmihalyi. 2006. *Experience Sampling Method: Measuring the Quality of Everyday Life.* Thousand Oaks, CA: Sage.

Helander, B. 1988. *The Slaughtered Camel: Coping with Fictitious Descent among the Hubeer of Southern Somalia.* Uppsala, Sweden: University of Uppsala.

———. 1991. Words, worlds and wishes: The aesthetics of Somali kinship. *Cultural Anthropology* 6 (1): 113–120.

Henderson, S., and M. Gilding. 2004. "I've never clicked so much with anyone in my life": Trust and hyperpersonal communication in online friendships. *New Media and Society* 6 (4): 487–506.

Hendrick, C., and S. S. Hendrick. 1994. Attachment theory and close adult relationships. *Psychological Inquiry* 5 (1): 38–41.

Hendrick, S. S., and C. Hendrick. 1987. Love and sexual attitudes, self-disclosure and sensation seeking. *Journal of Social and Personal Relation-ships* 4 (3): 281–297.

Hendrix, L. 1976. Kinship, social networks, and integration among Ozark resi-dents and out-migrants. *Journal of Marriage and Family* 38:97–104.

Hendry, L. B., and M. Reid. 2000. Social relationships and health: The meaning of social "connectedness" and how it relates to health concerns for rural Scottish adolescents. *Journal of Adolescence* 23 (6): 705–719.

Henrich, J., et al. 2006. Costly punishment across human societies. *Science* 312 (5781): 1767–1770.

Henrich, N., and J. Henrich. 2007. *Why Humans Cooperate.* New York: Oxford University Press.

Henry J. Kaiser Family Foundation. 2006. Trends and indicators in the chang-ing health care marketplace. Pub. No. 7031. http://www.kff.org.insurance/7031/, accessed March 22, 2010.

Herdt, G., and M. K. McClintock. 2000. The magical age of 10. *Archives of Sexual Behavior* 29 (6): 587–606.

Herrmann, E., J. Call, M. V. Hernandez-Lloreda, B. Hare, and M. Tomasello. 2008. Humans have evolved specialized skills of social cognition: The cul-tural intelligence hypothesis. *Science* 317:1360–1366.

Herskovitz, M. 1934. The "best friend" in Dahomey. In *Negro Anthology*, ed. N. Cunard. London: Wishart and Co.

Hess, B. 1972. Friendship. In *Aging and Society*. Vol. 3, *A Sociology of Age Stratification*, ed. M. W. Riley. New York: Russell Sage Foundation.

Hickman, J. M. 1963. The Aymara of Chinchera, Peru: Persistence and change in bicultural context. PhD diss., Cornell University, Ithaca, NY.

Hilger, M. I. 1951. *Chippewa Child Life and Its Cultural Background*. Washington, D.C.: Government Printing Office.

Hill, P. 1972. *Rural Hausa: A Village and Setting*. Cambridge: Cambridge University Press.

Hill, R. A., and R. I. M. Dunbar. 2003. Social network size in humans. *Human Nature* 14 (1): 53–72.

Hinde, R., and S. Atkinson. 1970. Assessing the roles of social partners in maintaining mutual proximity. *Animal Behaviour* 18:169–176.

Hinde, R. A. 1997. *Relationships: A Dialectical Perspective*. Hove, United Kingdom: Psychology Press.

Hinfelaar, H. F. 1994. *Bemba-Speaking Women in Zambia in a Century of Religious Change (1892–1992)*. Leiden: E. J. Brill.

Hirsch, F. 1976. *Social Limits to Growth*. Cambridge: Harvard University Press.

Hocart, A. M. 1929. *Lau Islands, Fiji*. Honolulu: Bernice P. Bishop Museum.

Hoff, E. V. 2004. A friend living inside me—the forms and functions of imaginary companions. *Imagination, Cognition and Personality* 24 (2): 151–189.

Hofferth, S., J. Boisjoly, and G. J. Duncan. 1999. The development of social capital. *Rationality and Society* 11 (1): 79–110.

Hogbin, I. 1939. Native land tenure in New Guinea. *Oceania* 10:113–165.

———. 1946. Puberty to marriage: A study of the sexual life of the natives of Wogeo, New Guinea. *Oceania* 16:185–209.

Hollis, A. C. 1905. *The Masai: Their Language and Folklore*. Oxford: Clarendon Press.

Honigmann, J. J. 1949. *Culture and Ethos of Kaska Society*. New Haven, CT: Yale University Press.

Hopkins, J. 2002. *Cultivating Compassion: A Buddhist Perspective*. New York: Broadway Books.

Hornik, J. 1992. Tactile stimulation and consumer response. *Journal of Consumer Research* 19:449–458.

Horowitz, L. M., and R. French. 1979. Interpersonal problems of people who describe themselves as lonely. *Journal of Consulting and Clinical Psychology* 47 (4): 762–764.

Horrocks, J. E., and G. G. Thompson. 1946. A study of the friendship fluctuations of rural boys and girls. *Journal of Genetic Psychology* 69:189–198.

Howard, D. J., C. E. Genglar, and A. Jain. 1997. The name remembrance effect: A test of alternative explanations. *Journal of Social Behaviour and Personality* 12:801–810.

Howell, P. P. 1954. *A Manual of Nuer Law: Being an Account of Custom-*

ary Law, Its Evolution and Development in the Courts Established by the Sudan Government. London: Oxford University Press.

Howell, W. 1908–1910. The Sea Dyak. *Sarawak Gazette*, 38–40.

Howes, C. 1983. Patterns of friendship. *Child Development* 54 (4): 1041–1053.

———. 1988. Peer interaction of young children. *Monographs of the Society for Research in Child Development* 53 (1): 1–88.

Howes, C., C. E. Hamilton, and L. C. Philipsen. 1998. Stability and continuity of child-caregiver and child-peer relationships. *Child Development* 69 (2): 418–426.

Hruschka, D. J. 2006. The diverse functions of friendship. PhD diss., Emory University, Atlanta.

———. 2009. Defining cultural competence in context: Dyadic norms of friendship among U.S. high school students. *Ethos* 37 (2): 205–224.

Hruschka, D. J., and J. Henrich. 2006. Friendship, cliquishness, and the emergence of cooperation. *Journal of Theoretical Biology* 239 (1): 1–15.

Huber, D., P. Veinante, and R. Stoop. 2005. Vasopressin and oxytocin excited distinct neuronal populations in the central amygdala. *Science* 308 (5719): 245–248.

Hugh-Jones, S. 1979. *The Palm and the Pleiades: Initiation and Cosmology in Northwest Amazonia.* Cambridge: Cambridge University Press.

Hungry Wolf, B. 1980. *The Ways of My Grandmothers.* New York: Morrow.

Hunt, G., and S. Satterlee. 1986. Cohesion and division: Drinking in an English village. *Man* 21 (3): 521–537.

Hunt, M. E. V. 1962. *The Dynamics of the Domestic Group in Two Tzeltal Villages: A Constrastive Comparison.* Chicago: University of Chicago Press.

Huntingford, G. W. B. 1953. *The Southern Nilo-Hamites.* London: International African Institute.

Hurd, J. P. 1983. Kin relatedness and church fissioning among the "Nebraska" Amish of Pennsylvania. *Social Biology* 30 (1): 59–66.

Hurlbert, H. B. 1910. The status of woman in Korea. *Korea Review* 1:529–534.

Hurlbert, J. S., and A. C. Acock. 1990. The effects of marital status on the form and composition of social networks. *Social Science Quarterly* 71:163–171.

Hutchinson, H. W. 1957. *Village and Plantation Life in Northeastern Brazil.* Seattle: University of Washington Press.

Hutereau, A. 1909. *Notes sur la vue familiale et juridique de quelques populations du Congo Belge.* Brussels: Musee du Congo Belge.

Hyde, J. S. 2005. The gender similarity hypothesis. *American Psychologist* 60 (6): 581–592.

Ibrahim, S. E. 1980. Anatomy of Egypt's militant Islamic groups: Methodological note and preliminary findings. *International Journal of Middle East Studies* 12 (4): 423–453.

Ikkink, K. K., and T. van Tilburg. 1999. Broken ties: Reciprocity and other factors affecting the termination of older adults' relationships. *Social Networks* 21 (2): 131–146.

Ingold, T. 1976. *The Skolt Lapps Today.* London: Cambridge University Press.

Ingram, P., and P. W. Roberts. 2000. Friendships among competitors in the Sydney hotel industry. *American Journal of Sociology* 106 (2): 387–423.

Ingram, P., and X. Zou. 2008. Business friendships. *Research in Organizational Behavior* 28:167–184.

Insel, T. R. 2003. Is social attachment an addictive disorder? *Physiology and Behavior* 79:351–357.

Insel, T. R., and R. D. Fernald. 2004. How the brain processes social information: Searching for the social brain. *Annual Review of Neuroscience* 27:697–722.

Insel, T. R., and L. J. Young. 2001. The neurobiology of attachment. *Nature Reviews Neuroscience* 2 (2): 129–136.

Irvine, J. T. 1974. *Caste and Communication in a Wolof Village.* Philadelphia: University of Pennsylvania.

Islam, A. K. M. A. 1974. *A Bangladesh Village: Conflict and Cohesion: An Anthropological Study of Politics.* Cambridge, MA: Schenkman Pub. Co.

Itkonen, T. I. 1948. *The Lapps in Finland Up to 1945.* Porvoo, Finland, and Helsinki: Werner Soderstrom Osakeyhtio.

Ivey, P. K. 2000. Cooperative reproduction in Ituri forest hunter-gatherers: Who cares for Efe infants? *Current Anthropology* 41 (5): 856–866.

Jackson, R. M. 1977. Social structure and process in friendship choice. In *Networks and Places: Social Relations in the Urban Setting,* ed. C. Fischer. New York: Free Press.

Jacobs, J. B. 1977. Local politics in rural Taiwan: A field study of kuan-hsi, face, and faction in Matsu township. PhD diss., Columbia University, New York.

Jacobson, D. 1973. *Itinerant Townsmen: Friendship and Social Order in Urban Uganda.* Menlo Park, CA: Cummings Publishing Co.

Jacobson-Widding, A. 1981. Friendship, trust, and social structure: A comparative analysis of bond friendship among two ethnic groups in Congo-Brazzaville. *Folk: Dansk ethnografisk tidsskrift Kobenhavn* 23:45–63.

Jaffee, S., and J. S. Hyde. 2000. Gender differences in moral orientation: A meta-analysis. *Psychological Bulletin* 126 (5): 703–726.

James, D., and J. Drakich. 1993. Understanding gender differences in amount of talk: A critical review of research. In *Gender and Conversational Interaction,* ed. D. Tannen. Oxford: Oxford University Press.

Janelli, R. L. 1982. *Ancestor Worship and Korean Society.* Stanford, CA: Stanford University Press.

Jankowiak, W. R., and E. F. Fischer. 1992. A cross-cultural perspective on romantic love. *Ethnology* 31 (2): 149–155.

Jay, R. 1969. *Javanese Villagers: Social Relations in Rural Modjokuto.* Cambridge: MIT Press.

Jenness, D. 1922. *The Life of the Copper Eskimos.* Ottawa: F. A. Acland.

———. 1959. *The People of the Twilight.* Chicago: University of Chicago Press.

Jerrome, D., and G. C. Wenger. 1999. Stability and change in late-life friendships. *Ageing and Society* 19:661–676.

Jochelson, W. 1908. *The Koryak.* Leiden: E. J. Brill.

Johnson, A. J. 2001. Examining the maintenance of friendships: Are there differences between geographically close and long-distance friends? *Communication Quarterly* 49 (4): 424–435.

Johnson, B. 2007. Twenty-eight people ask Hugh MacLeod to be their friend each day. What's so special about him? *The Guardian,* December 15.

Johnson, D. D. P., and O. Kruger. 2004. The good of wrath: Supernatural punishment and the evolution of cooperation. *Political Theology* 5 (2): 159–176.

Johnson, F. L., and E. J. Aries. 1983. Conversational patterns among same-sex pairs of late-adolescent close friends. *Journal of Genetic Psychology* 142 (2): 225–238.

Johnson, M. P., and L. Leslie. 1982. Couple involvement and network structure: A test of the dyadic withdrawal hypothesis. *Social Psychology Quarterly* 45:34–43.

Johnson, M. W., and W. K. Bickel. 2002. Within-subject comparison of real and hypothetical money rewards in delay discounting. *Journal of the Experimental Analysis of Behavior* 77 (2): 129–146.

Joiris, D. V. 2003. The framework of central African hunter-gatherers and neighboring societies. *African Study Monographs* suppl. 28:57–79.

Jones, B., and H. Rachlin. 2006. Social discounting. *Psychological Science* 17 (4): 283–286.

Jones, D. 2000. Group nepotism and human kinship. *Current Anthropology* 41 (5): 779–809.

Jones, D. C. 1985. Persuasive appeals and responses to appeals among friends and acquaintances. *Child Development* 56:757–763.

Jones, D. C., J. B. Newman, and S. Bautista. 2005. A three-factor model of teasing: The influence of friendship, gender, and topic on expected emotional reactions to teasing during early adolescence. *Social Development* 14 (3): 421–439.

Jones, D. C., and K. Vaughan. 1990. Close friendships among senior adults. *Psychology and Aging* 5 (3): 451–457.

Jones, S. B. 1973. Geographic mobility as seen by the wife and mother. *Journal of Marriage and Family* 35 (2): 210–218.

Juhasz, A. M. C. 1979. A concept of divorce. *Journal of Family and Economic Issues* 2 (4): 471–482.

Kalmijn, M. 2002. Sex segregation of friendship networks: Individual and structural determinants of having cross-sex friends. *European Sociological Review* 18:101–117.

———. 2003. Shared friendship networks and the life course: An analysis of survey data on married and cohabiting couples. *Social Networks* 25 (3): 231–249.

Kan, S. 1989. *Symbolic Immortality: The Tlingit Potlatch of the Nineteenth Century.* Washington, D.C.: Smithsonian Institution Press.

Kane, P. 1925. *Wanderings of an Artist among the Indians of North America from Canada to Vancouver's Island and Oregon through the Hudson's*

Bay Company's Territory and Back Again. Toronto: Radisson Society of Canada.

Kang, Y. 1931. *The Grass Roof.* New York: Scribner's.

Kanter, R. M. 1972. *Commitment and Community: Communes and Utopias in Sociological Perspective.* Cambridge: Harvard University Press.

Karp, I. 1980. Beer drinking and social experience in an African society: An essay in formal sociology. In *Explorations in African Systems of Thought,* ed. I. Karp and C. S. Bird. Bloomington: Indiana University Press.

Karremans, J. C., and H. Aarts. 2007. The role of automaticity in determining the inclination to forgive close others. *Journal of Experimental Social Psychology* 43 (6): 902–917.

Karremans, J. C., and P. A. M. Van Lange. 2008. The role of forgiveness in shifting from "me" to "we." *Self and Identity* 7:75–88.

Karsten, R. 1932. *Indian Tribes of the Argentine and Bolivian Chaco: Ethnological Studies.* Helsinki: Akademische Buchlandlung.

Kasarda, J. D., and M. Janowitz. 1974. Community attachment in mass society. *American Sociological Review* 47:427–433.

Katz, J. E., and P. Aspden. 1997. A nation of strangers? *Communications— ACM* 40 (12): 81–86.

Katz, J. E., and R. E. Rice. 2002. *Social Consequences of Internet Use: Access, Involvement, and Interaction.* Cambridge: MIT Press.

Kaufman, H. K. 1960. *Bangkhuad: A Community Study in Thailand.* Locust Valley, NY: Association for Asian Studies.

Kedit, P. M. 1991. Meanwhile, back home: Bejalai and its effects on Iban men and women. In *Female and Male in Borneo: Contributions and Challenges to Gender Studies,* ed. V. H. Sutlive. Williamsburg, VA: Borneo Research Council, Inc.

Keesing, R. 1972. Simple models of complexity: The lure of kinship. In *Kinship Studies in the Morgan Centennial Year,* ed. P. Reining. Washington, D.C.: Anthropological Society of Washington.

Keil, C. 1979. *Tiv Song.* Chicago: University of Chicago Press.

Keller, M. 2004. Self in relationship. In *Morality, Self and Identity: Essays in Honor of Augusto Blasi,* ed. D. K. Lapsley and D. Narvaez. Mahwah, NJ: Erlbaum.

Keller, M., W. Edelstein, C. Schmid, F. Fang, and G. Fang. 1998. Reasoning about the responsibilities and obligations in close relationships: A comparison of two cultures. *Developmental Psychology* 34:731–741.

Keller, M., and P. Wood. 1989. Development of friendship reasoning: A study of interindividual differences in intraindividual change. *Developmental Psychology* 25:820–826.

Kelley, H. H. 1986. Personal relationships: Their nature and significance. In *The Emerging Field of Personal Relationships,* ed. R. Gilmour and S. Duck. Mahwah, NJ: L. Erlbaum.

Keltner, D., L. Capps, A. M. Kring, R. C. Young, and E. A. Heerey. 2001. Just teasing: A conceptual analysis and empirical review. *Psychological Bulletin* 127 (2): 229–248.

Kemp, P. 1935. *Healing Ritual: Studies in the Technique and Tradition of the Southern Slavs*. London: Faber and Faber Limited.

Kennedy, G. F. 1977. The Korean kye: Maintaining human scale in a modernizing society. *Korean Studies* 1:197–222.

Kennedy, R. 1986. Women's friendships on Crete: A psychological perspective. In *Gender and Power in Rural Greece*, ed. J. Dubisch. Princeton, NJ: Princeton University Press.

Kenny, D. A., and L. LaVoie. 1982. Reciprocity of interpersonal interaction: A confirmed hypothesis. *Social Psychology Quarterly* 45:54–58.

Kenny, D. A., and W. Nasby. 1980. Splitting the reciprocity correlation. *Journal of Personality and Social Psychology* 38:249–256.

Kenrick, D. T., and M. R. Trost. 2000. An evolutionary perspective on human relationships. In *The Social Psychology of Personal Relationships*, ed. W. Ickes and S. Duck. New York: Wiley.

Kephart, W. M. 1967. Some correlates of romantic love. *Journal of Marriage and Family* 29 (3): 470–474.

Khuri, F. I. 1968. The etiquette of bargaining in the Middle East. *American Anthropologist* 70 (4): 698–706.

Kiefer, T. M. 1968. Institutionalized friendship and warfare among the Tausug of Jolo. *Ethnology* 7:225–224.

———. 1972. *The Tausug: Violence and Law in a Philippine Moslem Society*. New York: Holt, Rinehart and Winston.

Kim, C. S. 1992. *The Culture of Korean Industry: An Ethnography of Poongsan Corporation*. Tucson: University of Arizona Press.

Kim, H. J., and J. B. Stiff. 1991. Social networks and the development of close relationships. *Human Communication Research* 18 (1): 70–91.

Kim, H. S., D. K. Sherman, D. Ko, and S. E. Taylor. 2006. Pursuit of comfort and pursuit of harmony: Culture, relationships, and social support seeking. *Personality and Social Psychology Bulletin* 32 (12): 1595–1607.

Kim, K. H., and H. Yun. 2007. Cying for me, cying for us: Relational dialectics in a Korean social network site. *Journal of Computer Mediated Communication* 13 (1): 298–318.

Kirsch, P., C. Esslinger, Q. Chen, D. Mier, S. Lis, S. Siddhanti, H. Gruppe, V. S. Mattay, B. Gallhofer, and A. Meyer-Lindenberg. 2005. Oxytocin modulates neural circuitry for social cognition and fear in humans. *Journal of Neuroscience* 25 (49): 11489–11493.

Kito, M. 2005. Self-disclosure in romantic relationships and friendships among American and Japanese college students. *Journal of Social Psychology* 145 (2): 127–140.

Klass, M. 1978. *From Field to Factory: Community Structure and Industrialization in West Bengal*. Philadelphia: Institute for the Study of Human Issues.

Klohnen, E. C., J. A. Weller, S. Luo, and M. Choe. 2005. Organization and predictive power of general and relationship-specific attachment models: One for all, and all for one? *Personality and Social Psychology Bulletin* 31 (12): 1665–1682.

Knez, E. I. 1960. *Sam Jong Dong: A South Korean Village.* Ann Arbor: University Microfilms.

Knippenberg, B., and H. Steensma. 2003. Future interaction expectation and the use of soft and hard influence tactics. *Applied Psychology: An International Review* 52 (1): 55–67.

Kogan, L. 2006. Oprah and Gayle, uncensored. *O,* August 1.

Kohl, J. G. [1860] 1985. *Kitchi-Gami: Life among the Superior Ojibway.* St. Paul, MN: Borealis Books.

Kohlberg, L. 1969. Stage and sequence: The cognitive-developmental approach to socialization. In *Handbook of Socialization Theory and Research,* ed. D. Goslin. Chicago: Rand-McNally.

Kojetin, B. A. 1996. Adult attachment styles with romantic partners, friends, and parents. PhD diss., University of Minnesota, Minneapolis.

Komarovsky, M. 1962. *Blue-Collar Marriage.* New York: Random House.

Komter, A. E. 2004. *Social Solidarity and the Gift.* Cambridge: Cambridge University Press.

Komter, A. E., and W. Vollebergh. 2002. Solidarity in Dutch families: Family ties under strain? *Journal of Family Issues* 23 (2): 171–188.

Kon, I. S., and V. A. Losenkov. 1978. Friendship in adolescence: Values and behavior. *Journal of Marriage and Family* 40 (1): 143–155.

Konner, M. 1975. Relations among infants and juveniles in comparative perspective. In *The Origins of Behavior: Friendship and Peer Relations,* ed. M. Lewis and L. A. Rosenblum. New York: Wiley.

Korchmaros, J. D., and D. A. Kenny. 2001. Emotional closeness as a mediator of the effect of genetic relatedness on altruism. *Psychological Science* 12 (3): 262–265.

————. 2006. An evolutionary and close-relationship model of helping. *Journal of Social and Personal Relationships* 23 (1): 21–43.

Korpi, T. 2001. Good friends in bad times? Social networks and job search among the unemployed in Sweden. *Acta Sociologica* 44:157–170.

Kosfeld, M., M. Heinrichs, P. J. Zak, U. Fischbacher, and E. Fehr. 2005. Oxytocin increases trust in humans. *Nature* 435:673–676.

Kovacs, D. M., J. G. Parker, and L. W. Hoffman. 1996. Behavioral, affective, and social correlates of involvement in cross-sex friendship in elementary school. *Child Development* 67:2269–2286.

Kramer, K. 1978. Tebu. In *Muslim Peoples: A World Ethnographic Survey,* ed. R. V. Weekes. Westport, CT: Greenwood Press.

Kranton, R. E. 1996. Reciprocal exchange: A self-sustaining system. *American Economic Review* 86 (4): 830–851.

Krappmann, L. 1996. Amicitia, drujba, shin-yu, philia, freundschaft, friendship: On the cultural diversity of a human relationship. In *The Company*

They Keep: Friendships in Childhood and Adolescence, ed. W. M. Bukowski and A. F. Newcomb. Cambridge: Cambridge University Press.

Kremensek, S. 1983. On the fringe of the town. In *Urban life in Mediterranean Europe: Anthropological Perspectives*, ed. M. Kenny and D. I. Kertzer. Urbana: University of Illinois Press.

Kropotkin, P. 1902. *Mutual Aid: A Factor of Evolution*. London: William Heinemann.

Krueger, F., K. McCabe, J. Moll, N. Kriegeskorte, R. Zahn, M. Strenziok, and A. Heinecke, and J. Grafman. 2007. Neural correlates of trust. *PNAS* 104 (50): 20084–20089.

Kruger, D. J. 2001. Psychological aspects of adaptations for kin directed altruistic helping behaviors. *Social Behavior and Personality* 29 (4): 323–330.

———. 2003. Evolution and altruism: Combining psychological mediators with naturally selected tendencies. *Evolution and Human Behavior* 24:118–125.

Kunkel, A. W., and B. R. Burleson. 1998. Social support and emotional lives of men and women: An assessment of the different cultures perspective. In *Sex Differences and Similarities in Communication: Critical Essays and Empirical Investigations and Gender in Interaction*, ed. D. J. Canary and K. Dindia. Mahwah, NJ: Lawrence Erlbaum.

Kurcer, M. A., and E. Pehlivan. 2002. Hepatitis B seroprevalence and risk factors in urban areas of Malatya. *Turkish Journal of Gastroenterology* 13 (1): 1–5.

Kurdek, L. A. 2002. On being insecure about the assessment of attachment styles. *Journal of Social and Personal Relationships* 19 (6): 811–834.

Kurdek, L. L., and D. Krile. 1982. A developmental analysis of the relation between peer acceptance and both interpersonal understanding and perceived social self-competence. *Child Development* 53 (6): 1485–1491.

Kurth, S. B. 1970. Friendships and friendly relations. In *Social Relationships*, ed. G. J. McCall. Chicago: Aldine.

Kutsukake, N. 2006. The context and quality of social relationships affect vigilance behaviour in wild chimpanzees. *Ethology* 112 (6): 581–591.

La Freniere, P. J., and W. R. Charlesworth. 1987. Effects of friendship and dominance status on preschoolers' resource utilization in a cooperative/competitive situation. *International Journal of Behavioral Development* 10 (3): 345–358.

La Gaipa, J. J. 1977. Testing a multidimensional approach to friendship. In *Theory and Practice in Interpersonal Relationships*, ed. S. W. Duck. New York: Academic Press.

———. 1979. A developmental study of the meaning of friendship in adolescence. *Journal of Adolescence* 2 (3): 201–213.

Lagace, R. O. 1979. The HRAF probability sample: Retrospect and prospect. *Cross-cultural Research* 14 (3): 211–229.

Lagae, C. R. 1926. *Les Azande or Niam-Niam: Zande Organizations, Religious and Magical Beliefs*. Brussels: Vromant and Co.

Lakoff, G., and M. Johnson. 1999. *Philosophy in the Flesh: The Embodied Mind and Its Challenge to Western Thought.* New York: Basic Books.

Lamm, H., and T. Schwinger. 1980. Norms concerning distributive justice: Are needs taken into consideration in allocation decisions? *Social Psychology Quarterly* 43 (4): 425–429.

Lancaster, R. 1966. *Piegan: A Look from Within at the Life, Times, and Legacy of an American Indian Tribe.* Garden City, NY: Doubleday.

Landa, J. T. 1994. *Trust, Ethnicity, and Identity: Beyond the New Institutional Economics of Ethnic Trading Networks.* Ann Arbor: University of Michigan Press.

Landes, R. 1937. *Ojibwa Sociology.* New York: Columbia University Press.

Lang, O. 1946. *Chinese Family and Society.* New Haven, CT: Yale University Press.

Larken, P. M. 1927. Impressions of the Azande. *Sudan Notes and Records* 10:86–134.

Larson, R., and M. Csikszentmihalyi. 1983. The experience sampling method. *New Directions for Methodology of Social and Behavioral Science* 15:41–56.

Larson, R., M. H. Richards, G. Moneta, G. Holmbeck, and E. Duckette. 1996. Changes in adolescents: Daily interactions with their families from ages 10 to 18: Disengagement and transformation. *Developmental Psychology* 32 (4): 744–754.

Larson, R., and S. Verma. 1999. How children and adolescents spend time across the world: Work, play, and developmental opportunities. *Psychological Bulletin* 125 (6): 701–736.

Larson, R., S. Verma, and J. Dworkin. 2003. Adolescent without family disengagement: The daily family lives of Indian middle class teenagers. In *Cross-cultural Perspectives in Human Development,* ed. T. S. Saraswathi. Newbury Park, CA: Sage.

Larson, R. W., and N. Bradney. 1988. Precious moments with family members and friends. In *Families and Social Networks,* ed. R. M. Milardo. Thousand Oaks, CA: Sage.

Lasnet, D. 1900. *Une mission au Senegal.* Paris: Augustin Challamel.

Latour, F. 2000. Two 15-year-olds charged in T case: Girls face counts of attempted rape. *Boston Globe,* February 1, 2000.

Laumann, E. O. 1973. *Bonds of Pluralism: The Form and Substance of Urban Social Networks.* New York: Wiley.

Laursen, B. P., B. D. Finkelstein, and N. T. Betts. 2001. A developmental meta-analysis of peer conflict resolution. *Developmental Review* 21 (4): 423–449.

Leach, E. 1961. *Pul Eliya, a Village in Ceylon: A Study of Land Tenure and Kinship.* Cambridge: Cambridge University Press.

Leach, J. W., and E. R. Leach. 1983. *The Kula: New Perspectives on Massim Exchange.* Cambridge: Cambridge University Press.

Leblebicioglu, H., D. Turan, M. Sunbul, S. Esen, and C. Eroglu. 2003. Transmission of human immunodeficiency virus and hepatitis B virus by blood brotherhood rituals. *Scandinavian Journal of Infectious Disease* 35 (3): 210.

Ledbetter, A.M., E.M. Griffin, and G.G. Sparks. 2007. Forecasting "friends forever": Longitudinal investigations of sustained closeness between best friends. *Personal Relationships* 14 (2): 343–350.

Ledeneva, A.V. 1998. *Russia's Economy of Favours: Blat, Networking, and Informal Exchange*. Cambridge: Cambridge University Press.

Lederberg, A.R. 1987. Temporary and long-term friendships in hearing and deaf preschoolers. *Merrill-Palmer Quarterly* 33 (4): 515–533.

Leech, G., P. Rayson, and A. Wilson. 2001. *Word Frequencies in Written and Spoken English: Based on the British National Corpus*. London: Longman.

Leider, S., M.M. Mobius, T. Rosenblat, and Q. Do. 2009. Directed altruism and enforced reciprocity in social networks. *Quarterly Journal of Economics* (November): 1815–1851.

Lemche, E., V.P. Giampietro, S.A. Surguladze, E.J. Amaro, C.M. Andrew, S.C.R. Williams, M.J. Brammer, et al. 2006. Human attachment security is mediated by the amygdala: Evidence from combined fMRI and psychophysiological measures. *Human Brain Mapping* 27 (8): 623–635.

Lempers, J.D., and D.S. Clark-Lempers. 1992. Young, middle, and late adolescents' comparisons of the functional importance of five significant relationships. *Journal of Youth and Adolescence* 21 (1): 53–96.

Lenhart, A., M. Madden, and P. Hitlin. 2005. *Teens and Technology: Youth Are Leading the Transition to a Fully Wired and Mobile Nation*. Washington, D.C.: Pew Internet and American Life Project.

Lenoir, J.D. 1973. The Paramacca Maroons: A study in religious acculturation. PhD diss., New School for Social Research, New York.

Lesser, A. 1933. *The Pawnee Ghost Dance Hand Game*. New York: Columbia University Press.

Levine, D.N. 1965. *Wax and Gold: Tradition and Innovation in Ethiopian Culture*. Chicago: University of Chicago Press.

Levine, R. 2006. *The Power of Persuasion: How We're Bought and Sold*. Hoboken, N.J.: Wiley.

Levine, R., S. Sato, T. Hashimoto, and J. Verma. 1995. Love and marriage in eleven cultures. *Journal of Cross-cultural Psychology* 26 (5): 554–571.

Levy, M. 1949. *The Family Revolution in China*. Cambridge: Harvard University Press.

Lewis, C.S. 1960. *The Four Loves*. New York: Harcourt, Inc.

Lewis, O. 1951. *Life in a Mexican Village: Tepoztlan Restudied*. Urbana: University of Illinois Press.

Leyton, E. 1974a. Friends and "friends": The nexus of friendship, kinship and class in Aughnaboy. In *The Compact: Selected Dimensions of Friendship*, ed. E. Leyton. St. John's, Canada: Memorial University of Newfoundland.

———, ed. 1974b. *The Compact: Selected Dimensions of Friendship*. St. John's, Canada: Institute of Social and Economic Research, Memorial University of Newfoundland.

Li, H. Z. 2002. Culture, gender and self-close-other(s) connectedness in Canadian and Chinese samples. *European Journal of Social Psychology* 32 (1): 93–104.

———. 2003. Inter- and intra-cultural variations in self-other boundary: A qualitative-quantitative approach. *International Journal of Psychology* 38 (3): 138–150.

Li, H. Z, Z. Zhang, G. Bhatt, and Y. Yum. 2006. Rethinking culture and self-construal: China as middle land. *Journal of Social Psychology* 146:5, 591–605.

Li, N. P., R. A. Halterman, M. J. Cason, G. P. Knight, and J. K. Maner. 2008. The stress-affiliation paradigm revisited: Do people prefer the kindness of strangers or their attractiveness? *Personality and Individual Differences* 44 (2): 382–391.

Liberman, M. 2006. Sex on the brain. *Boston Globe*, September 24.

Lindenbaum, S. 1979. *Kuru Sorcery: Disease and Danger in the New Guinea Highlands.* Mountain View, CA: Mayfield Publishing Company.

Lindenbaum, S., and R. Glasse. 1969. Fore age mates. *Oceania* 39:165–183.

Lindholm, C. 1982. *Generosity and Jealousy: The Swat Pukhtun of Northern Pakistan.* New York: Columbia University Press.

Linnekin, J. 1985. *Children of the Land: Exchange and Status in a Hawaiian Community.* New Brunswick, NJ: Rutgers University Press.

Linton, R. 1933. *The Tanala: A Hill Tribe of Madagascar.* Chicago: Field Museum Press.

Litwak, E. 1960. Occupational mobility and extended family cohesion. *American Sociological Review* 25:9–21.

Litwak, E., and I. Szelenyi. 1969. Primary group structures and their functions: Kin, neighbors, and friends. *American Sociological Review* 34:465–481.

Livingstone, D., and C. Livingstone. 1865. *Narratives of an Expedition to the Zambesi.* London: John Murray.

Lodge, O. 1941. *Peasant Life in Jugoslavia.* London: Selley, Service & Co.

Loewenstein, G. F., and D. A. Small. 2007. The scarecrow and the tin man: The vicissitudes of human sympathy. *Review of General Psychology* 11 (2): 112–126.

Logan, J. R., and G. D. Spitze. 1996. *Family Ties: Enduring Relations between Parents and Their Grown Children.* Philadelphia: Temple University Press.

Loizos, P. 1991. Gender and kinship in marriage and alternative contexts. In *Contested Identities: Gender and Kinship in Modern Greece*, ed. P. Loizos and E. Papataxiarchis. Princeton, NJ: Princeton University Press.

Lonkila, M. 1997. Informal exchange in relations in post-Soviet Russia: A comparative perspective. *Sociological Research Online* 2 (2).

Low, B. 1988. Measures of polygyny in humans. *Current Anthropology* 29 (1): 189–194.

Low, H. B. 1892. The natives of Borneo. *Journal of the Anthropological Institute of Great Britain and Ireland* 21:110–137.

Lowe, C. A., and J. W. Goldstein. 1970. Reciprocal liking and attributions of ability: Mediating effects of perceived intent and personal involvement. *Journal of Personality and Social Psychology* 16 (2): 291–298.

Lowie, R.J. 1929. *Hopi Kinship*. New York: American Museum of Natural History.

Luce, R.D., and H. Raiffa. 1957. *Games and Decisions*. New York: Wiley.

Lydon, J.E., D.W. Jamieson, and J.G. Holmes. 1997. The meaning of social interactions in the transition from acquaintanceship to friendship. *Journal of Personality and Social Psychology* 73 (3): 536–548.

Lynd, R.S., and H.M. Lynd. 1956. *Middletown: A Study in Contemporary American Culture*. New York: Harcourt, Brace and Company.

Maccoby, E. 1990. Gender and relationships: A developmental account. *American Psychologist* 45 (4): 513–520.

———. 1998. *The Two Sexes: Growing Up Apart, Coming Together*. Cambridge: Harvard University Press.

MacDougall, J.J. 1981. Elite friendship ties and their political-organizational functions: The case of Indonesia. *Bijdragen tot de taal, land en volkenkunde* 137 (1): 61–89.

MacGeorge, E.L., A.R. Graves, B. Feng, S.J. Gillihan, and B.R. Burleson. 2004. The myth of gender cultures: Similarities outweigh differences in men's and women's provision of and responses to supportive communication. *Sex Roles* 50 (3–4): 143–175.

MacRae, H. 1996. Strong and enduring ties: Older women and their friends. *Canadian Journal of Aging* 15:374–392.

Madden, G.J., B.R. Raiff, C.H. Lagorio, A.M. Begotka, A.M. Mueller, D.J. Hehli, and A.A. Wegener. 2004. Part 1, Delay discounting of potentially real and hypothetical rewards. Part 2, Between- and within-subject comparisons. *Experimental and Clinical Psychopharmacology* 12 (4): 251–261.

Madsen, E.A., R.J. Tunney, G. Fieldman, and H.C. Plotkin. 2007. Kinship and altruism: A cross-cultural experimental study. *British Journal of Psychology* 98 (2): 339–359.

Magdol, L., and D.R. Bessel. 2003. Social capital, social currency, and portable assets: The impact of residential mobility on exchanges of social support. *Personal Relationships* 10:149–169.

Magel, E.A. 1984. *Folktales from Gambia: Wolof Fictional Narratives*. Washington, D.C.: Three Continents Press.

Mahony, F.J. 1971. A Trukese theory of medicine. PhD diss., Stanford University, Palo Alto, CA.

Maier, R.A., and P.J. Lavrakas. 1976. Lying behavior and evaluation of lies. *Perceptual and Motor Skills* 42:575–581.

Mailath, G.J., and L. Samuelson. 2006. *Repeated Games and Reputations: Long-Run Relationships*. Oxford: Oxford University Press.

Majolo, B., K. Ames, R. Brumpton, R. Garratt, K. Hall, and N. Wilson. 2006. Human friendship favours cooperation in the iterated prisoner's dilemma. *Behaviour* 143:1383–1395.

Malherbe, W. A. 1931. *Tiv-English Dictionary with Grammar Notes and Index*. Lagos: Government Printer.

Malinowski, B. 1922. *Argonauts of the West Pacific*. New York: Dutton.

———. 1926. *Crime and Custom in Savage Society*. London: Kegan Paul, Trench, Trubner and Company.

———. 1929. *The Sexual Life of Savages in North Western Melanesia*. New York: Horace Liveright.

———. 1935. *Coral Gardens and Their Magic: A Study of the Methods of Tilling the Soil and of Agricultural Rites in the Trobriand Islands*. New York: American Book Company.

Man, E. H. 1932. *On the Aboriginal Inhabitants of the Andaman Islands*. London: Royal Anthropological Institute of Great Britain and Ireland.

Mandelbaum, D. G. 1936. Friendship in North America. *Man* 36:205–2–6.

Maner, J. K., C. N. DeWall, R. F. Baumeister, and M. Schaller. 2007. Does social exclusion motivate interpersonal reconnection? Resolving the "porcupine problem." *Journal of Personality and Social Psychology* 92 (1): 42–55.

Maner, J. K., and M. T. Gailliot. 2006. Altruism and egoism: Prosocial motivations for helping depend on relationship context. *European Journal of Social Psychology* 37 (2): 347–358.

Maner, J. K., C. L. Luce, S. L. Neuberg, R. B. Cialdini, S. Brown, and B. J. Sagarin. 2002. The effects of perspective taking on motivations for helping: Still no evidence for altruism. *Personality and Social Psychology Bulletin* 28:1601–1610.

Manning, N. 1995. Social policy and the welfare state. In *Russia in Transition: Politics, Privatisation and Inequality*, ed. D. Lane. London: Longman.

Mannix, E. A., and G. F. Loewenstein. 1993. Managerial time horizons and interfirm mobility: An experimental investigation. *Organizational Behavioral and Human Decision Processes* 56 (2): 266–284.

Marak, K. R. 1997. *Traditions and Modernity in Matrilineal Tribal Society*. New Delhi: Inter-India Publications.

Marazziti, D., and D. Canale. 2004. Hormonal changes when falling in love. *Psychoneuroendocrinology* 29:931–936.

Markowitz, F. 1991. Russkaia Druzhba: Russian friendship in American and Israeli contexts. *Slavic Review* 50 (3): 637–645.

Marlowe, D., K. J. Gergen, and A. N. Doob. 1966. Opponent's personality, expectation of social interaction, and interpersonal bargaining. *Journal of Personality and Social Psychology* 3 (2): 206–213.

Marshall, D. S. 1950. Cuna folk: A conceptual scheme involving the dynamic factors of culture, as applied to Cuna Indians of Darien. PhD diss., Harvard University, Cambridge.

Marshall, M. 1977. The nature of nurture. *American Ethnologist* 4 (4): 643–662.

Martin, K. D., and N. C. Smith. 2008. Commercializing social interaction: The ethics of stealth marketing. *Journal of Public Policy and Marketing* 27 (1): 45–56.

Mashek, D. J., A. Aron, and M. Boncimino. 2003. Confusions of self with close others. *Personality and Social Psychology Bulletin* 29 (3): 382–392.

Masheter, C. 1991. Postdivorce relationships between ex-spouses: The roles of attachment and interpersonal conflict. *Journal of Marriage and Family* 53 (1): 103–110.

Masters, W. M. 1953. *Rowanduz: A Kurdish Administrative and Mercantile Center*. Ann Arbor, MI: University Microfilms.

Matsumoto, D., N. Haan, G. Yabrove, P. Theodorou, and C. Cooke Carney. 1986. Preschoolers' moral actions and emotions in prisoner's dilemma. *Developmental Psychology* 22 (5): 663–670.

Matthews, S. H. 1986. *Friendships through the life course*. Beverly Hills, CA: Sage.

Mauss, M. 1954. *The Gift: Forms and Functions of Exchange in Archaic Societies*. Translated by I. Cunnison. London: Cohen and West.

Mayer, A. C. 1966. *Caste and Kinship in Central India: A Village and Its Region*. Berkeley: University of California Press.

Maynard Smith, J., and E. Szathmary. 1995. *The Major Transitions in Evolution*. Oxford: Oxford University Press.

Mayseless, O., and M. Scharf. 2007. Adolescents' attachment representations and their capacity for intimacy in close relationsihps. *Journal of Research on Adolescence* 17 (1): 23–50.

McClintock, M. K., and G. Herdt. 1996. Rethinking puberty: The development of sexual attraction. *Current Directions in Psychological Science* 5 (6): 178–183.

McCollum, A. T. 1990. *The Trauma of Moving: Psychological Issues for Women*. Newbury Park, CA: Sage.

McCorkle, B. H., E. S. Rogers, E. C. Dunn, A. Lyass, and Y. M. Wan. 2008. Increasing social support for individuals with serious mental illness: Evaluating the Compeer model of intentional friendship. *Community Mental Health Journal* 44 (5): 359–366.

McCormack, A. P. 1964. Khasis. In *Ethnic Groups of Mainland Southeast Asia*, ed. F. M. Lebar, G. C. Hickey and J. K. Musgrave. New Haven, CT: HRAF Press.

McCreery, J. L. 1974. The symbolism of popular Taoist magic. PhD diss., Cornell University, Ithaca, NY.

McCullough, M. E. 2008. *Beyond Revenge: The Evolution of the Forgiveness Instinct*. San Francisco: Jossey Bass.

McCullough, M. E., K. C. Rachal, S. J. Sandage, and E. L. Worthington. 1998. Part 1, Interpersonal forgiving in close relationships. Part 2, Theoretical elaboration and measurement. *Journal of Personality and Social Psychology* 75 (6): 1586–1603.

McDougall, P., and S. Hymel. 2007. Same-gender versus cross-gender friendship conceptions. *Merrill-Palmer Quarterly* 53 (3): 347–380.

McElreath, R. 2004. Social learning and the maintenance of cultural variation:

An evolutionary model and data from East Africa. *American Anthropologist* 106 (2): 308–321.

McElreath, R., and R. Boyd. 2007. *Mathematical Models of Social Evolution: A Guide for the Perplexed.* Chicago: University of Chicago Press.

McFarland, C., R. Buehler, and L. MacKay. 2001. Affective responses to social comparisons with extremely close others. *Social Cognition* 19 (5): 547–586.

McGillicuddy-De Lisi, A. V., C. Watkins, and A. J. Vinchur. 1994. The effect of relationship on children's distributive justice reasoning. *Child Development* 65 (6): 1694–1700.

McGinn, K. L., and A. T. Keros. 2002. Improvisation and the logic of exchange in socially embedded transactions. *Administrative Science Quarterly* 47 (3): 442–473.

McGuire, B. P. 1988. *Friendship and Community: The Monastic Experience.* Kalamazoo, MI: Cistercian Publications.

McKenna, K. Y. A., A. S. Green, and M. E. J. Gleason. 2002. Relationship formation on the Internet: What's the big attraction? *Journal of Social Issues* 58 (1): 9–32.

McMichael, C., and L. Manderson. 2004. Somali women and well-being: Social networks and social capital among immigrant women in Australia. *Human Organization* 63 (1): 88–99.

McPherson, J. M., L. Smith-Lovin, and M. E. Brashears. 2006. Social isolation in America: Changes in core discussion networks over two decades. *American Sociological Review* 71 (3): 353–375.

McPherson, M., L. Smith-Lovin, and J. M. Cook. 2001. Birds of a feather: Homophily in social networks. *Annual Reviews in Sociology* 27:415–444.

Mead, M. 1937a. The Arapesh of New Guinea. In *Cooperation and Competition among Primitive Peoples,* ed. M. Mead. New York: McGraw-Hill.

———. 1937b. The Manus of the Admiralty Islands. In *Cooperation and Competition among Primitive Peoples,* ed. M. Mead. New York: McGraw-Hill.

Meggitt, M. J. 1965. *The Lineage System of the Mae Enga.* New York: Barnes and Noble.

Mehl, M. R., S. Vazire, N. Ramirez-Esparza, R. B. Slatcher, and J. W. Pennebaker. 2007. Are women really more talkative than men? *Science* 317 (5834): 82.

Mehu, M., K. Grammer, and R. I. M. Dunbar. 2007. Smiles when sharing. *Evolution and Human Behavior* 28:415–422.

Melville, H. 1846. *Typee: A Peep at Polynesian Life during a Four Months' Residence in a Valley of the Marquesas.* New York: Wiley and Putnam.

———. 1847. *Omoo: A Narrative of the South Seas.* London: John Murray.

Mendelson, M. J., and A. C. Kay. 2003. Positive feelings in friendship: Does imbalance in the relationship matter? *Journal of Social and Personal Relationships* 20 (1): 101–116.

Merker, M. 1910. *The Masai: Ethnographic Monograph of an East African Semite People.* Berlin: Dietrich Reimer.

Mesch, G., and I. Talmud. 2006. The quality of online and offline relationships:

The role of multiplexity and duration of social relationships. *The Information Society* 22:137–143.

Messman, S. J., D. J. Canary, and K. S. Hause. 2000. Motives to remain platonic, equity, and the use of maintenance strategies in opposite-sex friendships. *Journal of Social and Personal Relationships* 17 (1): 67–94.

Metts, S., and W. R. Cupach. 1995. Postdivorce relations. In *Explaining Family Interactions*, ed. M. A. Fitzpatrick and A. L. Vangelisti. Newbury Park, CA: Sage.

Metts, S., W. R. Cupach, and R. A. Bejlovec. 1989. "I love you too much to ever start liking you": Redefining romantic relationships. *Journal of Social and Personal Relationships* 6 (3): 259–274.

Metzger, D. 1964. *Interpretations of Drinking Performances in Aguacatenango.* Chicago: University of Chicago.

Meyerowitz, E. L.-R. 1974. *The Early History of the Akan States of Ghana.* London: Red Candle Press.

Michael, R. T., J. H. Gagnon, E. O. Laumann, and G. Kolata. 1994. *Sex in America: A Definitive Study.* Boston: Little, Brown and Company.

Mikulincer, M., and M. Selinger. 2001. The interplay between attachment and affiliation systems in adolescents' same-sex friendships: The role of attachment style. *Journal of Social and Personal Relationships* 18 (1): 81–106.

Mikulincer, M., and P. R. Shaver. 2007. *Attachment in Adulthood.* New York: Guilford.

Mikulincer, M., P. R. Shaver, O. Gillath, and R. A. Nitzberg. 2005. Attachment, caregiving, and altruism: Boosting attachment security increases compassion and helping. *Journal of Personality and Social Psychology* 89 (5): 817–839.

Milardo, R. M. 1982. Friendship networks in developing relationships: Converging and diverging social environments. *Social Psychology Quarterly* 45 (3): 162–172.

Milicic, B. 1998. The grapevine forest: Kinship, status and wealth in a Mediterranean community (Selo, Croatia). In *Kinship, Networks, and Exchange*, ed. T. Schweizer and D. R. White. Cambridge: Cambridge University Press.

Miller, D. T., J. S. Downs, and D. A. Prentice. 1998. Minimal conditions for the creation of a unit relationship: The social bond between birthdaymates. *European Journal of Social Psychology* 28 (3): 475–481.

Miller, L., P. Rozin, and A. P. Fiske. 1998. Food sharing and feeding another person suggest intimacy: Two studies of American college students. *European Journal of Social Psychology* 28 (3): 423–436.

Miller, R. J., and Y. Darlington. 2002. Who supports? The providers of social support to dual-parent families caring for young children. *Journal of Community Psychology* 30 (5): 461–473.

Mills, J., and M. S. Clark. 1982. Communal and exchange relationships. *Review of Personality and Social Psychology* 3:121–144.

Mitani, J. C. 2009. Male chimpanzees form enduring and equitable social bonds. *Animal Behaviour* 77:633–640.

Mitani, J.C., D.A. Merriwether, and C. Zhang. 2000. Male affiliation, cooperation and kinship in wild chimpanzees. *Animal Behaviour* 59 (4): 885–893.

Moffatt, M. 1986. The discourses of the dorm: Race, friendship and "culture" among college youth. In *Symbolizing America*, ed. H. Varenne. Lincoln: University of Nebraska Press.

Money, J. 1980. *Love and Love Sickness: The Science of Sex, Gender Difference, and Pair-Bonding.* Baltimore: Johns Hopkins University Press.

Monsour, N. 1992. Meaning of intimacy in cross- and same-sex friendships. *Journal of Social and Personal Relationships* 9 (2): 277–295.

Montaigne, M. de. 1993. *The Complete Essays.* Translated by M.A. Screech. London: Penguin.

Montaya, R.M., and C.A. Insko. 2008. Toward a complete understanding of the reciprocity of liking effect. *European Journal of Social Psychology* 38 (3): 477–498.

Moore, D. 1990. Drinking, the construction of ethnic identity and social process in a Western Australian youth subculture. *British Journal of Addiction* 85:1265–1278.

Moore, G. 1990. Structural determinants of men's and women's personal networks. *American Sociological Review* 55 (5): 726–735.

Moorhouse, G. 1972. *Calcutta.* New York: Harcourt Brace Jovanovich.

Morgan, C.J. 1979. Eskimo hunting groups, social kinship, and the possibility of kin selection in humans. *Evolution and Human Behavior* 1 (1): 83–91.

Morgan, L.H. 1877. *Ancient Society.* New York: H. Holt and Company.

Morgan, W.R., and J. Sawyer. 1967. Bargaining, expectations, and the preference for equality over equity. *Journal of Personality and Social Psychology* 6 (2): 139–149.

Morris, S.C., and S. Rosen. 1973. Effects of felt adequacy and opportunity to reciprocate on help seeking. *Journal of Experimental Social Psychology* 9 (3): 265–276.

Morry, M.M. 2005. Relationship satisfaction as a predictor a similarity rating: A test of the attraction-similarity hypothesis. *Journal of Social and Personal Relationships* 22 (4): 561–584.

———. 2007. The attraction-similarity hypothesis among cross-sex friends: Relationship satisfaction, perceived similarities, and self-serving perceptions. *Journal of Social and Personal Relationships* 24 (1): 117–138.

Morton, H. 1980. Who gets what, when and how? Housing in the Soviet Union. *Soviet Studies* 32:235–259.

Morton, J. 1992. Country, people, art: The Western Aranda, 1870–1990. In *The Heritage of Namatjira: The Watercolourists of Central Australia*, ed. J. Hardy, J.V.S. Megaw, and M.R. Megaw. Melbourne: William Heineman Australia.

Moss, H.B., G.L. Panzak, and R.E. Tarter. 1993. Sexual functioning of male anabolic steroid abusers. *Archives of Sexual Behavior* 22 (1): 1–12.

Mounts, N.S., and H.S. Kim. 2007. Parental goals regarding relationships and management of peers in a multiethnic sample. *New Directions for Child and Adolescent Development* 2007 (116): 17–33.

Muir, D. E., and E. A. Weinstein. 1962. The social debt: An investigation of lower-class and middle-class norms of social obligation. *American Sociological Review* 27 (4): 532–539.

Mukherjea, C. 1962. *The Santals.* Calcutta: A. Mukherjee & Co.

Mulder, N. 1996. *Inside Thai Society: Interpretations of Everyday Life.* Amsterdam: Pepin Press.

Munch, A., J. M. McPherson, and L. Smith-Lovin. 1997. Gender, children, and social contact: The effects of childrearing for men and women. *American Sociological Review* 62 (4): 509–520.

Murdock, G. 1949. *Social Structure.* New York: MacMillan, Co.

Murray, S. L., J. L. Derrick, S. Leder, and J. G. Holmes. 2008. Balancing connectedness and self-protection goals in close relationships: A levels-of-processing perspective on risk regulation. *Journal of Personality and Social Psychology* 94 (3): 429–459.

Murray, S. L., J. G. Holmes, G. Bellavia, D. W. Griffin, and D. Dolderman. 2002. Kindred spirits? The benefits of egocentrism in close relationships. *Journal of Personality and Social Psychology* 82 (4): 563–581.

Murray, S. L., J. G. Holmes, and N. L. Collins. 2006. Optimizing assurance: The risk regulation system in relationships. *Psychological Bulletin* 132 (5): 641–666.

Murray, S. O. 2004. Africa: Sub-Saharan, pre-independence. In *GLBTQ: An Encyclopedia of Gay, Lesbian, Bisexual, Transgender, and Queer Culture,* ed. C. J. Summers. Chicago: GLBTQ, Inc.

Naegele, K. D. 1958. Friendship and acquaintances: An exploration of some social distinctions. *Harvard Educational Review* 28:232–252.

Nakane, C. 1967. *Garo and Khasi: A Comparative Study in Matrilineal Systems.* Paris: Mouton.

Naroll, R. 1962. The activity bias of ethnography: A reply to Engelmann. *American Anthropologist* 64 (4): 833–835.

——. 1967. The proposed HRAF Probability Sample. *Cross-cultural Research* 2 (2): 70–80.

Naroll, R., and F. Naroll. 1963. On bias of exotic data. *Man* 63:24–26.

Nash, J. C. 1970. *In the Eyes of the Ancestors: Belief and Behavior in a Mayan Community.* New Haven, CT: Yale University Press.

National Alliance for Caregiving and AARP. 2004. *Caregiving in the U.S.*

Neckerman, H. J. 1996. The stability of social groups in childhood and adolescence: The role of the classroom social environment. *Social Development* 5 (2): 131–145.

Neilson, W. S. 1999. The economics of favors. *Journal of Economic Behavior and Organization* 39 (4): 387–397.

Newcomb, A., and C. L. Bagwell. 1995. Children's friendship relations: A meta-analytic review. *Psychological Bulletin* 117 (2): 306–347.

Newcomb, T. M. 1961. *The Acquaintance Process*. New York: Holt, Rinehart and Winston.

Niiya, Y., P. C. Ellsworth, and S. Yamaguchi. 2006. Amae in Japan and the United States: An exploration of a "culturally unique" emotion. *Emotion* 6:279–295.

Nimuendaju, C. 1946. *The Eastern Timbira*. Berkeley: University of California Press.

Niv, Y., D. Joel, and P. Dayan. 2006. A normative perspective on motivation. *Trends in Cognitive Science* 10 (8): 375–381.

Nix, J. C. L. 1999. Addressing friendship transitions in early adulthood: An inductive examination of how individuals manage romantic relationships and close friendship simultaneously. PhD diss., Ohio State University, Columbus.

Norbeck, E., and M. Norbeck. 1956. Child training in a Japanese fishing community. In *Personal Character and Cultural Milieu: A Collection of Readings*, ed. D. C. Haring. Syracuse, NY: Syracuse University Press.

Nordenskiold, E. 1938. *An Historical and Ethnological Survey of the Cuna Indians*. Göteborg: Göteborgs Museum.

Nowak, M. A., and K. Sigmund. 1992. Tit for tat in heterogeneous populations. *Nature* 355:250–253.

Nyakatura, J. 1970. *Aspects of Bunyoro Custom and Tradition*. Nairobi: East African Literature Bureau.

Oates, K., and M. Wilson. 2002. Nominal kinship cues. *Proceedings of the Royal Society B* 269 (1487): 105–109.

Oberg, K. 1973. *The Social Economy of the Tlingit Indians*. Seattle: University of Washington Press.

Obermeyer, G. J. 1968. *Structure and Authority in a Bedouin Tribe: The 'Aishabit of the Western Desert of Egypt*. Bloomington: Indiana University.

Odling-Smee, F. J., K. N. Laland, and M. W. Feldman. 2003. *Niche Construction: The Neglected Process in Evolution*. Princeton, NJ: Princeton University Press.

O'Connell, L. 1984. An exploration of exchange in three social relationships. *Journal of Social and Personal Relationships* 1:333–345.

O'Donovan, A., and B. M. Hughes. 2008. Factors that moderate the effect of laboratory-based social support on cardiovascular reactivity to stress. *International Journal of Psychology and Psychological Therapy* 8 (1): 85–102.

Oetzel, J. G., S. Ting-Toomey, Y. Yokochi, T. Masumoto, and J. Takai. 2000. A typology of facework behaviors in conflicts with best friends and relative strangers. *Communication Quarterly* 48 (4): 397–419.

Okada, F. E. 1973. Ritual brotherhood: A cohesive factor in Nepalese society. *Journal of Anthropological Research* 13:212–222.

Oliker, S. J. 1989. *Best Friends and Marriage: Exchange among Women*. Berkeley: University of California Press.

———. 1998. The modernization of friendship: Individualism, intimacy, and gender in the nineteenth century. In *Placing Friendship in Context*, ed. R. G. Adams and G. A. Allan. Cambridge: Cambridge University Press.

Olson, R. L. 1967. *Social Structure and Social Life of Tlingit in Alaska*. Berkeley: University of California Press.

Orans, M. 1965. *The Santal: A Tribe in Search of a Great Tradition*. Detroit: Wayne State University Press.

Ortigue, S., and F. Bianchi-Demicheli. 2008. Why is your spouse so predictable? Connecting mirror neuron system and self-expansion model of love. *Medical Hypotheses* 71 (6): 941–944.

Ortner, S. B. 1978. *Sherpas through Their Rituals*. Cambridge: Cambridge University Press.

———. 2003. *New Jersey Dreaming: Capital, Culture, and the Class of '58*. Durham, NC: Duke University Press.

Oschema, K. 2006. Blood-brothers: A ritual of friendship and the construction of the imagined barbarian in the middle ages. *Journal of Medieval History* 32 (3): 275–301.

Osgood, C. 1951. *The Koreans and Their Culture*. New York: Ronald Press Company.

Oswald, D. L., E. M. Clark, and C. M. Kelly. 2004. Friendship maintenance: An analysis of individual and dyad behaviors. *Journal of Social and Clinical Psychology* 23 (3): 413–441.

Otten, S., and K. Epstude. 2006. Overlapping mental representation of self, ingroup, and outgroup: Unraveling self-stereotyping and self-anchoring. *Personality and Social Psychology Bulletin* 32 (7): 957–969.

Ottenberg, S. 1989. *Boyhood Rituals in an African Society: An Interpretation*. Seattle: University of Washington Press.

Overall, N. C., G. J. O. Fletcher, and M. D. Friesen. 2003. Mapping the intimate relationship mind: Comparisons between three models of attachment representations. *Personality and Social Psychology Bulletin* 29 (12): 1479–1493.

Owens, R. A. 2003. Friendship features associated with college students' friendship maintenance and dissolution following problems. PhD diss., Eberly College, Morgantown, WV.

Packard, V. 1972. *A Nation of Strangers*. New York: David McKay.

Pahl, R. 2000. *On Friendship*. Malden, MA: Polity Press.

Pahl, R., and D. J. Pevalin. 2005. Between family and friends: A longitudinal study of friendship choice. *British Journal of Sociology* 56 (3): 433–450.

Paine, R. 1969. In search of friendship: An exploratory analysis in "middleclass" culture. *Man* 4 (4): 505–524.

———. 1974. Anthropological approaches to friendship. In *The Compact: Selected Dimensions of Friendship*, ed. E. Leyton. St. John's, Canada: Memorial University of Newfoundland.

Pakaluk, M. 1991. *Other Selves: Philosophers on Friendship*. Indianapolis: Hackett Pub. Co.

Palacio, J. O. 1982. Food and social relations in a Garifuna village. PhD diss., University of California, Berkeley.

Papataxiarchis, E. 1991. Friends of the heart: Male commensal solidarity, gender, and kinship in Aegean Greece. In *Contested Identities: Gender and Kinship in Modern Greece,* ed. P. Loizos and E. Papataxiarchis. Princeton, NJ: Princeton University Press.

Parin, P., F. Morgenthaler, and G. Parin-Matthey. 1963. *The Whites Think Too Much: Psychoanalytic Investigations among the Dogon.* Zurich: Atlantis Verlag.

Park, J. H., and M. Schaller. 2005. Does attitude similarity serve as a heuristic cue for kinship? Evidence of an implicit cognitive association. *Evolution and Human Behavior* 26:158–170.

Park, J. H., M. Schaller, and M. Van Vugt. 2008. The psychology of human kin recognition: Heuristic cues, erroneous inferences, and their implications. *Review of General Psychology* 12 (3): 215–235.

Park, R. E. 1928. Human migrations and the marginal man. *American Journal of Sociology* 33:881–893.

Parker, J. G., C. M. Low, A. R. Walker, and B. K. Gamm. 2005. Friendship jealousy in young adolescents: Individual differences and links to sex, self-esteem, aggression and social adjustment. *Developmental Psychology* 41 (1): 235–250.

Parker, S., and B. de Vries. 1993. Patterns of friendship for women and men in same and cross-sex relationships. *Journal of Social and Personal Relationships* 10 (4): 617–626.

Parkin, R. 1997. Kinship with trees. *Journal of the Royal Anthropological Institute* 3 (2): 374–376.

Parks, M. R., and K. Floyd. 1996. Making friends in cyberspace. *Journal of Computer Mediated Communication* 46 (1): 80–97.

Parks, M. R., and L. D. Roberts. 1998. Making Moosic: The development of personal relationships on line and a comparison to their off-line counterparts. *Journal of Social and Personal Relationships* 15 (4): 517–537.

Parlee, M. B. 1979. The friendship bond. *Psychology Today* 13:43–54.

Parman, S. 1972. *Sociocultural change in a Scottish crofting township.* Ann Arbor, MI: University Microfilms.

———. 1990. *Scottish Crofters: An Historical Ethnography of a Celtic Village.* New York: Holt, Rinehart and Winston.

Parsons, E. C. 1917. Ceremonial friendship at Zuni. *American Anthropologist* 19 (1): 1–8.

Parsons, T., and E. Shils. 1951. *Toward a General Theory of Action: Theoretical Foundations for the Social Sciences.* Cambridge: Harvard University Press.

Pasternak, B. 1972. *Kinship and Community in Two Chinese Villages.* Stanford, CA: Stanford University Press.

Pataki, S. P., C. Shapiro, and M. S. Clark. 1994. Children's acquisition of appropriate norms for friendships and acquaintances. *Journal of Social and Personal Relationships* 11 (3): 427–442.

Patterson, B. R., L. Bettini, and J. F. Nussbaum. 1993. The meaning of friendship across the life-span: Two studies. *Communication Quarterly* 41:145–160.

Paulme, D. 1940. *Social Organization of the Dogon*. Paris: Domat-Montchrestian.

Pavlovic, J. M. 1973. *Folk Life and Customs in the Kragujevac Region of the Jasenica in Sumdaija*. New Haven, CT: HRAF.

Pehrson, R. N. 1957. *The Bilateral Network of Social Relations in Konkama Lapp District*. Bloomington: Indiana University Press.

Peletz, M. G. 1995. Kinship studies in late twentieth-century anthropology. *Annual Review of Anthropology* 24 (1): 343–372.

Pellegrini, D. S. 1986. Variability in children's level of reasoning about friendship. *Journal of Applied Developmental Psychology* 7:341–354.

Pelto, P. J. 1962. *Individualism in Skolt Lapp Society*. Helsinki: Finnish Antiquities Society.

Pemberton, M, and C. Sedikides. 2001. When do individuals help close others improve? The role of information diagnosticity. *Journal of Personality and Social Psychology* 81 (2): 234–246.

Peng, Y. 2004. Kinship networks and entrepreneurs in China's transitional economy. *American Journal of Sociology* 109 (5): 1045–1074.

Penning, M. J., and N. L. Chappell. 1987. Ethnicity and informal supports among older adults. *Journal of Aging Studies* 1 (2): 145–160.

Peshkin, A. 1972. *Kanuri Schoolchildren: Education and Social Mobilization in Nigeria*. New York: Holt, Rinehart and Winston.

Peter, Prince of Greece. 1963. *A Study of Polyandry*. The Hague: Mouton.

Peters, E. L. 1990. *The Bedouin of Cyrenaica: Studies in Personal and Corporate Power*. Cambridge: Press Syndicate of the University of Cambridge.

Pettit, B. 2004. Moving and children's social connections: Neighborhood context and the consequences of moving for low-income families. *Sociological Forum* 19 (2): 285–311.

Pfeffer, J., C. T. Fong, R. B. Cialdini, and R. R. Portnoy. 2006. Overcoming the self-promotion dilemma: Interpersonal attraction and extra help as a consequence of who sings one's praises. *Personality and Social Psychology Bulletin* 32 (10): 1362–1374.

Phelps, E. A., and J. E. LeDoux. 2005. Contributions of the amygdala to emotion processing: From animal models to human behavior. *Neuron* 48:175–187.

Phillips, H. P. 1966. *Thai Peasant Personality: The Patterning of Interpersonal Behavior in the Village of Bang Chan*. Berkeley: University of California Press.

Pieters, R., and H. Robben. 1999. Consumer evaluation of money as a gift: A two-utility model and an empirical test. *Kyklos* 52:173–200.

Piker, S. 1968. Friendship to the death in rural Thai society. *Human Organization* 27:200–204.

Pimentel, E. E. 2000. Just how do I love thee? Marital relations in urban China. *Journal of Marriage and Family* 62 (1): 32–47.

Pinson, A. 1979. Kinship and economy in modern Iceland: A study in social continuity. *Ethnology* 18:183–197.

———. 1985. The institution of friendship and drinking patterns in Iceland. *Anthropological Quarterly* 58 (2): 75–82.

Pitt-Rivers, J. 1954. *The People of the Sierra.* Oxford: Oxford University Press.

———. 1968. Pseudo-kinship. In *International Encyclopedia of the Social Sciences,* ed. D. L. Sills and R. K. Merton. London: Free Press.

———. 1973. The kith and the kin. In *The Character of Kinship,* ed. J. Goody and M. Fortes. Cambridge: Cambridge University Press.

Planalp, S., and K. Garvin-Doxas. 1994. Using mutual knowledge in conversations: Friends as experts on each other. In *Dynamics of relationships,* ed. S. W. Duck. Newbury Park, CA: Sage.

Plato. 1892. *Meno.* Translated by B. Jowett. London: Oxford University Press.

Polzer, J. T., M. A. Neale, and P. O. Glenn. 1993. The effects of relationships and justification in an interdependent allocation task. *Group Decision and Negotiation* 2:135–148.

Porges, S. W. 2003. The polyvagal theory: Phylogenetic contributions to social behavior. *Physiology and Behavior* 79 (3): 503–513.

Pospisil, L. J. 1978. *The Kapauka Papuans of West New Guinea.* New York: Holt, Rinehart, and Winston.

Powdthavee, N. 2008. Putting a price tag on friends, relatives and neighbors: Using surveys of life satisfaction to value social relationships. *Journal of Socio-economics* 37:1459–1480.

Precourt, W. E. 1979. Ideological bias: A data quality control factor for cross-cultural research. *Cross-cultural Research* 14 (2): 115–131.

Preecha, K. 1980. *Marketing in North-central Thailand: A Study of Socio-economic Organization in a Thai Market Town.* Bangkok: Chulalongkorn University Social Research Institute.

Price, D. H. 2004. *Atlas of World Cultures: A Geographical Guide to Ethnographic Literature.* Caldwell, NJ: Blackburn Press.

Price, R. 1991. *Two Evenings in Saramaka.* Chicago: University of Chicago Press.

Priel, B., D. Mitrany, and G. Shahar. 1998. Closeness, support and reciprocity: A study of attachment styles in adolescence. *Personality and Individual Differences* 25 (6): 1183–1198.

Pryde, D. 1972. *Nunaga: My Land, My Country.* Edmonton: M. G. Hurtig Ltd.

Pulakos, J. 1989. Young adult relationships: Siblings and friends. *Journal of Psychology* 123:237–244.

Putnam, P. 1948. The Pygmies of the Ituri Forest. In *A Reader in General Anthropology,* ed. C. S. Coon. New York: Henry Holt and Company.

Putnam, R. D. 2000. *Bowling Alone: The Collapse and Revival of American Community.* New York: Simon & Schuster.

Quirin, M., J. C. Pruessner, and J. Kuhl. 2008. HPA system regulation and adult attachment anxiety: Individual differences in reactive and awakening cortisol. *Psychoneuroendocrinology* 33:581–590.

Rachlin, H., and B. A. Jones. 2008. Altruism among relatives and non-relatives. *Behaviour Processes* 79:120–123.

Radcliffe-Brown, A. R. 1922. *The Andaman Islanders: A Study in Social Anthropology*. Cambridge: Cambridge University Press.

———. 1940. On joking relationships. *Africa* 12 (3): 195–210.

———. 1949. A further note on joking relationships. *Africa* 19 (2): 113–140.

———. 1952. *Structure and Function in Primitive Society*. New York: Free Press.

Rao, N., and S. M. Stewart. 1999. Cultural influences on sharer and recipient behavior: Sharing in Chinese and Indian preschool. *Journal of Cross-cultural Psychology* 30 (2): 219–241.

Rattray, R. S. 1927. *Religion and Art in Ashanti*. Oxford: Clarendon Press.

Rawlins, W. K. 1992. *Friendship Matters*. Hawthorne, NY: Aldine Transactions.

Read, M. 1956. *The Ngoni of Nyasaland*. Oxford: Oxford University Press.

Reay, M. 1953. Social control amongst the Orokaiva. *Oceania* 24:110–118.

Reddy, A. M. 1973. On the role of ritual friendship in the mobility of wealth in the Visakhapatnam Agency. *Man in India* 53 (3): 243–255.

Reed, A. D. E. 2003. *Papua New Guinea's Last Place: Experiences of Constraint in a Postcolonial Prison*. New York: Berghahn Books.

Reed, R. 1995. *Prophets of Agroforestry: Guarani Communities and Commercial Gathering*. Austin: University of Texas Press.

Reed-Danahay, D. E. 1999. Friendship, kinship, and the life course in rural Auvergne. In *The Anthropology of Friendship*, ed. S. M. Bell and S. Coleman. Oxford, UK: Berg.

Reeder, H. M. 2000. "I like you as a friend": The role of attraction in cross-sex friendship. *Journal of Social and Personal Relationships* 17 (3): 329–348.

Reeve, H. K. 1998. Kinship and reciprocity with some new twists. In *Handbook of Evolutionary Psychology*, ed. C. B. Crawford and D. Krebs. Mahwah, NJ: Lawrence Erlbaum.

Regan, P. C. 1998. Of lust and love: Beliefs about the role of sexual desire in romantic relationships. *Personal Relationships* 5 (2): 139–157.

———. 2004. Sex and the attraction process: Lessons from science (and Shakespeare) on lust, love, chastity, and fidelity. In *The Handbook of Sexuality in Close Relationships*, ed. J. H. Harvey, A. Wenzel, and S. Sprecher. Mahwah, NJ: Lawrence Erlbaum.

Regan, P. C., and E. Berscheid. 1995. Gender differences in beliefs about the causes of male and female sexual desire. *Personal Relationships* 2 (4): 345–358.

Reichel-Dolmatoff, G. 1951. *The Kogi: A Tribe of the Sierra Nevada de Santa Marta, Colombia*. Bogota: Editorial Iqueima.

Reina, R. E. 1959. Two patterns of friendship in a Guatemalan community. *American Anthropologist* 61 (1): 44–50.

Reis, H. T. 1998. Gender differences in intimacy and related behaviors: Context and process. In *Sex Differences in Similarities in Communication*, ed. D. J. Canary and K. Dindia. Mahwah, NJ: Lawrence Erlbaum.

Reisman, J. M. 1979. *Anatomy of Friendship*. New York: Irvington Publishers.

———. 1990. Intimacy in same-sex friendships. *Sex Roles* 23 (1–2): 65–82.

Rezende, C. 1999. Building affinity through friendship. In *The Anthropology of Friendship*, ed. S. Bell and S. Coleman. Oxford: Berg Publishers.

Rheingold, H. 1994. *The Virtual Community: Finding Connection in a Computerized World*. London: Secker & Warburg.

Rholes, W., and J. A. Simpson. 2004. *Adult Attachment: Theory, Research and Clinical Implications*. New York: Guilford.

Rich, G. W. 1980. Kinship and friendship in Iceland. *Ethnology* 19 (4): 475–493.

———. 1989. Problems and prospects in the study of Icelandic kinship. In *The Anthropology of Iceland*, ed. E. P. Durrenberger and G. Palsson. Iowa City: University of Iowa Press.

Richards, A. I. 1939. *Land, Labour and Diet in Northern Rhodesia: An Economic Study of the Bemba Tribe*. London: Oxford University Press.

Richards, M. H., P. A. Crowe, R. Larson, and A. Swarr. 1998. Developmental patterns of gender differences in the experience of peer companionship during adolescence. *Child Development* 69 (1): 154–163.

Richerson, P. J., R. T. Boyd, and J. Henrich. 2003. Cultural evolution of human cooperation. In *Genetic and Cultural Evolution of Cooperation*, ed. P. Hammerstein. Cambridge: MIT Press.

Rick, S., and G. F. Loewenstein. 2008. Intangibility in intertemporal choice. *Philosophical Transactions of the Royal Society B* 363:3813–3824.

Rilling, J. K. 2008. Neuroscientific approaches and applications within anthropology. *American Journal of Physical Anthropology* 137 (S47): 2–32.

Rilling, J. K., D. R. Goldsmith, A. L. Glenn, M. R. Jairam, H. A. Elfenbein, J. E. Dagenais, C. D. Murdock, and G. Pagnoni. 2008. The neural correlates of the affective response to unreciprocated cooperation. *Neuropsychologia* 46 (5): 1256–1266.

Rilling, J. K., B. King-Casas, and A. G. Sanfey. 2008. The neurobiology of social decision-making. *Current Opinions in Neurobiology* 18 (2): 159–165.

Rivers, W. H. R. 1900. A genealogical method of collecting social and vital statistics. *Journal of the Anthropological Institute of Great Britain and Ireland* 30:74–82.

———. 1915. *Kinship and Social Organization*. Cambridge, United Kingdom: Constable & Co. Ltd.

Rizzo, T. A. 1989. *Friendship Development among Children in School*. Norwood, NJ: Ablex.

Rizzo, T. A., and W. A. Corsaro. 1988. Toward a better understanding of Vygotsky's process of internalization: Its role in the development of the concept of friend. *Developmental Review* 8:219–237.

Robben, H. S. J., and T. M. M. Verhallen. 1994. Behavioral costs as determinants of cost perception and preference formation for gifts to receive and gifts to give. *Journal of Economic Psychology* 15 (2): 333–350.

Robbins, M. C. 1979. Problem-drinking and the integration of alcohol in rural

Buganda. In *Beliefs, Behaviors, and Alcoholic Beverages*, ed. M. Marshall. Ann Arbor: University of Michigan.

Robbins, M. C., and P. L. Kilbride. 1987. Microtechnology in rural Buganda. In *Technology and Social Change*, ed. H. R. Bernard and P. J. Pelto. Prospect Heights, IL: Waveland Press.

Roberto, K. A. 1996. Friendships between older women: Interactions and reactions. In *Relationships between Women in Later Life*, ed. K. A. Roberto. Binghamton, NY: Haworth Press.

———. 1999. Qualities of older women's friendships: Stable or volatile. *International Journal of Aging and Human Development* 48 (1): 81–83.

Roberto, K. A., and J. P. Scott. 1986. Friendships of older men and women: Exchanged patterns and satisfaction. *Psychology and Aging* 1 (2): 103–109.

Roberts, G., and T. N. Sherratt. 1998. Development of cooperative relationships through increasing investment. *Nature* 6689:175–178.

Roberts, J. M. 1965. Kinsmen and friends in Zuni culture: A terminological note. *El Palacio* 72 (4): 38–43.

Roe, S. K. 2007. A brief history of an ethnographic database: The HRAF collection of ethnography. *Behavioral and Social Sciences Librarian* 25 (2): 47–77.

Roheim, G. 1945. *The Eternal Ones of the Dream: A Psychoanalytic Interpretation of Australian Myth and Ritual*. New York: International Universities Press.

Rohner, R. P., B. R. DeWalt, and R. C. Ness. 1973. Ethnographer bias in cross-cultural research: An empirical study. *Cross-cultural Research* 8 (4): 275–308.

Rook, K. S. 1987. Reciprocity of social exchange and social satisfaction among older women. *Journal of Personality and Social Psychology* 52 (1): 145–154.

Roschelle, A. R. 1997. *No More Kin: Exploring Race, Class, and Gender in Family Networks*. Newbury Park, CA: Sage.

Roscoe, J. 1911. *The Baganda*. London: Macmillan.

Rose, A., W. Carlson, and E. M. Waller. 2007. Prospective associations of co-rumination with friendship and emotional adjustment. *Developmental Psychology* 43 (4): 1019–1031.

Rose, A., and K. D. Rudolph. 2006. A review of sex differences in peer relationships processes: Potential trade-offs for the emotional and behavioral development of girls and boys. *Psychological Bulletin* 132 (1): 98–131.

Rose, R. 1995. Russia as an hourglass society: A constitution without citizens. *East European Constitutional Review* 4 (3): 34–42.

Rose, S. M. 1984. How friendships end: Patterns among young adults. *Journal of Social and Personal Relationships* 1:267–277.

———. 1985. Same- and cross-sex friendships and the psychology of homosociality. *Sex Roles* 12 (1–2): 63–74.

Rose, S. M., and F. C. Serafica. 1986. Keeping and ending casual, close and best friendships. *Journal of Social and Personal Relationships* 3 (3): 275–288.

Rosenblatt, A., and J. Greenberg. 1991. Examining the world of the depressed:

Do depressed people prefer others who are depressed? *Journal of Personality and Social Psychology* 60 (4): 620–629.

Roshchin, E. 2006. The concept of friendship: From princes to states. *European Journal of International Relations* 12 (4): 599–624.

Rosnow, R. L., R. Rosenthal, and D. B. Rubin. 2000. Contrasts and correlations in effect-size estimation. *Psychological Science* 11:446–453.

Ross, H. S., and S. P. Lollis. 1989. A social relations analysis of toddler peer relationships. *Child Development* 60 (5): 1082–1091.

Ross, L. 1995. Reactive devaluation in negotiation and conflict resolution. In *Barriers to Conflict Resolution*, ed. K. J. Arrow, R. H. Mnookin, L. Ross, A. Tversky and R. B. Wilson. New York: W. W. Norton and Company.

Rossi, P. H., and A. B. Anderson, eds. 1982. *Measuring Social Judgements: The Factorial Survey Approach*. Beverly Hills, CA: Sage.

Roth, M. A., and J. G. Parker. 2001. Affective and behavioral responses to friends who neglect their friends for dating partners: Influences of gender, jealousy and perspective. *Journal of Adolescence* 24 (3): 281–296.

Rotundo, E. A. 1989. Romantic friendship: Male intimacy and middle-class youth in the northern United States, 1800–1900. *Journal of Social History* 23:1–25.

Royal Anthropological Institute of Great Britain and Ireland. [1874] 1951. *Notes and Queries on Anthropology*. London: Routledge & Kegan Paul.

Ruan, D. 1998. The content of the General Social Survey discussion networks. *Social Networks* 20:247–264.

Rubin, L. B. 1985. *Just Friends: The Role of Friendship in Our Lives*. New York: Harper and Row.

Rubin, R. L., H. Yang, and M. Porte. 2000. A comparison of self-reported self-disclosure among Chinese and North Americans. In *Balancing the Secrets of Private Disclosures*, ed. S. Petronio. Mahwah, NJ: Lawrence Erlbaum.

Rubin, Z., and S. Shenker. 1978. Friendship, proximity, and self-disclosure. *Journal of Personality* 46 (1): 1–22.

Rushton, J. P. 1989. Genetic similarity, human altruism, and group selection. *Behavioral and Brain Sciences* 12:503–559.

Rust, F. 1969. *Nama Worterbuch*. Pietermaritzburg, South Africa: University of Natal Press.

Rybak, A., and F. T. McAndrew. 2006. How do we decide whom our friends are? Defining levels of friendship in Poland and the United States. *Journal of Social Psychology* 146 (2): 147–163.

Rynkiewich, M. 1976. Adoption and land tenure among Arno Marshallese. In *Transactions in Kinship: Adoption and Fosterage in Oceania*, ed. I. P. Brady. Honolulu: University of Hawaii Press.

Sagan, K., M. Pondel, and M. A. Wittig. 1981. The effect of anticipated future interaction on reward allocation in same- and opposite-sex dyads. *Journal of Personality* 49 (4): 438–449.

Sahli, N. 1979. Smashing: Women's relationships before the fall. *Chrysalis* 8:17–27.

Sahlins, M. 1972. *Stone Age Economics.* New York: Aldine de Gruyter.

———. 1976. *The Use and Abuse of Biology: An Anthropological Critique of Sociobiology.* Ann Arbor: University of Michigan Press.

St. John, D. P. 1981. The dream-vision experience of the Iroquois. PhD diss., Fordham University, New York.

Salamone, F. A. 1974. *Gods and Goods in Africa: Persistence and Change in Ethnic Religious Identity in Yauri Emirate, North-western State, Nigeria.* New Haven, CT: HRAF.

Salisbury, R. F. 1962. *From Stone to Steel: Economic Consequences of Technological Change in New Guinea.* Melbourne: Melbourne University Press.

Salmi, A. M. 2000. Bonds, bottles, blat and banquets: Birthdays and networks in Russia. *Ethnologia Europaea* 30 (1): 31–44.

Sampson, R. J. 1988. Local friendship ties and community attachment in mass society: A multilevel systemic model. *American Sociological Review* 53:766–779.

Sandin, B. 1980. *Iban Adat and Augury.* Penang: Penerbit Universiti Sains Malaysia.

Sanford, S., and D. Eder. 1984. Adolescent humor during peer interaction. *Social Psychology Quarterly* 47 (3): 235–243.

Santos-Granero, F. 2007. Of fear and friendship: Amazonian sociality beyond kinship and affinity. *Journal of the Royal Anthropological Institute* 13:1–18.

Sapadin, L. A. 1988. Friendship and gender: Perspectives of professional men and women. *Journal of Social and Personal Relationships* 5 (4): 387–403.

Saraswathi, T. S., and R. Dutta. 1988. *Invisible Boundaries: Grooming for Adult Roles.* New Delhi: Northern Book Centre.

Sarpong, P. 1977. *Girls' Nubility Rites on Ashanti.* Tema, Ghana: Ghana Pub. Corp.

Schapera, I. 1930. *Khoisan Peoples of Southern Africa.* London: Routledge.

Schefold, R. 1980. The sacrifices of Sakuddei (Mentawai Archipelago, Western Indonesia): An attempt at classification. In *Man, Meaning and History,* ed. R. Schefold, W. Schoorl, and J. Tennekes. The Hague: Martinus Nijhoff.

Schino, G., and F. Aureli. 2008. Grooming reciprocation among female primates: A meta-analysis. *Biology Letters* 4 (1): 9–11.

Schlegel, A. 1989. Fathers, daughters, and kachina dolls. *European Review of Native American Studies* 3 (1): 7–10.

Schlegel, A., and H. Barry. 1991. *Adolescence: An Anthropological Inquiry.* New York: Free Press.

Schneider, B. H., K. Dixon, and S. Udvari. 2007. Closeness and competition in the inter-ethnic and co-ethnic friendships of early adolescents in Toronto and Montreal. *Journal of Early Adolescence* 27 (1): 115–138.

Schneider, B. H., G. Tomada, S. Normand, E. Tonci, and P. Domini. 2008. Social support as a predictor of school bonding and academic motivation following the transition to Italian middle school. *Journal of Social and Personal Relationships* 25 (2): 287–310.

Schneider, C. S., and D. A. Kenny. 2000. Cross-sex friends who were once romantic partners: Are they platonic friends now? *Journal of Social and Personal Relationships* 17 (3): 451–466.

Schneider, D. 1968. *American Kinship: A Cultural Account.* Chicago: University of Chicago Press.

Schubert, T. W., and S. Otten. 2002. Overlap of self, ingroup, and outgroup: Pictorial measures of self-categorization. *Self and Identity* 1 (4): 353–376.

Schultz, J. W. 1930. *The Sun God's Children.* Boston: Houghton Mifflin.

Schultze, L. 1907. *Aus Namaland und Kalahari.* Jena, Germany: Gustav Fischer Press.

Schwartz, R. 1974. The crowd: Friendship groups in a Newfoundland outport. In *The Compact: Selected Dimensions of Friendship,* ed. E. Leyton. St John's, Canada: Memorial University of Newfoundland. .

Schwimmer, E. G. 1974. Friendship and kinship: An attempt to relate two anthropological concepts. In *The Compact: Selected Dimensions of Friendship,* ed. E. Leyton. St. John's, Canada: Memorial University of Newfoundland.

Schwinger, T., and H. Lamm. 1981. Justice norms in allocation decisions: Need consideration as a function of resource adequacy for complete need satisfaction, recipients' contributions, and recipients' interpersonal attraction. *Social Behavior and Personality* 9 (2): 235–241.

Scollon, C. N., C. Kim-Prieto, and E. Diener. 2003. Experience sampling: Promises and pitfalls, strengths and weaknesses. *Journal of Happiness Studies* 4 (1): 5–34.

Scribner, S., and M. Cole. 1973. Cognitive consequences of formal and informal education. *Science* 182 (4112): 553–559.

Secord, P. F., and C. W. Backman. 1964. Interpersonal congruency, perceived similarity, and friendship. *Sociometry* 27:115–127.

Sedikides, C., W. K. Campbell, G. D. Reeder, and A. J. Elliott. 1998. The self-serving bias in relational context. *Journal of Personality and Social Psychology* 74 (2): 378–386.

Seiter, J. S. 2007. Ingratiation and gratuity: The effect of complimenting customers on tipping behavior in restaurants. *Journal of Applied Social Psychology* 37 (3): 478–485.

Seiter, J. S., J. Bruschke, and C. Bai. 2002. The acceptability of deception as a function of perceivers' culture, deceiver's intention and deceiver-deceived relationship. *Western Journal of Communication* 66:158–180.

Sekaquaptewa, H. 1969. *Me and Mine: The Life Story of Helen Sekaquaptewa.* Tucson: University of Arizona Press.

Seki, K., D. Matsumoto, and T. T. Imahori. 2002. The conceptualization and expression of intimacy in Japan and the United States. *Journal of Cross-cultural Psychology* 33 (3): 303–319.

Seligman, C. G., and B. S. Seligman. 1932. *The Azande: Pagan Tribes of the Nilotic Sudan.* London: George Routledge & Sons.

Selman, R. 1980. *The Growth of Interpersonal Understanding.* New York: Academic Press.

Serbin, L. A., L. C. Moller, J. Gulko, K. K. Powlishta, and K. A. Colburne. 1994. The emergence of gender segregation in toddler playgroups. *New Directions for Child and Adolescent Development* 65:7–17.

Shack, W. 1963. Religious ideas and social action in Gurage bond-friendship. *Africa* 33:198–208.

Shackelford, T. K., and D. M. Buss. 1996. Betrayal in mateships, friendships, and coalitions. *Personality and Social Psychology Bulletin* 22 (11): 1151–1164.

Shah, P. P., K. T. Dirk, and N. Chervany. 2006. The multiple pathways of high performing groups: The interaction of social networks and group performance. *Journal of Organizational Behavior* 27 (3): 299–317.

Shakespeare, W. 1974. *The Complete Works of William Shakespeare.* London: Abbey Library.

Shapiro, E. G. 1975. Effect of expectations of future interaction on reward allocations in dyads: Equity and equality. *Journal of Personality and Social Psychology* 31 (5): 873–880.

Sharabany, R., R. Gershoni, and J. Hofman. 1981. Girlfriend, boyfriend: Age and sex differences in intimate friendship. *Developmental Psychology* 17 (6): 800–808.

Sharp, L. 1978. *Bang Chan: Social History of a Rural Community in Thailand.* Ithaca, NY: Cornell University Press.

Sharp, L., and S. Rapaport, eds. 1966. *Stability and Change in Thai Culture.* Ithaca, NY: Cornell University.

Sharpley, C. F., and J. Rodd. 1985. The effects of real versus hypothetical stimuli upon preschool children's helping behavior. *Early Child Development and Care* 22 (4): 303–313.

Shavit, Y., C. Fischer, and Y. Koresh. 1994. Kin and nonkin under collective threat: Israeli networks during the Gulf War. *Social Forces* 72:1197–1215.

Shearn, D., L. Spellman, B. Straley, J. Meirick, and K. Stryker. 1999. Empathic blushing in friends and strangers. *Motivation and Emotion* 23 (4): 307–316.

Sheets, V. L., and R. Lugar. 2005a. Friendship and gender in Russia and the United States. *Sex Roles* 52 (1–2): 131–140.

———. 2005b. Sources of conflict between friends in Russia and the United States. *Cross-cultural Research* 39 (4): 380–398.

Sherry, J. F. 1983. Gift giving in anthropological perspective. *Journal of Consumer Research* 10:157–168.

Sherzer, J. 1983. *Kuna Ways of Speaking: An Ethnographic Perspective.* Austin: University of Texas Press.

Shimony, A. 1961. *Conservatism among the Iroquois at the Six Nations Reserve.* New Haven, CT: Department of Anthropology, Yale University.

Shlapentokh, V. 1989. *Private and Public Life of the Soviet People: Changing Values in Post-Stalin Russia.* New York: Oxford University Press.

Shostak, M. 1981. *Nisa: The Life and Words of a !Kung Woman.* Cambridge: Harvard University Press.

Shrauger, J. S., and S. C. Jones. 1968. Social validation and interpersonal evaluations. *Journal of Experimental Social Psychology* 4:315–323.

Shulman, S., J. Elicker, and L. A. Sroufe. 1994. Stages of friendship growth in preadolescence as related to attachment history. *Journal of Social and Personal Relationships* 11 (3): 341–361.

Shumaker, S. A., and J. S. Jackson. 1979. The aversive effects of nonreciprocated benefits. *Social Psychology Quarterly* 42 (2): 148–158.

Sias, P. M., R. G. Heath, T. Perry, D. Silva, and B. Fix. 2004. Narratives of workplace friendship deterioration. *Journal of Social and Personal Relationships* 21 (3): 321–340.

Sieroszewski, W. 1896. *The Yakut: An Experiment in Ethnographic Research.* Moscow: HRAF translation.

Sik, E., and B. Wellman. 1999. Network capital in capitalist, communist, and postcommunist countries. In *Networks in the Global Village*, ed. B. Wellman. Boulder, CO: Westview Press.

Silk, J. B. 1980. Adoption and kinship in Oceania. *American Anthropologist* 82 (4): 799–820.

———. 1990. Human adoption in evolutionary perspective. *Human Nature* 1:25–52.

———. 2002. Using the "F"-word in primatology. *Behaviour* 139:421–446.

———. 2003. Cooperation without counting: The puzzle of friendship. In *Genetic and Cultural Evolution of Cooperation*, ed. P. Hammerstein. Cambridge: MIT Press.

Silva, P. 1990. Buddhist psychology: A review of theory and practice. *Current Psychology* 9 (3): 236–254.

Silver, A. 1990. Friendship in commercial society: Eighteenth-century social theory and modern sociology. *American Journal of Sociology* 95 (6): 1474–1504.

Sime, J. D. 1983. Affiliative behavior during escape to building exits. *Journal of Environmental Psychology* 3 (1): 21–41.

Simic, A. 1973. *The Peasant Urbanites: A Study of Rural-Urban Mobility in Serbia.* New York: Seminar Press.

Simpson, J. A. 1987. The dissolution of romantic relationships: Factors involved in relationship stability and emotional distress. *Journal of Personality and Social Psychology* 53 (4): 683–692.

Simpson, J. A., B. Campbell, and E. Berscheid. 1986. The association between romantic love and marriage: Kephart (1967) twice revisited. *Personality and Social Psychology Bulletin* 12 (3): 363–372.

Singer, E., and J. Doornenball. 2006. Learning morality in peer conflict: A study of schoolchildren's narratives about being betrayed by a friend. *Childhood* 13:225–245.

Skrefsrud, L. O. 1942. *Traditions and Institutions of the Santals.* New Delhi: Gyan.

Slingerland, E. G. 2003. *Confucius Analects, with Selections from Traditional Commentaries.* Indianapolis: Hackett Publishing Company.

Smaniotto, R. C. 2004. *"You Scratch My Back, I Scratch Yours" versus "Love Thy Neighbour": Two Proximate Mechanisms or Reciprocal Altruism?* Groningen, the Netherlands: University of Groningen.

Smith, A. 2002. *The Theory of Moral Sentiments.* Translated by K. Haakonssen. Cambridge Texts in the History of Philosophy. Cambridge and New York: Cambridge University Press.

Smith, D. R. 1983. *Palauan Social Structure.* New Brunswick, NJ: Rutgers University Press.

Smith, M. F. 1954. *Baba of Karo, a Woman of the Muslim Hausa.* London: Faber and Faber.

Smith, M. G. 1957. Cooperation in Hausa society. *Information* 11:1–20.

Smuts, B. B. 1985. *Sex and Friendship in Baboons.* New York: Aldine Pub. Co.

Song, S. H., and D. Olshfski. 2008. Friends at work: A comparative study of work attitudes in Seoul City government and New Jersey State. *Administration and Society* 40 (2): 147–169.

Soueif, A. 2000. *The Map of Love.* New York: Random House.

South, S. J., and D. L. Haynie. 2004. Friendship networks of mobile adolescents. *Social Forces* 83 (1): 315–350.

Southern, T. 1962. An interview with Stanley Kubrick: Director of *Lolita.* http://terrysouthern.com/archive/skint.htm, accessed May 2009.

Southwold, M. 1965. The Ganda of Uganda. In *Peoples of Africa,* ed. J. L. Gibbs. New York: Holt, Rinehart, and Winston.

———. 1971. Meanings of kinship. In *Rethinking Kinship and Marriage,* ed. R. Needham. London: Tavistock Publications.

Sozou, P. D., and R. M. Seymour. 2005. Costly but worthless gifts facilitate courtship. *Proceedings of the Royal Society B* 272:1877–1884.

Spencer, B. 1927. *The Arunta: A Study of a Stone Age People.* London: Macmillan and Co.

Spencer, P. 1988. *The Maasai of Matapato: A Study of Rituals of Rebellion.* Manchester, United Kingdom: Manchester University Press.

Spencer, R. F. 1956. Exhortation and the Klamath ethos. *Proceedings of the American Philosophical Society* 100 (1): 77–86.

Spillius, J. 1957. Natural disaster and political crisis in a Polynesian society: An exploration of operational research. *Human Relations* 10:3–27, 113–125.

Spoehr, A. 1941. "Friends" among the Seminole. *Chronicles of Oklahoma* 19 (3): 252.

Sprecher, S., A. Aron, E. Hatfield, A. Cortese, E. Potapova, and A. Levitskaya. 1994. Love: American style, Russian style, and Japanese style. *Personal Relationships* 1 (4): 349–369.

Sprecher, S., and R. Chandak. 1992. Attitudes about arranged marriages and dating among men and women from India. *Free Inquiry in Creative Sociology* 20:59–69.

Sprecher, S., and B. Fehr. 2006. Compassionate love for close others and humanity. *Journal of Social and Personal Relationships* 22 (5): 629–651.

Sprecher, S., and P. C. Regan. 1998. Passionate and companionate love in court-ing and young married couples. *Sociological Inquiry* 68 (2): 163–185.

Sproull, L., and S. Kiesler. 1986. Reducing social context cues: Electronic mail in organizational communication. *Management Science* 32 (11): 1492–1512.

Stack, C. B. 1974. *All Our Kin: Strategies for Survival.* New York: Harper.

Stark, R., and W. S. Bainbridge. 1980. Networks of faith: Interpersonal bonds and recruitment to cults and sects. *American Journal of Sociology* 85 (6): 1376–1395.

Staub, E., and L. Sherk. 1970. Need for approval, children's sharing behavior, and reciprocity in sharing. *Child Development* 41 (1): 243–252.

Stegmiller, P. F., and E. Knight. 1925. Arrow shooting and hunting of Khasi. *Anthropos* 20:607–623.

Stephen, A. M. 1969. *Hopi Journal of Alexander M. Stephen.* New York: AMS Press.

Stephenson, J. B. 1984. *Ford, a Village in the West Highlands of Scotland: A Case Study of Repopulation and Social Change in a Small Community.* Lexington: University of Kentucky Press.

Stern, T. 1963. Ideal and expected behavior as seen in Klamath mythology. *Journal of American Folklore* 76 (299): 21–30.

Sternberg, R. J. 1996. Construct validation of a triangular love scale. *European Journal of Social Psychology* 27 (3): 313–335.

Stewart-Williams, S. 2007. Altruism among kin and non-kin: Effects of cost of helping and reciprocal exchange. *Evolution and Human Behavior* 28 (3): 193–198.

———. 2008. Human beings as evolved nepotists: Exceptions to the rule and the effects of the cost of help. *Human Nature* 19 (4): 414–425.

Stinson, L., and W. Ickes. 1992. Empathic accuracy in the interactions of male friends vs. male strangers. *Journal of Personality and Social Psychology* 62 (5): 787–797.

Stouffer, S. A., and J. Toby. 1951. Role conflict and personality. *American Journal of Sociology* 56 (5): 305–407.

Stout, D. B. 1947. *San Blas Cuna Acculturation: An Introduction.* New York: Viking Fund.

Strathern, A. 1971. *The Rope of Moka: Big-Men and Ceremonial Exchange in Mount Hagen, New Guinea.* Cambridge: Cambridge University Press.

Strayer, F. F., and J. M. Noel. 1986. The prosocial and antisocial functions of preschool agression: An ethological study of triadic conflict among young children. In *Altruism and Aggression,* ed. C. Zahn-Waxler, E. M. Cummings, and R. Iannotti. Cambridge: Cambridge University Press.

Strehlow, T. G. H. 1947. *Aranda Traditions.* Melbourne: Melbourne University Press.

Suizzo, M. A. 2007. Parents' goals and values for children: Dimensions of inde-pendence and interdependence across four U.S. ethnic groups. *Journal of Cross-cultural Psychology* 38 (4): 506–530.

Sullivan, H.S. 1953. *The Interpersonal Theory of Psychiatry*. New York: Norton.

Surra, C.A. 1985. Courtship types: Variation in interdependence between partners and social networks. *Journal of Personality and Social Psychology* 49 (2): 357–375.

Sutlive, V.H. 1973. *From Longhouse to PASAR: Urbanization in Sarawak, East Malaysia*. Ann Arbor, MI: University Microfilms.

Suttles, G.D. 1970. Friendship as a social institution. In *Social Relationships*, ed. G.J. McCall. Chicago: Aldine.

Sutton, A.J., K.R. Abrams, D.R. Jones, T.A. Sheldon, and F. Song. 2000. *Methods for Meta-analysis in Medical Research*. New York: Wiley.

Suzuki, L.K., and P.M. Greenfield. 2002. The construction of everyday sacrifice in Asian Americans and European Americans: The roles of ethnicity and acculturation. *Cross-cultural Research* 36 (3): 200–228.

Sverdrup, H.U. 1938. *With the People of the Tundra*. Oslo: Gyldendal Norsk Forlag.

Swingle, P.G. 1966. Effects of emotional relationship between protagonists in a two-person game. *Journal of Personality and Social Psychology* 4 (3): 270–279.

Swingle, P.G., and J.S. Gillis. 1968. Effect of the emotional relationship between protagonists in the prisoner's dilemma. *Journal of Personality and Social Psychology* 8 (2): 160–165.

Szanton, M.C.B. 1972. *A Right to Survive: Subsistence Marketing in a Lowland Philippine Town*. University Park: Pennsylvania State University Press.

Tagiuri, R., R.R. Blake, and J.S. Bruner. 1953. Some determinants of the perception of positive and negative feelings in others. *Journal of Abnormal Psychology* 48 (4): 585–592.

Talayesva, D.C. 1942. *Sun Chief: The Autobiography of a Hopi Indian*. New Haven, CT: Yale University Press.

Tambiah, S.J. 1965. Kinship fact and fiction in relation to the Kandyan Sinhalese. *Journal of the Anthropological Institute of Great Britain and Ireland* 95:131–173.

Tanaka, I. 1998. Social diffusion of modified louse egg-handling techniques during grooming in free-ranging Japanese macaques. *Animal Behaviour* 56 (5): 1229–1236.

Tannen, D. 1990. *You Just Don't Understand: Men and Women in Conversation*. New York: Morrow.

Tanner, J. 1830. *A Narrative of the Captivity and Adventures of John Tanner*. New York: Baldwin & Cradock.

Taylor, M. 1999. *Imaginary Companions and the Children Who Create Them*. New York: Oxford University Press.

Taylor, R.J., L.M. Chatters, and V.M. Mays. 1988. Parents, children, in-laws, and non-kin as sources of emergency assistance to elderly black Americans. *Family Relations* 37 (3): 298–304.

Taylor, S. E., L. C. Klein, B. P. Lewis, T. L. Gruenewald, R. A. R. Gurung, and J. A. Updegraff. 2000. Biobehavioral responses to stress in females: Tend-and-befriend, not fight-or-flight. *Psychological Review* 107 (3): 411–429.

Taylor, S. E., W. T. Welch, H. S. Kim, and D. K. Sherman. 2007. Cultural differences in the impact of social support on psychological and biological stress response. *Psychological Science* 18 (9): 831–837.

Tegnaeus, H. 1952. *Blood Brothers: An Ethno-sociological Study of Institutions of Blood-Brotherhood.* Stockholm: Ethnographical Museum of Sweden.

Teigen, K. H., M. V. G. Olsen, and O. E. Solas. 2005. Giver-receiver asymmetries in gift preferences. *British Journal of Social Psychology* 44 (1): 125–144.

Tenbrunsel, A. E., K. A. Wade-Benzoni, J. Moag, and M. H. Bazerman. 1999. The negotiation matching process: Relationship and partner selection. *Organizational Behavior and Human Decision Processes* 80 (3): 252–283.

Tennov, D. 1979. *Love and Limerance: The Experience of Being in Love.* New York: Scarborough House.

Terwiel, B. J. 1975. *Monks and Magic: An Analysis of Religious Ceremonies in Central Thailand.* Lund, Sweden: Studentlitteratur.

Tesch, S. A., and R. R. Martin. 1983. Friendship concepts of young adults in two age groups. *Journal of Psychology* 115 (11): 7–12.

Tesser, A., R. Gatewood, and M. Driver. 1968. Some determinants of gratitude. *Journal of Personality and Social Psychology* 9 (3): 233–236.

Tesser, A., and J. Smith. 1980. Some effects of task relevance and friendship on helping: You don't always help the one you like. *Journal of Experimental Social Psychology* 16 (6): 582–590.

Theriault, M. K. 1992. *Moose to Moccasins: The Story of Ka Kita Wa Pa No Kwe.* Toronto: Natural Heritage.

Thibaut, J., and H. H. Kelley. 1959. *The Social Psychology of Groups.* New York: Wiley.

Thompson, G. G., and J. E. Horrocks. 1947. A study of the friendship fluctuations of urban boys and girls. *Journal of Genetic Psychology* 70:53–63.

Thompson, L. 1940. *Fijian Frontier.* San Francisco: American Council, Institute of Pacific Relations.

Thye, S. R., J. Yoon, and E. J. Lawler. 2002. A theory of relational cohesion: Review of a research program. In *Group Cohesion, Trust and Solidarity,* ed. S. R. Thye and E. J. Lawler. New York: JAI.

Tiger, L. 1969. *Men in Groups.* New York: Random House.

Tilly, C., and C. H. Brown. 1967. On uprooting, kinship, and the auspices of migration. *International Journal of Comparative Sociology* 8 (2): 139–164.

Titiev, M. 1967. The Hopi use of kinship terms for expressing sociocultural values. *Anthropological Linguistics* 9 (6): 44–49.

———. 1972. *The Hopi Indians of Old Oraibi: Change and Continuity.* Ann Arbor: University of Michigan Press.

Tixier, V. 1940. *Travels on the Osage Prairies.* Norman: University of Oklahoma Press.

Tomasello, M. 2001. *The Cultural Origins of Human Cognition.* Cambridge: Harvard University Press.

Tooby, J., and L. Cosmides. 1996. Friendship and the banker's paradox: Other pathways to the evolution of adaptations for altruism. *Proceedings of the British Academy* 88:119–143.

Tooker, E. 1970. *The Iroquois Ceremonial of Midwinter.* Syracuse, NY: Syracuse University Press.

Traupmann, J., and E. Hatfield. 1981. Love and its effect on mental and physical health. In *Aging: Stability and Change in the Family,* ed. R. Fogel, E. Hatfield, S. Kiesler, and E. Shanas. New York: Academic Press.

Trief, D. 2004. Models of altruism as applied to human blood and organ donation. *Penn Science* 2 (2).

Trivers, R. 1971. The evolution of reciprocal altruism. *Quarterly Review of Biology* 46:35–57.

———. 2006. Reciprocal altruism: 30 years later. In *Cooperation in Humans and Primates,* ed. P. M. Kappeler and C. van Schaik. Berlin: Springer.

Trompenaars, A., and C. Hampden-Turner. 1998. *Riding the Waves of Culture: Understanding Cultural Diversity in Global Business.* New York: McGraw-Hill.

Tropp, L. R., and S. C. Wright. 2001. Ingroup identification as the inclusion of ingroup in the self. *Personality and Social Psychology Bulletin* 27 (5): 585–600.

Tsang, J. 2006. The effects of helper intention on gratitude and indebtedness. *Motivation and Emotion* 30 (3): 198–204.

Tschopik, H. 1951. *The Aymara of Chucuito, Peru.* Vol. 1, *Magic.* New York: American Museum of Natural History.

Tulchin, A. A. 2007. Same-sex couples creating households in old regime France: The uses of the *affrerement. Journal of Modern History* 79:613–647.

Turkle, S. 1996. Virtuality and its discontents. *American Prospect* 24:50–58.

Turnbull, C. M. 1962. *The Forest People.* New York: Simon and Schuster.

———. 1965. *Wayward Servants: The Two Worlds of the African Pygmies.* Garden City, NY: Natural History Press.

———. 1972. *The Mountain People.* New York: Simon and Schuster.

Turner, V. W. 1952. *The Lozi Peoples of North-western Rhodesia.* London: International African Institute.

Twain, M. 1894. *The Tragedy of Pudd'nhead Wilson.* Hartford, CT: American Publishing Company.

Tylor, E. B. 1889. On a method of investigating the development of institutions: Applied to laws of marriage and descent. *Journal of the Anthropological Institute of Great Britain and Ireland* 18:245–272.

Uhl, S. 1991. Forbidden friends: Cultural veils of female friendship in Andalusia. *American Ethnologist* 18 (1): 90–105.

Urry, J. 1972. *Notes and Queries on Anthropology* and the development of field methods in British anthropology, 1870–1920. *Proceedings of the Royal Anthropological Institute of Great Britain and Ireland* 1972:45–57.

Valkenberg, P. M., and J. Peter. 2007. Online communication and adolescent well-being: Testing the stimulation versus the displacement hypothesis. *Journal of Computer Mediated Communication* 12:1169–1182.

Vallee, F. G. 1955. Burial and mourning customs in a Hebridean community. *Journal of the Royal Anthropological Institute of Great Britain and Ireland* 85:119–130.

Valley, K. L., J. Moag, and M. H. Bazerman. 1998. A matter of trust: Effects of communication on the efficiency and distribution of outcomes. *Journal of Economic Behavior and Organization* 34 (2): 211–238.

Van der Poel, W. H. M. 1993. *Personal Networks: A Rational-Choice Explanation of Their Size and Composition.* Lisse, the Netherlands: Swets and Zeitlinger.

Van Horn, K. R., and J. C. Marques. 2000. Interpersonal relationships in Brazilian adolescents. *International Journal of Behavioral Development* 24 (2): 199–203.

Van Overwalle, F. 2009. Social cognition and the brain: A meta-analysis. *Human Brain Mapping* 30:829–859.

Vandell, D. L., and E. C. Mueller. 1980. Peer play and friendships during the first two years. In *Friendship and Social Relations in Children,* ed. H. C. Foot, A. J. Chapman and J. R. Smith. London: Wiley.

Vaquera, E., and G. Kao. 2007. Do you like me as much as I like you? Friendship reciprocity and its effects on school outcomes among adolescents. *Social Science Research* 37 (1): 2006.

Vaughan, G. M., H. Tajfel, and J. Williams. 1981. Bias in reward allocation in an intergroup and an interpersonal context. *Social Psychology Quarterly* 44 (1): 37–42.

Vaughn, B. E., M. R. Azria, L. Krzysik, L. R. Caya, K. K. Bost, W. Newell, and K. L. Kazura. 2000. Friendship and social competence in a sample of preschool children attending Head Start. *Developmental Psychology* 36 (3): 326–338.

Venema, L. B. 1978. *The Wolof of Saloum: Social Structure and Rural Development in Senegal.* Wageningen, the Netherlands: Centre for Agricultural Publication and Documentation.

Vennum, T. 1988. *Wild Rice and the Ojibway People.* St. Paul: Minnesota Historical Society Press.

Verbrugge, L. M. 1979. Multiplexity in adult friendships. *Social Forces* 57 (4): 1286–1309.

Verkuyten, M., and K. Masson. 1996. Culture and gender differences in the perception of friendship by adolescents. *International Journal of Psychology* 31 (5): 207–217.

Vespo, J. E., and M. Caplan. 1993. Preschoolers' differential conflict behavior with friends and acquaintances. *Early Education and Development* 4 (1): 45–53.

Vigil, J. M. 2007. Asymmetries in friendship preferences and social styles of men and women. *Human Nature* 18 (2): 143–161.

Villa Rojas, A. 1969. The Tzeltal. In *Handbook of Middle American Indians*, ed. R. Wauchope. Austin: University of Texas Press.

Villaverde, J. 1909. The Ifugaos of Quiangan and vicinity. *Philippine Journal of Science* 4:237–262.

Virginia, L. 2006. Genealogy as theory, genealogy as tool: Aspects of Somali "clanship." *Social Identities* 12 (4): 471–485.

von Grumbkow, J., E. Deen, H. Steensma, and H. Wilke. 1976. The effect of future interaction on the distribution of rewards. *European Journal of Social Psychology* 6 (1): 119–123.

Vrticka, P., F. Andersson, D. Grandjean, D. Sander, and P. Vuilleumier. 2008. Individual attachment style modulates human amygdala and striatum activation during social appraisal. *PLoS One* 3 (8): e2868.

Wagner, G. 1949. *The Bantu of Kavirondo*. London: Oxford University Press.

Waite, L. J., and S. C. Harrison. 1992. Keeping in touch: How women in mid-life allocate social contacts with kith and kin. *Social Forces* 70 (3): 637–655.

Waldfogel, J. 1993. The deadweight loss of Christmas. *American Economic Review* 83 (5): 1328–1336.

Walker, K. 1994. Men, women, and friendship: What they say, what they do. *Gender and Society* 8:246–265.

———. 1995. "Always there for me": Friendship patterns and expectations among middle- and working-class men and women. *Sociological Forum* 10 (2): 273–296.

Walker, M. K. 1974. Social constraints, individuals, and social decisions in a Scottish rural community. PhD diss., University of Illinois, Urbana-Champaign.

Wall, S. M., S. M. Pickert, and L. M. Paradise. 1984. American men's friendships: Self-reports on meaning and closeness. *Journal of Psychology* 116 (2): 179–186.

Wallace, A. F. C. 1950. Some psychological characteristics of the Delaware Indians during the 17th and 18th centuries. *Pennsylvania Archaeologist* 20:33–39.

———. 1969. *The Death and Rebirth of the Seneca*. New York: Vintage Books.

Wallman, S. 1974. Kinship, a-kinship, anti-kinship: Variations in the logic of kinship situations. In *The Compact: Selected Dimensions of Friendship*, ed. E. Leyton. St. John's, Canada: Memorial University of Newfoundland.

Walshe, Maurice. 1995. *The Long Discourses of the Buddha: A Translation of the Digha Nikaya*. Somerville, MA: Wisdom Publications.

Walster, E., G. W. Walster, and E. Berscheid. 1978. *Equity Theory and Research*. Boston: Allyn and Bacon.

Walters, M., and E. A. Stow. 2001. *Darwin's Mentor: John Stevens Henslow, 1796–1861*. Cambridge and New York: Cambridge University Press.

Walther, J. B., B. Van Der Hiede, S. Y. Kim, and D. Westerman. 2008. The role of friends' appearance and behavior on evaluations of individuals on Facebook: Are we known by the company we keep? *Human Communication Research* 34 (1): 28–49.

Warren, D. M. 1975. *The Techiman-Bobo of Ghana: An Ethnography of Akan Society*. Dubuque, IA: Kendall/Hunt Pub. Co.

Watson, J. 2002. Starting small and commitment. *Games and Economic Behavior* 38:176–199.

Watson, J. B. 1952. Cayua culture change. *Memoirs of the American Anthropological Association* 73:1–144.

Watson, V. D. 1944. Notes on the kinship system of the Cayua Indians. *Sociologia* 6:1–23.

Watts, D. P. 2002. Reciprocity and interchange in the social relationships of wild male chimpanzees. *Behaviour* 138:343–370.

Way, N. 1998. *Everyday Courage: The Lives and Stories of Urban Teenagers*. New York: New York University Press.

Way, N., and L. Chen. 2000. Close and general friendships among African American, Latino, and Asian American adolescents from low-income families. *Journal of Adolescent Research* 15 (2): 274–301.

Way, N., and M. L. Greene. 2006. Trajectories of perceived friendship quality during adolescence: The patterns and contextual predictors. *Journal of Research on Adolescence* 16 (2): 293–320.

Weaver, S. M. 1972. *Medicine and Politics among the Grand River Iroquois*. Ottawa: National Museums of Canada.

Webster, G. D. 2003. Prosocial behavior in families: Moderators of resource sharing. *Journal of Experimental Social Psychology* 39 (6): 644–652.

Wedgewood, C. H. 1942. Notes on the Marshall Islands. *Oceania* 13:1–23.

Weimer, B. L., K. A. Kerns, and C. M. Oldenburg. 2004. Adolescents' interactions with best friends: Associations with attachment style. *Journal of Experimental Child Psychology* 88 (1): 102–120.

Weiner, A. B. 1988. *The Trobrianders of Papua New Guinea*. New York: Holt, Rinehart and Winston.

———. 1992. *Inalienable Possessions: The Paradox of Keeping-While-Giving*. Berkeley: University of California Press.

Weinstock, J. S., and L. A. Bond. 2000. Conceptions of conflict in close friendships and ways of knowing among young college students. *Journal of Social and Personal Relationships* 17 (4–5): 687–696.

Weiss, L., and M. F. Lowenthal. 1975. Life-course perspectives on friendship. In *Four Stages of Life*, ed. M. F. Lowenthal. San Francisco: Jossey-Bass.

Weisz, C., and L. F. Wood. 2005. Social identity support and friendship outcomes: A longitudinal study predicting who will be friends and best friends 4 years later. *Journal of Social and Personal Relationships* 22 (3): 416–432.

Wellman, B. 1979. The community question: The intimate networks of East Yorkers. *American Journal of Sociology* 84:1201–1231.

Wellman, B., R. Y. Wong, D. Tindall, and N. Nazer. 1997. A decade of network change: Turnover, persistence and stability in personal communities. *Social Networks* 19 (1): 27–50.

Wellman, B., and S. Wortley. 1990. Different strokes for different folks: Community ties and social support. *American Journal of Sociology* 96 (3): 558–588.

Weltfish, G. 1965. *The Lost Universe.* New York: Basic Books.

Werking, K. 1997. *We're Just Good Friends: Women and Men in Nonromantic Relationships.* New York: Guilford Press.

Wharton, E. 1934. *A Backward Glance: Reminiscences.* New York: Appleton & Co.

Wheeler, L., H.T. Reis, and M.H. Bond. 1989. Collectivism-individualism in everyday social life: The middle kingdom and the melting pot. *Journal of Personality and Social Psychology* 57 (1): 79–86.

Whitaker, I.W. 1955. *Social Relations in a Nomadic Lappish Community.* Oslo: Norsk Folkmuseum.

Whitesell, N.R., and S. Harter. 1996. The interpersonal context of emotion: Anger with close friends and classmates. *Child Development* 67 (4): 1345–1359.

Whiting, B.B., and C.P. Edwards. 1988. *Children of Different Worlds.* Cambridge: Harvard University Press.

Whiting, J. 1941. *Becoming Kwoma.* New Haven, CT: Yale University Press.

Whitman, W., and M.W. Whitman. 1947. *The Pueblo Indians of San Ildefonso: A Changing Culture.* New York: Columbia University Press.

Wierzbicka, Anna. 1997. *Understanding Cultures through Their Key Words.* Oxford: Oxford University Press.

Wieselquist, J. 2007. Commitment and trust in young adult friendships. *Interpersona* 1 (2): 209–220.

Wieselquist, J., C.E. Rusbult, C.A. Foster, and C.R. Agnew. 1999. Commitment, pro-relationship behavior, and trust in close relationships. *Journal of Personality and Social Psychology* 77 (5): 942–966.

Wiessner, P.W. 1977. Hxaro: A regional system of reciprocity for reducing risk among the !Kung San. PhD diss., University of Michigan, Ann Arbor.

———. 1982. Risk, reciprocity and social influences in !Kung San economics. In *Politics and History in Band Societies,* ed. E. Leacock and R. Lee. Cambridge: Cambridge University Press.

———. 1994. The pathways of the past: !Kung San hxaro exchange and history. In *Uberlebensstrategien in Afrika,* ed. M. Bollig and F. Klees. Cologne: Heinrich-Barth-Institut.

———. 1998. Indoctrinability and the evolution of socially defined kinship. In *Ethnic Conflict and Indoctrination,* ed. I. Eibl-Eibesfeldt and F.K. Salter. New York: Berghahn Books.

———. 2002a. Hunting, healing, and hxaro exchange—a long-term perspective on !Kung (Ju/'hoansi) large-game hunting. *Evolution and Human Behavior* 23 (6): 407–436.

———. 2002b. Taking the risk out of risky transactions: A forager's dilemma. In *Risk Transactions,* ed. F.K. Salter. Oxford: Berghahn Books.

Wilkins, J.F., and S. Thurner. nd. The Jerusalem Game: Cultural evolution of the golden rule. Working paper, Santa Fe Institute.

Wilkins, R., and E. Gareis. 2006. Emotion expression and the locution "I love you": A cross-cultural study. *International Journal of Intercultural Relations* 30 (1): 51–75.

Wilkinson, G. S. 1984. Reciprocal food sharing in the vampire bat. *Nature* 308:181–184.

———. 1985. The social organization of the common vampire bat. Part 1, Pattern and cause of association. *Behavioral Ecology and Sociobiology* 17:111–121.

Wilkinson, R. J. 1925. Malay proverbs on Malay character. In *Papers on Malay Subjects: Malay Literature,* by R. J. Wilkinson, pp. 1–19. Kuala Lumpur: F. M. S. Government Press.

Williams, K. D., C. K. T. Cheung, and W. Choi. 2000. Cyberostracism: Effects of being ignored over the Internet. *Journal of Personality and Social Psychology* 79 (5): 748–762.

Williams, K. D., and K. L. Sommer. 1997. Social ostracism by coworkers: Does rejection lead to loafing or compensation? *Personality and Social Psychology Bulletin* 23 (7): 693–706.

Winn, K. I., D. W. Crawford, and J. L. Fischer. 1991. Equity and commitment in romance versus friendship. *Journal of Social Behaviour and Personality* 6:301–314.

Winslow, J., and T. R. Insel. 2002. The social deficits of the oxytocin knockout mice. *Neuropeptides* 36 (2–3): 221–229.

Wiseman, D. B., and I. P. Levin. 1996. Comparing risky decision making under conditions of real and hypothetical consequences. *Organizational Behavior and Human Decision Processes* 66 (3): 241–250.

Wiseman, J. P. 1986. Friendship: Bonds and binds in a voluntary relationship. *Journal of Social and Personal Relationships* 3 (2): 191–211.

Wissler, C. 1911. *The Social Life of the Blackfoot Indians.* New York: American Museum of Natural History.

Witherspoon, G. 1975. *The Central Concepts of Navajo World View.* Lisse, the Netherlands: Peter de Ridder Press.

Wolak, J., K. J. Mitchell, and D. Finkelhor. 2002. Close online relationships in a national sample of adolescents. *Adolescence* 37:441–456.

Wolf, E. 1966. Kinship, friendship, and patron-client relations in complex societies. In *The Social Anthropology of Complex Societies,* ed. M. Banton. London: Tavistock Publications.

Wolf, M. 1978. *Child Training and the Chinese Family.* Stanford, CA: Stanford University Press.

Wolff, K., E. M. Tsapakis, A. R. Winstock, D. Hartley, D. Holt, M. L. Forsling, and K. J. Aitchison. 2006. Vasopressin and oxytocin secretion in response to the consumption of Ecstasy in a clubbing population. *Journal of Psychopharmacology* 20 (3): 400–410.

Won-Doornink, M. 1985. Self-disclosure and reciprocity in conversation: A cross-national study. *Social Psychology Quarterly* 48 (22): 97–107.

———. 1991. Self-disclosure and reciprocity in South Korean and US male dyads. In *Cross-cultural Interpersonal Communication*, ed. S. Ting-Toomey and F. Korzenny. Thousand Oaks, CA: Sage.

Wong, S. C. H., and M. H. Bond. 1999. Personality, self-disclosure and friendship between Chinese university roommates. *Asian Journal of Social Psychology* 2 (2): 201–214.

Wood, J. G. 1868. *Natural History of Man: Africa*. New York: George Routledge and Sons.

World Values Survey 2005 Official Data File. 2009. World Values Survey Association. www.worldvaluessurvey.org.

Worldbank. 2009. *Governance Matters 8: Governance Indicators for 1996–2008*. Washington, D.C.: Worldbank.

Wright, P. A. 1986. Language shift and the redefinition of social boundaries among the Caribs of Belize. PhD diss., City University of New York, New York.

Wright, P. H. 1974. The delineation and measurement of some key variables in the study of friendship. *Representative Research in Social Psychology* 5:93–96.

———. 1978. Toward a theory of friendship based on a conception of self. *Human Communication Research* 4 (3): 196–207.

———. 1982. Men's friendships, women's friendships and the alleged inferiority of the latter. *Sex Roles* 8 (1): 1–20.

———. 1998. Toward an expanded orientation to the study of sex differences in friendship. In *Sex Differences and Similarities in Communication*, ed. D. J. Canary and K. Dindia. Mahwah, NJ: Lawrence Erlbaum.

Xiaohe, X., and M. K. Whyte. 1990. Love matches and arranged marriages: A Chinese replication. *Journal of Marriage and Family* 52 (3): 709–722.

Xie, B. 2008. Multimodal computer-mediated communication and social support among older Chinese Internet users. *Journal of Computer Mediated Communication* 13:728–750.

Yakubovich, V., and I. Kozina. 2000. The changing significance of ties: An exploration of the hiring channels in the Russian transitional labor market. *International Sociology* 15 (3): 479–500.

Yalman, N. 1971. *Under the Bo Tree: Studies of Caste, Kinship and Marriage in the Interior of Ceylon*. Berkeley: University of California Press.

Yamagishi, T., and N. Mifune. 2008. Does shared group membership promote altruism? *Rationality and Society* 20 (1): 5–30.

Yamagishi, T., S. Terai, T. Kiyonari, N. Mifune, and S. Kanazawa. 2007. The social exchange heuristic: Managing errors in social exchange. *Rationality and Society* 19 (3): 259–291.

Yamamoto, K., and N. Suzuki. 2006. The effects of social interaction and personal relationships on facial expressions. *Journal of Nonverbal Behavior* 30 (4): 167–179.

Yanagisako, S. 1979. Family and household: The analysis of domestic groups. *Annual Review of Anthropology* 8 (1): 161–205.

Ye, J. 2006. Maintaining online friendship: Cross-cultural analyses of links among relational maintenance strategies, relational factors, and channel-related factors. PhD diss., Georgia State University, Athens.

———. 2007. Attachment style differences in online relationship involvement: An examination of interaction characteristics and relationship satisfaction. *Cyberpsychology and Behavior* 10 (4): 605–607.

Yi, K. 1975. *Kinship System in Korea.* New Haven, CT: HRAF.

You, H. S., and K. Malley-Morrison. 2000. Young adult attachment styles and intimate relationships with close friends: A cross-cultural study of Koreans and Caucasians. *Journal of Cross-cultural Gerontology* 31 (4): 528–534.

Young, A. L. 1975. Magic as a "quasi-profession": The organization of magic and magical healing among Amhara. *Ethnology* 14 (3): 245–265.

Young, M., and P. Willmott. 1962. *Family and Kinship in East London.* New York: Penguin.

Youniss, J. 1980. *Parents and Peers in Social Development.* Chicago: University of Chicago Press.

Youniss, J., and J. Smollar. 1985. *Adolescents' Relations with Mothers, Fathers, and Friends.* Chicago: University of Chicago Press.

Yum, Y., and K. Hara. 2005. Computer-mediated relationship development: A cross-cultural comparison. *Journal of Computer Mediated Communication* 11 (1): 133–152.

Zak, P. J., R. Kurzban, and W. T. Matzner. 2005. Oxytocin is associated with human trustworthiness. *Hormones and Behavior* 48 (5): 522–527.

Zamma, K. 2002. Grooming site preferences determined by lice infection among Japanese macaques. *Primates* 43 (1): 41–50.

Zeisberger, D. 1910. *David Zeisberger's History of Northern American Indians.* Columbus, OH: Fred H. Heer.

Zimmer, C., and H. Aldrich. 1987. Resource mobilization through ethnic networks: Kinship and friendship ties of shopkeepers in England. *Sociological Perspectives* 30:422–445.

Zvoch, K. 1999. Family type and investment in education: A comparison of genetic and stepparent families. *Evolution and Human Behavior* 20 (6): 453–464.

Index

First names of authors are given only when mentioned in the text; otherwise, initials are used. References with multiple authors are listed by the first author's name, et al.

371

Text: 10/13 Aldus
Display: Aldus
Compositor: BookMatters, Berkeley
Printer and binder: Thomson-Shore, Inc.